T0214123

Cancer Bioinformatics

Ying Xu • Juan Cui • David Puett

Cancer Bioinformatics

 Springer

Ying Xu
Department of Biochemistry
and Molecular Biology
University of Georgia
Athens, GA, USA

Juan Cui
Department of Computer Science
and Engineering
University of Nebraska
Lincoln, NE, USA

David Puett
Department of Biochemistry
and Molecular Biology
University of Georgia
Athens, GA, USA

ISBN 978-1-4939-4303-6   ISBN 978-1-4939-1381-7 (eBook)
DOI 10.1007/978-1-4939-1381-7
Springer New York Heidelberg Dordrecht London

Printed on acid-free paper

Springer is part of Springer Science+Business Media (www.springer.com)

Preface

In his superb exposition, *The Emperor of All Maladies: A Biography of Cancer*, Mukherjee attributes the earliest documentation of cancer to the brilliant Egyptian, Imhotep, who some 4,500 years ago clearly described a case of breast cancer (Mukherjee 2010). Roughly two millennia later (ca. 400 BC), the Greek physician Hippocrates named the disease *karkinos* (the Greek word for crab), which has now come down to us as cancer. Some five to six centuries later while practicing in Rome (ca. 130–200 AD), the Greek physician, Claudius Galen, who was influenced by the four humors constituting the human body as proposed by the Hippocratic school, i.e., blood, phlegm, yellow bile, and black bile, attributed cancer to an excess of black bile. It took centuries before Vesalius (sixteenth century) and Baillie (eighteenth century) put the black bile hypothesis to rest, thus indirectly encouraging surgeons to begin resection of solid tumors. (Surgical procedures had been done earlier by some fearless surgeons, but few patients survived the ordeal and infection that likely followed.) The later introduction of anesthesia and antibiotics in the nineteenth to twentieth centuries, as well as more sterile operating environments, thrust surgery (and later radiation therapy) as a major treatment of this disease, an approach that is still used whenever possible. In the middle of the twentieth century and continuing today, chemotherapy and hormonal therapy emerged as a complement to, and sometimes instead of, surgery and radiation therapy to treat cancer.

A number of theories have been proposed regarding those factors that may drive and facilitate a cancer to initiate, develop, and metastasize, and these have guided cancer studies in the past few decades. An insightful speculation was made by Otto Warburg following his seminal work in the 1920s: "Cancer … has countless secondary causes. But … there is only one prime cause, [which] is the replacement of respiration of oxygen in normal body cells by a fermentation of sugar" (Warburg 1969).

The first discovery of oncogenes and tumor suppressor genes about 40 years ago marked another major milestone in our understanding of cancer development, which has profoundly influenced research in this area during the past three decades. It has become a widely held belief that cancer is ultimately a disease caused by genomic mutations. Aided by the rapidly increasing pool of a variety of *omic* data such as genomic, transcriptomic, epigenomic, metabolomic, glycomic, lipidomic, and pharmacogenomic data collected on both cell lines and cancer tissues, spectacular progress has been made in the past two decades in our understanding of cancer, particularly in terms of how the microenvironment and the immune system contribute to the whole process of neoplasm formation and survival.

In spite of the considerable progress made, however, a number of salient questions remain to be answered. The authors posit that a considerable amount of information needed to address and answer many of these questions already exists in the available *omic* databases, and much of these data are substantially undermined and underutilized. Among the many possible reasons, a key one, we believe, is that computational cancer biologists, as a community, have yet to sufficiently develop their independent thinking about the overall biology of cancer. The thinking should be quite different from the reductionist approaches that have been widely used in experimental studies of cancer in the past century and should enable them to address fundamental questions about cancer in a holistic manner as an evolving system. Many fundamental issues concerning cancer are intrinsically holistic by nature. Thus, when examining cancer as an evolutionary problem, its microenvironment, including the extracellular matrix and the immune and other stromal cells, must be considered as an integral part of the system. This strongly suggests that cell culture-based or animal model-based cancer studies must be complemented by cancer tissue-based studies in order to gain a full understanding of cancer. The *omic* data collected on cancer tissue samples, covering different developmental stages, is likely to contain the information on the interplay between cancer cells and their environment, and particularly how such interactions may drive the evolution in specific directions. Hence, we posit that mining such *omic* data for information discovery will, in the future, represent an essential component of cancer research, complementary to the current more reductionist-oriented approaches.

The goals of this book are to provide an overview of cancer biology from an informatics perspective and to demonstrate how *omic* data can be mined to generate new insights and a more comprehensive understanding that is needed to address a wide range of fundamental cancer biology questions. Throughout this book, the authors have attempted to establish the following key points: (1) cancer is a process of cell survival in an increasingly more stressful and difficult microenvironment, which co-evolves with the diseased cells; (2) cell proliferation is a cancer's way to reduce the stresses imposed on them for survival; (3) the challenges that the evolving cells must overcome are not only at the cell level, but more importantly at the tissue level, hence making cancer dominantly a tissue rather than a cell-only problem; (4) the survival pathway for each cancer is not created 'on the fly' through its selection of molecular malfunctions or genetic mutations, instead it is largely determined by substantial cellular programs encoded in the human genome, which originally evolved for other purposes; (5) subpopulations of cancer cells have managed to create the conditions needed to trigger such cellular program-guided survival pathways; (6) as the stresses become increasingly more challenging, cancer cells utilize increasingly less reversible stress-responses for their survival, thus making the disease progressively more malignant; (7) genomic mutations in sporadic cancers probably serve mainly as permanent replacements for ongoing functions to provide efficiency and sustainability for survival; in contrast, mutations in hereditary cancers dominantly play driver roles of cancer initiation, but in a sense different from driver mutations as defined in the current literature; (8) there is a fundamental difference between cell proliferation in primary *versus* metastatic cancers as the former is essential in overcoming the encountered stress(es) while the latter is simply a side product of a stress-response process, suggesting that their treatment regiments should be different; and (9) cancer survives and proliferates by continually evolving with natural selection having a major part in deciding which cells remain and which must perish.

For each chapter, the authors present the main topic by placing cancer in an evolutionary context, for example by raising and addressing questions such as: *What pressures are the evolving neoplastic cells currently under*, and *How have the cells responded to adapt to the pressures*? In addition, the authors also demonstrate through examples how to derive the desired information from the available *omic* data by asking questions and then addressing them using a hypothesis-driven data mining approach. An example could be as follows: *What is the difference between the main driving forces of primary versus metastatic cancer*? This can be addressed by identifying genes that are up-regulated consistently across all metastatic cancers *versus* their matching primary cancer tissues, and then delineating the particular pathways that are enriched by these genes.

This 14-chapter book consists of the following clusters of chapters. Chapters 1 and 2 introduce the basic biology and biochemistry of cancer and the available cancer *omic* data, as well as the type of information derivable from such data. Chapter 3 serves as an introduction to the use of *omic* data to address cancer-related problems, written for someone with only a limited knowledge of cancer; and Chap. 12 serves a similar purpose but for someone who has a general understanding about cancer at the molecular and cellular levels, e.g., having read a substantial portion of this book. Chapter 4 is a transition chapter, serving as an introduction to both information that can be derived from cancer genomes and elucidation of cancer mechanisms using such information. Chapters 5 through 9 represent the core of the book: elucidation of novel information and how to gain a new and better understanding about the fundamental biology of primary cancer, in which cancer is treated as an evolving system driven by specific pressures and assisted by certain facilitators at different developmental stages. A common theme is used when tackling a series of cancer-related key issues across these five chapters: *What stresses do the cancer cells need to overcome at a specific stage*, and *how do such cells utilize encoded stress-response systems to ensure their survival*? Chapters 10 and 11 extend this discussion to metastatic cancer, which, somewhat surprisingly, represents a different type of disease from primary cancers with fundamentally different drivers. Chapter 13 provides some general information to those new to the field about how to conduct meaningful data mining-based cancer research. Chapter 14 presents our perspectives about cancer research using a more holistic approach than is generally done.

The authors hope that this book will help in bridging the gap between experimental cancer biologists and computational biologists in their joint efforts to uncover the enormous wealth of information hidden in the cancer *omic* data. Success in this endeavor will lead to a better understanding of cancer, as well as assist computational biologists to develop independent thinking when tackling these complex problems. This approach will probably be less detail-oriented but more holistic and will likely span the entire range of cancer evolution, thus making it different from but complementary to those of their experimental peers. It is the authors' contention that more qualitative and quantitative utilization of the *omic* data will improve our overall understanding of cancer biology, hence leading to improved capabilities in early detection, development of more effective cancer treatments, and improvement in the quality of the patient lives.

The authors welcome any feedback from the reader regarding errors that need correcting and areas where the book could be improved. Such information will be highly valuable, particularly if there is a decision to write a future edition of the book.

Athens, GA, USA	Ying Xu
Lincoln, NE, USA	Juan Cui
Chapel Hill, NC, USA	David Puett

References

Mukherjee, S. (2010). The emperor of all maladies: a biography of cancer, Scribner.
Warburg, O. H. (1969). The prime cause and prevention of cancer, K. Triltsch.

Acknowledgments

The authors thank their many trainees, postdoctoral fellows, collaborators, and colleagues who have contributed enormously to this book. We wish to particularly acknowledge the following for their invaluable assistance and contributions. Mr. Chi Zhang and Ms. Sha Cao, two brilliant and dedicated doctoral students at the University of Georgia, have provided tremendous assistance in generating all of the case studies along with the figures showing the results of data analysis, in proofreading most chapters, and in coordinating the efforts in reference collection and figure generation. Dr. Ying Li and Dr. Wei Du, two young faculty members at the Jilin University College of Computer Science and Technology, have spent long hours and tireless efforts in collecting the majority of the references used throughout this book. Mr. Liang Chen and Ms. Yanjiao Ren, two Ph.D. students, also at the Jilin University College of Computer Science and Technology, have carried out the artistic design and the generation of all the non-boxplot figures used in this book. We particularly want to thank Mr. Liang Chen for his superb illustrations and figures, including the book cover image, that have clearly enhanced the presentation of the contents covered. Mr. Xin Chen, a visiting graduate student at the University of Georgia, has also contributed to improving the presentation of some of the figures. In addition, Ms. Sha Cao and Dr. Wei Du have organized journal clubs to present various chapters of the book at the University of Georgia and Jilin University, respectively. Feedback from these two journal clubs has been very useful in guiding us in the revision of the earlier drafts. A number of colleagues have read the early drafts of some chapters and provided invaluable comments and suggestions for further revision, including Professor Shaying Zhao of the University of Georgia, Professor Dong Xu of the University of Missouri, and Professor Yuan Yuan of the Chinese Medical University in Shenyang. We thank them profusely for their valuable comments and suggestions that clearly improved the quality of the book. Suffice it to say that any errors in the book are solely the responsibility of the authors.

We also wish to thank all the cancer biologists and funding agencies that have supported the publicly available cancer *omic* data, thus making our research as well as this book possible. It is a pleasure to thank Mr. Gilbert Miller for his financial contribution to our sequencing project of gastric cancer genomes, some of the results of which are used in Chap. 4. The continuous support from the Georgia Research Alliance and the University of Georgia Research Foundation over the years provided the financial stability in our cancer research.

We also take this opportunity to thank Springer US for encouraging us to write and publish this book. We particularly acknowledge our editor, Ms. Susan Lagerstrom-Fife, for her understanding and patience with us in completing this book project. Also, the assistance of Ms. Jennifer Malat of Springer US in formatting and developing consistency in the manuscript is greatly appreciated.

Last, but clearly not the least, we extend our heartfelt thanks to our families for their patience and understanding during the long work hours, including the weekends, required to complete this book project during the past year.

2014   YX, JC, DP

Contents

Abbreviations

The following abbreviations are used throughout the book.

A2AR	Adenosine A2a receptor
ABC	ATP-binding cascade
ABL1	ABL oncogene
ACTA	Actin, aortic smooth muscle
ACVR	Activin A receptor
AIDA	Axin interactor, dorsalization associated
AIF	Apoptosis-induced factor
AKT	V-AKT murine thymoma viral oncogene homolog
ALK	Anaplastic lymphoma receptor tyrosine kinase
ALL	Acute lymphoblastic leukemia
AML	Acute myelogenous leukemia
APAF	Adaptor protein apoptotic protease-activating factor
APO	Tumor necrosis factor
APC	Adenomatous polyposis coli
ARID	AT-rich interactive domain
ARSE	Arylsulfatase
ASXL	Additional sex combs like
ATM	Ataxia telangiectasia mutated
ATP	Adenosine triphosphate
ATP	ATPase, H^+/K^+ transporting
ATR	Ataxia telangiectasia and Rad3 related
BAD	BCL2-associated agonist of cell death
BAK1	BCL2-antagonist/killer
BARD1	BRCA1-associated ring domain protein
BAX	BCL2-associated X protein
BCC	Basal cell carcinoma
BCL2	B-cell lymphoma 2
BCL3	B-cell lymphoma 3
BCL2L	BCL2-like protein

BCR	Breakpoint cluster region
BID	BH3 interacting domain death agonist
BIM	BCL2-like 11
BOK	BCL2-related ovarian killer
BRAF	*B*-RAF proto-oncogene serine/threonine-protein kinase
BRCA	Breast cancer
CAS	CRK-associated substrate
CAD	Carbamoyl-phosphate synthetase
CCL	Chemokine ligand
CDC	Cell division control protein
CDK	Cyclin-dependent kinase
CDKN	Cyclin-dependent kinase inhibitor
CEACAM	Carcino-embryonic antigen-related cell adhesion molecule
CFD	Complement factor D
CHRM3	Cholinergic receptor, muscarinic
CLDN1	Senescence-associated epithelial membrane protein
CLL	Chronic lymphoblastic leukemia
CML	Chronic myelogenous leukemia
CNN1	Calponin
CNV	Copy number variation
COL	Collagen
COX	Cyclooxygenase
CPS	Carbamoyl-phosphate synthase
CRC	Colorectal cancer
CREB	CAMP-response element-binding protein
CSPG	Chondroitin sulfate proteoglycan
CTC	Circulating tumor cells
CTNN	Catenin (cadherin-associated protein)
CTSB	Cathepsin B
CYP	Cytochrome P450
DCC	Deleted in colorectal cancer
DEFA	Defensin, alpha
DES	Desmin
DHPS	Deoxyhypusine synthase
DMN	Synemin, intermediate filament protein
DNAPK	Protein kinase, catalytic polypeptide
DNM2	Dynamin
DNMT	DNA methyltransferase
DPT	Dermatopontin
DR	Tumor necrosis factor receptor
E2F	Transcription factor
EC	Enzyme classification
ECM	Extracellular matrix
ECT2L	Epithelial cell transforming sequence
EED	Embryonic ectoderm development

EFNA	Ephrin-A
EGR	Early growth response protein
EGF	Epidermal growth factor
EGFR	Epidermal growth factor receptor
EIF1A	Eukaryotic translation initiation factor
EMMPRIN	Extracellular matrix metalloproteinase inducer
EMT	Epithelial-mesenchymal transition
EP300	E1A-binding protein P300
EP	Prostaglandin E receptor
ER	Endoplasmic reticulum (subcellular compartment); or estrogen receptor (protein)
ERBB2	The same as HER2
ERK	Mitogen-activated protein kinase
EZH2	Enhancer of zeste homolog
FADD	FAS-associated protein with death domain
FAK	Focal adhesion kinase
FAM26D	Family with sequence similarity 26
FAS	A member of the tumor necrosis factor family
FGF	Fibroblast growth factor
FGFR	Fibroblast growth factor receptor
FH	Fumarate hydratase
FHL	Four and a half LIM domains
FLT3	fms-related tyrosine kinase
FN	Fibronectins
FOS	FBJ murine osteosarcoma viral oncogene homolog
FXR	Farnesoid X receptor
GAD	Glutamate decarboxylase
GADD	Growth arrest and DNA damage
GATA3	GATA-binding protein
GCLC	Glutamate-cysteine ligase, catalytic subunit
GCLM	Glutamate-cysteine ligase, modifier subunit
GDT	Growth-to-differentiation transition
GFRA	GDNF family receptor alpha
GKN1	Gastrokine-1 with strong anticancer activity
GNAS	Guanine nucleotide-binding protein
GNPDA	Glutamine-fructose-6-phosphate transaminase
GNPNAT	Glucosamine phosphate N-acetyltransferase
GP	Phosphoglucose isomerase
GPR	G-protein coupled receptor
GPX	Glutathione peroxidase
GSR	Glutathione reductase
GSTP	Glutathione S-transferase P
HARE	Hyaluronic acid receptor for endocytosis
HAS	Hyaluronic acid synthase
HDAC	Histone deacetylase

HDGF	Hepatoma-derived growth factor
HER2	V-ERB-B2 avian erythroblastic leukemia viral oncogene homolog
HGF	Hepatocyte growth factor
HIF	Hypoxia-induced factor
HLAA	Major histocompatibility complex A
HMOX1	Heme oxygenase
HNRPAB	Heterogeneous nuclear ribonucleoprotein
HO1	Heme oxygenase
HPV16L	Human papillomavirus 16 oncogene
HSF	Heat shock factor
HSPA2	Heat shock 70 kDa protein
HSPB1	Heat shock 27 kDa protein
HYAL	Hyaluronidase
IAP	Apoptosis inhibitor
ICAD	DNA fragmentation factor
ICGC	International Cancer Genome Consortium
IDH	Isocitrate dehydrogenase
IFITM2	Interferon-induced transmembrane protein
IGF	Insulin-like growth factor
IGFBP	Insulin-like growth factor-binding protein
IKK	Inducible I kappa-B kinase
IKZF	IKAROS family zinc finger
IL1B	Interleukin 1 beta
IL7R	Interleukin 7 receptor
IFIT	Interferon-induced protein
IFNR	Interferon gamma receptor
ING	Inhibitor of growth family
INMT	Indolethylamine *N*-methyltransferase
IRF	Interferon regulatory factor
JAK	Janus kinase
JNK1	Mitogen-activated protein kinase
JUN	Jun proto-oncogene
KAT	K (lysine) acetyltransferase
KC	Keratinocyte chemo-attractant
KCNN	Potassium intermediate/small conductance calcium-activated channel
KEAP	Kelch-like ECH-associated protein
KGF	Keratinocyte growth factor
KIT	V-kit Hardy-Zuckerman 4 feline
KRAS	Kirsten rat sarcoma viral oncogene homolog
LANCL	Lanc lantibiotic synthetase component c-like protein
LDLR	Low density lipoprotein receptor
LIPF	Gastric lipase with lipid-binding and retinyl-palmitate esterase activity
LMAN	Lectin, mannose-binding
LMNB	Laminin

LMOD	Leiomodin
LYVE1	Lymphatic vessel endothelial hyaluronic acid receptor
MAO	Monoamine oxidase
MAP	Mitogen-activated protein
MAPK	Mitogen-activated protein kinase
MCL	Myeloid cell leukemia
MCP	Macrophage chemo-attractant protein
MDK	Neurite growth-promoting factor
MDM2	Human homolog of mouse double minute 2
MEK1	Mitogen-activated protein kinase
MET	Hepatocyte growth factor receptor
MFAP	Microfibrillar-associated protein
MIP	Macrophage inflammatory protein
MLH	DNA mismatch repair protein
MLL	Myeloid/lymphoid or mixed-lineage leukemia
MMEJ	Microhomology-mediated end joining
MMP7	Matrix metalloproteinase-7
MS	Mass spectrometry
MSH	mutS homolog
mTORC	Mammalian target of rapamycin
MT3	Metallothionein
MUC	Mucin
MXI	MAX-interacting protein
MYC	Avian myelocytomatosis viral oncogene
MYD88	Myeloid differentiation primary response gene
MYH11	Myosin, heavy chain
MYLK	Myosin light chain kinase
NADH	Nicotinamide adenine dinucleotide
NANOG	A transcription factor critical to self-renewal of stem cells
NAV	Neuron navigator
NF	Neurofibromatosis
NFE2L	Nuclear factor, erythroid 2-like
NFκB	Nuclear factor of κ light polypeptide gene enhancer
NGF	Nerve growth factor
NHE	Na^+-H^+ exchanger
NMR	Nuclear magnetic resonance
NOXA	Phorbol-12-myristate-13-acetate-induced protein
NQO1	NAD(P)H dehydrogenase, quinone
NOTCH	Notch protein
NOX	Cell proliferation-inducing gene
NRAS	Neuroblastoma RAS
NUP160	Nucleoporin 160 kDa
OGT	O-linked N-acetylglucosamine
OR	Olfactory receptor
OCT	POU class 5 homeobox

OX40	A member of TNFR superfamily of receptors
P53	Tumor protein 53
PAI	Plasminogen activator inhibitor
PAK1	P21 protein (CDC42/RAC)-activated kinase
PAPPA	Pregnancy-associated plasma protein
PAR	Protease activated receptor
PAX5	Paired box protein
PCDH	Proto-cadherin
PDGFR	Platelet-derived growth factor receptor
PDK	*Phosphoinositide-dependent kinase*
PGE2	Prostaglandin E2
PGM3	Phosphoacetyl glucosamine mutase
PI3K	Phosphoinositide-3-kinase, regulatory subunit
PIK3CA	Phosphatidylinositol 4,5-bisphosphate 3-kinase
PKB	V-AKT murine thymoma viral oncogene homolog
PKC	Protein kinase
PKM2	Pyruvate kinase isozymes M2
POF1B	Premature ovarian failure 1β
PPA	Inorganic pyrophosphatase
PRDX	Peroxiredoxin
PRKCE	Protein kinase C epsilon type
PRK	Phosphoribulokinase
PRLTS	Platelet-derived growth factor receptor-like
PRRG1	Proline rich Gla
PSCA	Prostate stem cell antigen
PTEN	Phosphatase and tensin homolog
PTTG1	Pituitary tumor-transforming gene-1
PUMA	P53 up-regulated modulator of apoptosis
RAC	RAS-related C3 botulinum toxin
RAF	V-RAF-1 murine leukemia viral
RAS	The same as KRAS
RB	Retinoblastoma
RBM5	RNA-binding motif protein
REEP	Receptor accessory protein
RET	Proto-oncogene
RFE	Recursive feature elimination
RHAMM	Hyaluronic acid-mediated motility receptor
RHOA	RAS homolog gene
RIP	Receptor-interacting serine/threonine protein
RNS	Reactive nitrogen species
ROS	Reactive oxygen species
RPL22	Ribosomal protein
RTKN	Rhotekin
RUNX1	Runt-related transcription factor

S100	Calcium-binding protein
SATB	Special AT-rich sequence-binding protein
SETD	SET domain containing
SH2B3	Signal transduction protein link
SLC5	A glutamate transporter
SMAC	Second mitochondrial activator of caspases
SMAD4	Deleted in pancreatic carcinoma locus
SNP	Single-nucleotide polymorphism
SNRP	Small nuclear ribonucleoprotein polypeptide
SOD	Superoxide dismutase
SOX9	SRY (sex determining region Y)-box
SP	Specificity protein
SPI	Transcription factor PU
SRB1	Scavenger receptor B
STAT	Signal transducer and activator of transcription
SUZ12	Polycomb repressive complex
SVM	Support vector machines
TAL	T-cell acute lymphocytic leukemia protein
TAM	Tumor-associated macrophages
TCA cycle	Tricarboxylic acid cycle
TCF7L	Transcription factor like
TCGA	The Cancer Genome Atlas
TERT	Telomerase
TGF	Transforming growth factor
TGFβR	Transforming growth factor β receptor
TIMP	Tissue inhibitor of metalloproteinase
TLR	Toll-like receptor
TMED6	Transmembrane emp24 protein transport domain
TNF	Transforming necrosis factor
TNF	Transforming necrosis factor receptor
TRADD	Tumor necrosis factor receptor type 1-associated death domain
TRAF	Tumor necrosis factor receptor-associated factor
TRAIL	Transforming necrosis factor-related apoptosis-inducing ligand
TRIP	Thyroid hormone receptor interactor
TRK	Tyrosine receptor kinase
TSP	Thrombospondin
TTN	Titin
TXNL	Thioredoxin-like
UAP1	UDP-N acetylglucosamine pyrophosphorylase
UBFD	Ubiquitin family domain
UGDH	UDP-glucose dehydrogenase
UGP	UDP-glucose pyrophosphorylase
ULK	Unc-51 like autophagy activating kinase
UPA	Urokinase

VEGF	Vascular endothelial growth factor
VEGFR	Vascular endothelial growth factor receptor
VHL	Von Hippel-Lindau tumor suppressor
VLDLR	Very low density lipoprotein receptor
WISP	WNT-inducible signaling protein
WNT	Wingless-type MMTV integration site family
WT	Wilms tumor
ZNF367	Zinc finger protein 367 involved in transcriptional activation of erythroid genes

Chapter 1
Basic Cancer Biology

1.1 Overview of Cancer

Cancer has been recognized since early times, but treatment protocols and medications have lagged, by millennia, the initial observations of the disease. The tragic cases of childhood and teenage cancer notwithstanding, most cancers develop in the aging population, consistent with the nature of metabolic, genetic and other alterations discussed below and in various chapters. Epidemiological data show that, behind heart disease, cancer is the second leading cause of death worldwide, and many expect that in time cancer will overtake heart disease as the leading cause of mortality. Some 150 years ago it was demonstrated that cancer is composed of cells with morphology differing from that of normal cells. With information becoming available from numerous areas in biology and medicine, and capitalizing on major advances in technology, great strides were made in the twentieth century in unraveling many of the complexities of cancer, work that is continuing at an accelerating pace in the twenty-first century. It is now recognized that by far the majority of all cancers arises from environmental factors, metabolic disturbances, somatic mutations, and other pathophysiological processes (discussed throughout the book), while the remaining ones are attributable to germline mutations and are thus inheritable (familial).

In the early development of vertebrates, the embryonic stem cells undergo differentiation into the three primary cell layers, ectoderm, endoderm, and mesoderm. These, in turn, differentiate to give the 200-plus cell types of the human body comprising the myriad organs and supporting structures. The tissues can be categorized into four main groups, the epithelium, mesenchyme, nervous system and reticuloendothelial system, which in time can become subject to the development of cancer. It is also believed that normal cells throughout the body are continually in the process of undergoing changes that can result in cancer; fortunately, these events are spread over many years. From this it follows that, while one may die from cancer, individuals will often die from other causes before the cancer develops sufficiently to cause death. Clearly, the changes alluded to, as well as their rate of formation, depend on

© Springer Science+Business Media New York 2014
Y. Xu et al., *Cancer Bioinformatics*, DOI 10.1007/978-1-4939-1381-7_1

many variables such as genetic background, diet, environmental factors, etc. With tobacco smoke as the best documented example, one can convincingly argue for the importance of one's lifestyle in enhancing or diminishing the possible development of cancer.

Cancer has been considered by many investigators as a genetic disease, generally involving sequential random mutations and epigenetic changes. There is, however, now a school of thought being actively pursued by many scientists that the origins of cancer lie in cellular and micro-environmental perturbations that, in turn, can lead to genetic alterations or selection of such alterations. Indeed, cancer is now recognized as a very heterogeneous disease, even within the same type of cancer, and it may emerge that its origins can be attributable to a number of causes.

As discussed below and throughout the book, there are many metabolic/cellular micro-environmental disturbances and combinations of genomic alterations that can lead to cell transformation. Once established, or when being established, many other mutations accumulate in the tumor cells, each giving rise to clonal expansion. Regardless of the initiating cause(s) of cancer, there will be in time genetic alterations, e.g., mutations, amplifications, deletions and translocations, that facilitate growth, inhibit apoptosis (programmed cell death) and escape from immune destruction. The cells harboring metabolic alterations, micro-environmental changes and mutations that provide a growth advantage and best meet the other requirements for continued tumor survival will prevail, and the processes of natural selection and survival of the fittest and most adaptable become crucial for these cells. Thus, while Darwinian principles were originally proposed to explain the evolution of organisms, a similar rationale appears to underlie tumor progression. These events may lead to cellular heterogeneity, particularly since new mutations can arise due to loss-of-function of negative cell cycle regulators such as *P53* and perhaps even by gain-of-function of positive cell cycle regulators such as *RAS*, leading to persistent cell division and a statistical chance of errors in replication.

The following quotation (Eifert and Powers 2012) nicely summarizes the current thinking on the genetic component and alludes to the challenges ahead. "*Diversity and complexity are hallmarks of cancer genomes. Even cancers that arise from the same cell type can harbor a range of different genetic alterations that facilitate their unrestrained expansion and eventual metastasis. As a result, the behaviour of individual tumours—how they progress and eventually respond to therapy—can be varied and difficult to predict.*" Cancer development, survival and growth are, however, also heavily influenced, if not caused, by many of the aberrations in cancer metabolism and the microenvironment in which the tumor is located. Indeed, as alluded to above and discussed later in this book, some of these non-genetic alterations may become driving forces for the possible formation and/or survival of cancer. Another quotation is germane to a more holistic perspective of cancer (Nakajima and Van Houten 2013). "*The tumor must be recognized as an evolving ecosystem, adapting constantly to oxygen and nutrient availability*".

Large scale cancer genome sequencing is occurring at a rapid pace, and already the data are showing the extraordinary genomic complexity of tumors. It is common to find thousands, tens of thousands, or even hundreds of thousands of mutations and other genetic changes in a typical epithelial tumor. A working hypothesis was that only a limited number of the genetic alterations are necessary to initiate and/or propagate tumor formation in a single cell and that this genetically altered cell undergoes clonal expansion with increasing genetic changes. The few early key alterations are said to be "driver mutations" that confer a growth and survival advantage, in effect leading to the conversion of a normal cell, or one that is on the road to transformation from non-genetic causes, to one that is transformed and capable of sustained growth. The multitude of additional mutations are denoted as "passenger mutations" that are not required for tumor growth or survival. As discussed later, the driver mutations, at least for certain cancers, may occur sequentially, but whether there is any order to the process, whether there are many genes that can participate and how the genetic changes relate to phenotypic changes are not known (Ashworth et al. 2011).

The remainder of this chapter is focused on a succinct review of some of the aspects of cancer that are deemed important in its formation and growth. These sections will set the stage for the chapters dealing with *omics*-based cancer studies elsewhere in this book.

1.2 Hallmarks of Cancer

In 2000 Hanahan and Weinberg (2000) proposed six hallmarks of cancer to provide a framework for a better understanding of the basic molecular and cellular principles responsible for the development and maintenance of neoplasia, hallmarks that were extended in 2011 to a total of eight (Hanahan and Weinberg 2011). It is worthwhile to briefly review these hallmarks since they offer a rational understanding of the necessary changes that are required of normal cells to make the transition to a state of perpetual growth and survival. Suffice it to mention at this point that most of the following alterations can be attributed to one or a combination of the following: metabolic changes, hypoxia, extracellular matrix (ECM) alterations, epigenomic changes or somatic mutations, including chromosomal rearrangements, of key players in or regulators of the growth promoting or cell cycle pathways.

1.2.1 Sustained Proliferative Signaling

Unlike normal cells that tightly regulate their cell division, transformed cells have the ability to perpetuate growth-promoting signals and become refractory to growth-inhibiting processes. A variety of molecular mechanisms can contribute to sustained signaling for cell division, including the following examples: hyaluronic acid fragments (see Chap. 6), a constant supply of growth-promoting signals originally designed for tissue repair, constitutively activated (gain-of-function) growth factor receptors, a constitutively activated component of the cellular pathway for cell division, and the constitutive inactivation (loss-of-function) of growth-inhibiting components of the pathway for cell division A.

1.2.2   Evasion of Growth Suppressors

There are a number of negative regulators of the cell cycle, e.g., *RB* (retinoblastoma) and *P53* (tumor protein of 53 kDa) being two of the best known and studied, that must be overcome or evaded to ensure continued division of the aberrant cells. These two so-called tumor suppressors function in large part in responding to extra-cellular and intracellular signals, respectively. These important suppressors of growth are part of larger complex networks that in some manner serve to introduce redundancy in the regulation. In this vein, it should be mentioned that the ECM is important in modulating the balance of growth factors and growth suppressors. For example, when the ECM is altered from a highly elastic state to a one that is stiffer, the efficaciousness of growth factors can increase by 100-fold (see Chap. 4).

1.2.3   Resisting Cell Death

The cellular process of apoptosis (cell death or cell suicide) serves to rid the body of damaged or aged cells and is a powerful barrier to the development of cancer. *BAX* and *BAK* are two important mitochondrial membrane proteins that act to begin the process of apoptosis by disrupting the mitochondrial membrane and thus releasing cytochrome c; this in turn leads to the activation of caspases, a family of proteases key in releasing the apoptotic effectors. In opposition to this pathway are anti-apoptotic members of the *BCL2* family of proteins such as *BCL2*, *BCLB* and *MCL1*. Tumor cells have developed several mechanisms for overcoming the apoptotic pathway including the loss of *P53* function (a common alteration in cancer cells) and others that are actively being studied.

1.2.4   Enabling Replicative Immortality

Located on the ends of chromosomes, telomeres, composed of hexanucleotide repeats, are shortened as cells undergo progressive divisions. In time, after multiple divisions the telomeres become sufficiently shortened that cells are no longer viable, leading to senescence and eventual cell death. This seems to be the major reason that non-immortalized cells have a finite number of divisions and thus a finite life span. Telomerase is the enzyme responsible for adding these protective repeat segments of DNA to chromosomes, but it is present at progressively lower levels as cells divide. In contrast, cancer cells maintain relatively high levels of telomerase, thus ensuring that telomere shortening is minimized. In addition to the maintenance of telomere length, telomerase is now believed to also have other cellular functions related to growth.

1.2.5 Activation of Invasion and Metastasis

Carcinoma, the most common form of cancer and the main focus of this book, arise from epithelial cells that are engaged with neighboring cells and with the ECM. The protein E-cadherin is a well characterized cell-cell adhesion molecule, while interactions between cells and the ECM are regulated by other proteins (see Chap. 10). The processes of invasion and metastasis require several steps. First, the transformed cells must become disengaged from their interactions with other cells and with the ECM. This involves down-regulation of E-cadherin accompanied by metalloproteinases and cysteine cathepsin proteases, many of these being supplied by immune cells near the primary tumor. In addition, stromal cells neighboring the tumor, in response to signals from the cancer cells, secrete proteins facilitating invasiveness. This set of events is termed the *epithelial-mesenchymal transition* and also includes the ability of cancer cells to inhibit apoptosis. Second, the now loosely attached transformed cells undergo intravasation into blood and lymphatic vessels in their vicinity. Third, colonization to a distant site(s) then requires successful travel via the blood or lymph followed by the process of extravasation. Finally, growth of the cancer cell(s) at the new site completes the process of metastasis. Each of these processes requires many alterations in cell function that are systematically being investigated (see Chaps. 10 and 11).

1.2.6 Induction of Angiogenesis

The high energy requirements of tumors, both primary and secondary, necessitate a good blood supply for continuing availability of oxygen, nutrients and precursors for fuel-generating metabolic pathways. Angiogenesis refers to the sprouting of new blood vessels from existing ones, i.e. those produced during embryogenesis. This process is regulated by the protein, vascular endothelial growth factor-A (*VEGFA*), which acts through tyrosine kinase receptors to ensure the continued biosynthesis of new vessels. Except in a few physiological and pathological states, e.g. cancer, angiogenesis is quiescent in the adult, being inhibited in large part by thrombospondin-1.

1.2.7 Evasion of Immune Destruction

During evolution humans have developed a most sophisticated immune system, often discussed in two categories, the innate and the adaptive. The immune system is believed to be highly effective in protecting the body from the growth of transformed cells, both virally and non-virally induced. From this argument, one can argue that the cancers that do emerge have, in some manner, escaped immune surveillance or have developed the ability to counter an immune attack, particularly from T helper cells and natural killer cells, as discussed in details in Chap. 8.

1.2.8   Reprogramming Energy Metabolism

In the 1920s Otto Warburg reported that cancer cells increase their rate of glycolysis many fold over that of non-cancer cells. This reprogramming event occurs even in the presence of an ample supply of oxygen that would normally dictate that the end-product of glycolysis, pyruvate, would be converted to acetyl-CoA that in turn would enter the tricarboxylic acid (TCA) cycle (also known as the citric acid or Krebs cycle), eventually accounting for the conversion of oxygen to carbon dioxide and the generation of ATP. The putative regulatory factors responsible for this altered course of glucose metabolism will be discussed later, one hypothesis to account for the Warburg effect being that intermediates in glycolysis can be shuttled into other metabolic pathways for the biosynthesis of amino acids and nucleosides, required components for protein and nucleic acid synthesis, respectively. The important role of the glucosaminoglycan, hyaluronic acid, cannot be overlooked in cancer metabolism. This topic is briefly mentioned in Sect. 1.10 below and greatly elaborated on in Chap. 6.

1.2.9   Other Considerations

In addition to these delineated eight hallmarks of cancer, Hanahan and Weinberg also discussed processes defined as enabling characteristics of cancer: (a) genome instability and mutation, and (b) tumor-promoting inflammation. They concluded that the reduced cellular efficiency in genome maintenance and repair ultimately increases the rate of developing viable phenotypes of the cancer cells. The presence of immune cells in tumors prompted studies into their possible functions. Tantalizing results show, paradoxically, that the immune cells, normally charged with protecting the body, can aid tumor growth by secreting growth factors, pro-angiogenic factors, survival factors, and others that contribute positively to the survivability and growth of the tumor as discussed in detail in Chap. 7.

1.3   Proto-oncogenes, Oncogenes and Tumor Suppressor Genes

As discussed earlier, the cancer genome tends to contain numerous mutations and genomic rearrangements, but a central question is: *Are these causal for cancer or important for cancer growth and survival?* The introduction of the concept of an oncogene in the 1960s clearly represented a major breakthrough in defining an intellectual framework for studying cancer. It has provided useful guiding information in elucidating cancer mechanisms, particularly cancer drivers. However, this well-accepted concept seems, unfortunately, to have also restricted the thinking of cancer researchers somewhat since it requires that an oncogene must be the mutated or overexpressed form of a proto-oncogene, which is defined as genes involved in cell growth and differentiation. Originally attributed as being responsible for the origin of cancer, recent thinking by many has shifted the role of oncogenes from that of the originator to genetic alterations that arise during cancer evolution and selection of mutations that permit continued proliferation and survival.

1.3.1 The Rous Sarcoma Virus

The story begins with the elucidation of an avian retroviral oncogene prompted by the studies of Peyton Rous in the early 1900s at the Rockefeller Institute (now the Rockefeller University) in New York City. Interested in avian cancer, Rous was given a chicken harboring a sarcoma by an upstate chicken farmer who had read of his research at Rockefeller. Rous excised the tumor, then ground and filtered it to remove the cartilaginous residue. He found that upon injecting the soluble filtrate into certain strains of tumor-free chickens a sarcoma would develop. This represented a major breakthrough, demonstrating for the first time that this form of cancer was transmissible in chickens.

1.3.2 Proto-oncogenes and Oncogenes

Following many years of intense research by numerous investigators, the transmissible agent was identified as the (appropriately named) Rous sarcoma virus (*RSV*). Of interest to us in this section was the recognition that the oncogenic element in the retroviral genome was a mutated version of a highly conserved and essential gene in human cells, *SRC*. This gene encodes a tyrosine kinase that functions in a cellular growth pathway; the mutation of the gene in the retroviral genome renders the gene product constitutively active, thus the explanation for tumorigenicity in infected chickens. It appears that during a cycle of infection some time ago, *RSV* commandeered the normal cellular *SRC* gene, i.e. a proto-oncogene (also referred to as a cellular oncogene), from the infected bird and incorporated it into its genome. A subsequent mutation in the *SRC* gene was sufficient to render the protein constitutively active such that proliferation signaling occurred in the absence of proper growth signals. The mutation was responsible for the conversion of the proto-oncogene to an oncogene. To date, over 30 retroviral oncogenes have been identified, most of them being in rodent and avian viruses (Vogt 2012). [*N.B. While we do not know the exact constituents of the filtrate that were injected into the chickens, it surely contained some macromolecular constituents and probably cells associated with the sarcoma. Later studies demonstrated, however, that it was the presence of the viral SRC gene that produced the tumorigenicity.*]

In normal cellular function, the gene encoding almost any regulatory protein involved in cell growth or survival can undergo the proto-oncogene-to-oncogene conversion by certain mutations or amplification that result in constitutive activity promoting, say, cell division without the requirement of external or even internal growth signals. Even the growth factors or their receptors can be considered oncogenic if, for example, there are mutations in their genes that increase their expression. In addition to mutations, genomic rearrangements can also produce oncogenes if the proto-oncogene is translocated to make a fusion gene that is no longer regulated and possibly gives a constitutively active fusion protein. This is the basis of the Philadelphia chromosome that is responsible for many cases of chronic myelogenous leukemia (CML). With this definition one can consider hundreds of proto-oncogenes that have the potential to become oncogenes. Suffice it to say that the proto-oncogene-to-oncogene conversion does not necessarily lead to cancer; rather, it can be considered a signal.

1.3.3   Conversion of Proto-oncogenes to Oncogenes

There are several genetic alterations that convert proto-oncogenes to oncogenes, most of which were mentioned or alluded to above.

Mutations, often single base changes (point mutations), leading to gain-of-function of positive regulators of the cell cycle, e.g., growth factor receptors or *SRC*.

Chromosomal instability such as loss of portions of chromosomes or rearrangements, e.g., inversions, translocations, deletions and insertions, resulting in a gain-of-function of positive regulators. An example of such a translocation is the fusion of the *ABL* gene on chromosome 9 to chromosome 22 where it is fused to the *BCR* gene yielding a fusion protein of *BCR-ABL* where *ABL*, normally highly regulated, exhibits constitutive activity.

Gene amplification resulting in abnormally high expression of growth factor receptors (or growth factors) that function in a pathway leading to cell division, e.g. the *HER2* receptor in breast cancer.

Viral infection/insertion may also contribute to some forms of cancer, e.g., the human papilloma virus (HPV) and cervical cancer.

1.3.4   Tumor Suppressor Genes

We now turn our attention to the topic of tumor suppressor genes. These genes and their protein products refer to ones that function to prevent the progression of the cell cycle if conditions at some checkpoints are not met, e.g., DNA damage is detected and not repaired. For a tumor suppressor gene to lose its function, it

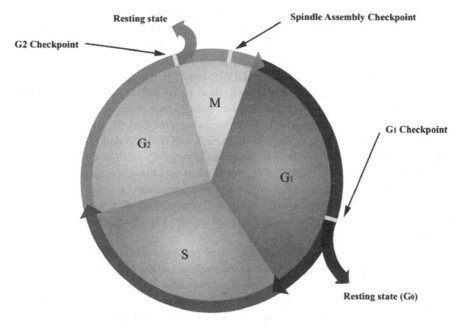

Fig. 1.1 A schematic view of the cell cycle showing the resting state (G_o), the first gap phase (G_1), the synthesis phase (S), the second gap phase (G_2) and the phase where mitosis occurs (M). A complete cycle can require some 18–24 h, although some cancer cells complete the cell cycle in less time

requires the loss of both copies of the gene while the loss of one copy increases the risk of cancer development. Within this class are the familiar *BRAC1* and *BRAC2* genes involved in a familial form of breast and ovarian cancers, and the *APC* (adenomatous polyposis coli) gene responsible for most cases of familial colorectal cancer. Of the many other tumor suppressor genes in the human genome, we will next discuss two well-studied examples that function in the cell cycle (Fig. 1.1).

Some cells in the body are dividing frequently, but most are in a resting or quiescent state denoted as G_o. A signal for cell division, such as growth factors initiating an intracellular signaling cascade or even the presence of a constitutively active oncogene in the signaling pathway, begins a process that takes the cells from the quiescent state to the first gap phase (G_1). Cyclin D family members are expressed and the proteins interact with cyclin-dependent kinase (*CDK*) complexes. The retinoblastoma (*RB*) gene, a tumor suppressor gene, encodes a nuclear protein that is a negative regulator of cell division, constantly maintaining cells in G_1 provided it is associated with another nuclear protein, the transcription factor *E2F*. The action of the *CDK* complex is to hyperphosphorylate *RB*, leading to dissociation of the *RB-E2F* complex. Freed of the inhibitory effects of *RB*, *E2F* acts to up-regulate itself, another cyclin, and enzymes required to carry out replication of the genomic DNA. These events, along with others, will lead to the progression from G_1 into the

S or synthesis phase of the cell cycle where DNA synthesis occurs. This progression is limited, however, by another protein *P53* (or TP53, tumor protein 53) that oversees DNA fidelity, along with other roles to be discussed later. Among its many actions *P53* can induce the activation of genes for DNA repair, cause cell cycle arrest or send the cell into apoptosis if DNA repair is not successful. Successful entry into the S phase results in the replication of ~3 billion base pairs of DNA accomplished with a variety of enzymes including ones capable of proofreading and repair of errors. As summarized, estimates suggest that the mutation rate in cells is some 10^{-12} to 10^{-9} per nucleotide in each cell division and that, of the 10^{14} cells comprising the average human, there are about 10^{16} division cycles during a lifetime (Duesberg 1987; Loeb 1989). Needless to say, these estimates are dependent on many factors and assumptions, and a wide range can be found in the literature. Although not discussed here, there are checkpoints and inhibitors at the $S \rightarrow G_2$ and the $G_2 \rightarrow M$ boundaries (G_2 is the second gap phase and M is mitosis, i.e., cell division).

RB and *P53* are, in effect, gatekeepers that prevent cells from dividing unless signaling pathways for growth impact on the nucleus and the cells have high fidelity, e.g., no damaged DNA. These negative regulators of cell division are thus critical components of cell oversight. From this perspective it is not surprising that mutations interfering with *RB* and *P53* functions could allow continual cell division and transcription of faulty DNA. Suffice it to say that the *RB* and/or *P53* genes are frequently mutated in various cancers.

1.4 Emerging Results on Cancer Genomes, Tumor Heterogeneity and Cancer Evolution

The emergence of rapid deep sequencing technology has provided an unprecedented opportunity to sequence large numbers of cancers for comparison with DNA sequences obtained from normal controls. In an interesting twist of fate, DNA sequencing is no longer the rate-limiting step in cancer genomics. Rather, it is the ability to analyze the copious amounts of data that are forthcoming from many laboratories and factory-like sequencing centers. From this perspective, the timing is good for bioinformaticians to enter cancer research with the possibility of adding substantively to our knowledge relating genomic changes to phenotypic changes in cancer patients.

1.4.1 What Is Being Learned?

The results from cancer genome sequencing are providing considerable information on mutations and the myriad other genomic changes, e.g., chromosomal gains, losses and rearrangements, present in most cancers (Stratton et al. 2009;

Pleasance et al. 2010; Garraway and Lander 2013; Alexandrov and Stratton 2014; The Cancer Genome Atlas Research Network 2011a, b, 2012a, b, c, 2013a, b, c, 2014; Alexandrov et al. 2013; Kandoth et al. 2013; Vogelstein et al. 2013). In one study 3,281 tumors from 12 different types of cancer (11 solid tumors plus acute myeloid leukemia) were analyzed for point mutations and small insertions and deletions (Kandoth et al. 2013). In this sampling 617,354 somatic mutations were identified: 398,750 missense; 145,488 silent and smaller numbers each of non-sense, splice site, non-coding RNA, non-stop read-through, frame-shift insertions/deletions (indels) and in-frame indels. *P53* was found to be the one most frequently mutated, and the lipid kinase gene, *PIK3CA* (phosphatidylinositol-4,5-bisphosphate 3-kinase, catalytic subunit alpha), was the second. Not surprisingly, many mutations appeared in genes encoding transcription factors; cell cycle regulators; signaling pathways, including receptor tyrosine kinase, *MAPK, PI3K, TGFβ* and *WNT*/β-catenin; and ECM related genes as detailed in Chap. 4.

Genome sequencing has also provided some surprising observations, but ones that are consistent with the emerging view that cancer is not just a disease of the genome. Sequencing studies by two labs (Mack et al. 2014; Parker et al. 2014) on three subtypes of ependymoma brain tumor found the following. One subtype had an intrachromosomal translocation yielding what appears to be a 'driver mutation' for cancer, and another subtype had abnormal epigenetic alterations. Of particular interest, however, was the finding that another subtype was devoid of gene mutations and aberrant epigenetic changes. These results emphasize the complexity of cancer and importantly the role of non-genomic changes driving cancer formation.

1.4.2 Driver and Passenger Mutations

In recent years there has been considerable interest in identifying the 'driver' mutations and separating them from the 'passenger' mutations. Although as discussed later (Chap. 5), there is a current movement to consider those crucial mutations as ones that were selected as necessary to maintain proliferation and survival of the developing cancer cell(s) and may not necessarily be causal to cancer. This information will of course direct many of the treatment modalities for specific cancers. Many mutations, particularly in older individuals, are known to exist before the occurrence of cancer and are believed to have nothing to do with the onset or continuation of cancer (Tomasetti et al. 2013). These innocuous mutations arise from the high number of cell divisions and the inherent errors that occur in proofreading and repair, as well as mutations from environmental causes that do not produce 'drivers' of cancer. One estimate is that there are about 140 genes that, with appropriate mutations, can become drivers (Vogelstein et al. 2013).

1.4.3   Major Findings

Some of the major findings from genomic sequencing of various tumors have been delineated and provide much insight into, if not cancer initiation, then at least its progression (Vogelstein et al. 2013). Some of these principles are listed below; however, we have qualified them as more likely being responsible, through natural selection, for unlimited growth and survival, not necessarily causal.

a. Solid tumors have an average of 33–66 somatic nonsynonymous mutations, predominantly single-base changes that are expected to alter the resulting proteins; however, a limited number of mutations are capable of sustaining cancer proliferation and survival. [*N.B. Vogelstein et al. (2013) claim that the majority of human cancers result from two to eight sequential mutations occurring over 20–30 years, each of which confers about a 0.4 % growth advantage.*]
b. There are about 140 such genes that if mutated can contribute to cancer, either via initiation, proliferation or survival.
c. Three cellular processes are regulated by these essential genes: cell fate determination, cell survival and genome maintenance.
d. Although the pathways altered by key mutations in different tumors are similar, each individual tumor is distinct.
e. Heterogeneity exists in the cells of tumors, and this can affect therapeutic effectiveness.

1.4.4   Metastasis

There is considerable interest in the delineation of the various changes that can drive a primary tumor to metastasis (see Chaps. 10 and 11), the cause of over 90 % of cancer mortality (Irmisch and Huelsken 2013). This is an important aspect of cancer research that bioinformatics can address when more data are available from the ongoing sequencing of cancer genomes and transcriptomes. In addition to the genetic changes referred to above, many alterations in metabolism, hypoxia, and other cellular processes exist that tend to drive cancer growth. These are covered in depth in Chap. 10.

1.4.5   Cancer Heterogeneity

It is important that the role of cancer heterogeneity be pointed out. Pathologists and clinicians have known for years that solid tumors are heterogeneous with regard to cellular morphology and patient responses to treatment. Thus, cancer heterogeneity, first proposed several decades ago (Nowell 1976), is an important aspect of cancer and an area that is being addressed (Meacham and Morrison 2013;

Burrell et al. 2013; Vogelstein et al. 2013). It is now appreciated that considerable heterogeneity exists in any given cancer, both at the molecular and cellular level. Cancer is clearly many diseases, and even individual tumors within similar types of cancer may be unique. A given tumor is likely composed of a dominant clone and several subclones, each of which may grow at different rates and respond differently to treatment(s). This intratumor heterogeneity impacts on the evolution of cancer and the natural selection of clones more favorable for sustained growth, survival and ability to colonize distant sites (extravasation and metastasis). Cancer heterogeneity, evolution and natural selection are emerging as significant features in our understanding of cancer growth and control (Klein 2013; Burrell and Swanton 2014; Lawrence et al. 2013) and are areas in which bioinformatics can provide considerable insight as more data become available.

1.5   An Early Sequential Model of Cancer Development

One of the early models to explain the development of cancer strictly from genetic changes is referred to as a "sequential model" based on a series of mutations. It is now well recognized that this model may be overly simplistic, but it is presented to introduce the concept and several genes that, when mutated, can function to aid propagation of cancer. Based on extensive studies of benign and malignant colorectal cancer (hereafter referred to simply as colon cancer), Fearson and Vogelstein proposed a sequential pathway for the development of malignancy, a pathway often referred to as the canonical pathway (Fearson and Vogelstein 1990). Colon cancer can be categorized into two forms, sporadic and familial, having respective frequencies of about 80 and 20 %. Sporadic colon cancer can be further divided into a form arising from mutations and/or chromosomal instability and a form attributable to microsatellite instability. These two forms of sporadic colon cancer exhibit frequencies of approximately 80–85 % and 15–20 %, respectively.

The canonical pathway was proposed to arise from a sequential or linear set of genetic alterations. In the majority of cases the *APC* gene, located on chromosome 5q, was found to undergo a mutation that reduced or abolished its activity and contributed to the formation of a benign lesion or early adenoma. While the protein encoded by *APC* has a number of biological actions including its role in the *WNT* pathway, cell adhesion (via E-cadherin), mitosis and cytoskeletal regulation, it is the former that has attracted most attention. Forming part of a complex with glycogen synthase kinase-3β and axin, loss of *APC* activity results in β-catenin escaping degradation and thus constitutively activating the *WNT* pathway that regulates numerous genes, some of which are involved in the cell cycle. We mention in passing that mutations in *APC* have been identified in many if not most cases of familial colon cancer.

Mutations to the oncogene *KRAS* on chromosome 12p12.1 have been frequently identified in intermediate adenomas, mutations that result in producing a defective GTP-binding protein that is involved in the mitogen-activated protein kinase

(*MAPK*) and other growth-promoting pathways. Again, constitutive activation of the protein *KRAS*, e.g., by inactivating the intrinsic GTPase activity that converts GTP to GDP leading to *KRAS* inactivation, results in a constant enhancement of growth-promoting pathways as well as a loss of cell polarity that could reduce cell adhesion.

Other allelic losses (via mutations or chromosomal loss) are often found in late adenomas and carcinomas with the *P53* and *DCC* (deleted in colorectal cancer) genes on chromosomes 17p13.1 and 18q21.1, respectively. *P53* is considered a tumor suppressor gene and serves as a gatekeeper for cells exiting G_1 to the S phase of the cell cycle. It promotes repair of damaged DNA, e.g. from errors in replication or environmental stress, and if repair is not successful terminates cell cycle progression and leads to apoptosis. Clearly, inactivation of this key cell cycle regulator could have major detrimental effects on the normal fidelity expected in the cell cycle. The *DCC* protein is a transmembrane protein that serves as a receptor for proteins involved in regulating axon guidance in the nervous system and also seems to participate in cell motility, signaling and overcoming apoptosis.

Although attractive in its simplicity, the sequential model is now believed by many investigators to function more as developing cancer cells are undergoing an "evolution and natural selection" phase to obtain a genomic background that perpetuates growth and evades apoptosis and immune destruction.

1.6 Epigenetics and Cancer

Most of the research on cancer has heretofore dealt with the role of genetic changes, i.e. alterations in the sequence of DNA, that lead to changes in normal cellular functions that regulate proliferation, survival, angiogenesis, metastasis and others. Recent studies have, however, documented that epigenetic changes are also important in the initiation and progression of cancer (Beck et al. 2012; Shen and Laird 2013; Timp and Feinberg 2013; Waldman and Schneider 2013; Suva et al. 2013). Such changes are attributable to modifications of chromatin and chromatin packaging, as emphasized by the appearance of mutations in genes involved in DNA methylation, histone modification and chromatin remodeling, with a number of mutations found to be tumor-specific.

Composed of nucleic acids and proteins, there are potentially many possibilities for epigenomic changes in chromatin. The protein core around which genomic DNA is wrapped is composed of a histone octamer with two copies each of four distinct histones, forming a nucleosome; these, in turn, form a helical arrangement. As summarized (Shen and Laird 2013), there are multiple sites for alterations that control the level of transcriptional activity, including DNA methylation, histone modifications and variants, interacting proteins, noncoding RNAs and nucleosome positioning.

Certainly one of the more prevalent alterations is that of DNA methylation, catalyzed and maintained by several DNA methyltransferases yielding primarily cytosine-5 methylation of CpG dinucleotides. Such enzymes can be considered as 'writers' since they in effect make the epigenetic mark. Another chemical change is that of histone modification, including methylation, acetylation and phosphorylation. The histone modifications are catalyzed by histone methyltransferases and demethylases, acetyltransferases and deacetylases, and kinases and phosphorylases. Those enzymes that remove the covalent tag are referred to as 'erasers'. Also, numerous histone variants have been identified. Another mode by which the epigenome can be altered is that of nucleosome positioning and remodeling, accomplished by sequence-specific binding proteins. These are important in selecting the form of chromatin, euchromatin (open form) and heterochromatin (a more closed form), thus enhancing or inhibiting the availability of readers, writers, erasers and other chromatin-binding proteins.

From this abbreviated overview, mentioning most but certainly not all of the factors responsible for defining and maintaining the epigenome, sequencing and functional studies have shown that mutations can occur in essentially all of the genes required to form and maintain the epigenome (Fullgrabe et al. 2011; Shen and Laird 2013; Timp and Feinberg 2013). Importantly, many of these mutations have been documented to be related, or at least correlated, at one level or another to tumorigenesis. This area will undoubtedly emerge as an important component of cancer as more results become available.

1.7 Cancer Cell Metabolism

1.7.1 Meeting the Energetic Requirement of Cells

Of the many types of foods ingested, the body uses three major classes as fuels for its energetic needs: carbohydrates, lipids (fats) and proteins. The chemical compositions and structures of these vastly different biomolecules, ranging from simple sugars to complex polysaccharides, fatty acids to triacylglycerols (triglycerides) and peptides to high molecular weight proteins. Yet, many of the different metabolic pathways converge at a common intermediate, acetyl-coenzyme A (acetyl-CoA) or a downstream intermediate, leading to the biosynthesis of adenosine triphosphate (ATP), an important source for cellular energetic needs.

The average sedentary adult requires about 2,000 Calories (Cal) per day to meet the normal requirements to maintain overall homeostasis, i.e., for heart, brain, lung, kidney and other organs to function. This daily requirement for any given sedentary individual can vary as much as ±400 Cal since it is influenced by age, gender and metabolic factors. Over 80 kg of ATP are required to meet this daily basal caloric need; however, the body contains only about 0.25 kg (Tymoczko et al. 2013). Thus, ATP is constantly being utilized and resynthesized to meet daily needs. For someone who is physically active, the caloric requirement rises dramatically, and consequently ATP biosynthesis must increase as well. [*N.B. The calorie, more specifically the gram or small calorie, is defined as the energy required to increase the temperature of 1 g of water 1 °C at standard atmospheric pressure (this corresponds to about 4.2 J). Biochemists and nutritionists, on the other hand, use a "large" or "kilogram" calorie, i.e. the Cal, that is equivalent to 1,000 "small" calories (about 4.2 kJ).*]

Metabolism of carbohydrates, proteins and lipids yields approximately 4 Cal/g, 4 Cal/g and 9 Cal/g, respectively. Thus, per unit weight ingested, lipids provide more than twice the Cal (or energy) than do carbohydrates and proteins; however, we rely on all three for energetic needs, particularly lipids and carbohydrates (and certainly circulating glucose for minute-to-minute cellular needs). Glucose can be metabolized anaerobically (absence of oxygen) to yield small amounts of ATP and aerobically (presence of oxygen) to obtain greater amounts. Fatty acids, most being obtained from lipolysis of triacylglycerols, can undergo β-oxidation, giving acetyl-CoA that can enter the TCA cycle for ATP production, and glycerol that can enter the hepatic gluconeogenic pathway and be converted to glucose for metabolism. Proteins are constantly turning over, and some of the amino acids so derived can serve as precursors for glucose synthesis (gluconeogenesis) or for synthesis of pyruvate or intermediates in the TCA cycle (see below).

Cancer cells also utilize carbohydrates, lipids and proteins to generate the ATP that is required to meet the energetic needs for proliferation, metastasis and survival. In order to appreciate the metabolic derangements unique to cancer, it is first necessary to understand, even if superficially, the pathways of normal metabolism, which are briefly treated in the following section with reference to the changes that occur in cancer.

1.7.2 Glucose Metabolism

Glucose metabolism will be discussed first since it is quite distinct in cancer cells compared to normal cells. For both normal and cancer cells, circulating glucose in the bloodstream enters cells via one or more glucose transporter proteins (*GLUTs*) and is then rapidly phosphorylated to glucose 6-phosphate by either of two ATP-dependent enzymes, hexokinase or glucokinase, thus ensuring retention within the cell (Reaction 1.1).

$$glucose + ATP \rightarrow glucose\,6 - phosphate + ADP + H^+ \tag{1.1}$$

There are three metabolic paths for glucose 6-phosphate within the cell. If the cell does not require ATP, glucose 6-phosphate can be metabolized to a high molecular weight polysaccharide of repeating glucose units, glycogen (Reaction 1.2). The three subsequent reactions are catalyzed by the enzymes phosphoglucomutase, UDP-glucose pyrophosphorylase and glycogen synthase, respectively, where UDP is uridine-diphosphate.

$$glucose\,6 - phosphate \rightarrow glucose\,1 - phosphate \rightarrow UDP - glucose \rightarrow glycogen \tag{1.2}$$

The two other metabolic paths for glucose 6-phosphate are of most interest here: the glycolytic pathway and the pentose phosphate pathway. Glycolysis represents a series of enzymatic reactions that convert the phosphorylated 6-carbon glucose into two molecules of pyruvate, a 3-carbon product. The pathway yields, in addition to two molecules of pyruvate, two molecules each of ATP, NADH and other reaction products for each molecule of glucose entering the pathway (Reaction 1.3 and Fig. 1.2).

$$\text{glucose} + 2ADP + 2NAD^+ + 2P_i \rightarrow 2\text{pyruvate} + 2ATP + 2NADH + 2H^+ + 2H_2O \quad (1.3)$$

Pyruvate can be converted to acetyl-CoA (Reaction 1.4), the main substrate for the TCA cycle, to lactate (Reaction 1.5) or to oxaloacetate (Reaction 1.6). The enzymes catalyzing these reactions are, respectively, pyruvate dehydrogenase, lactate dehydrogenase and pyruvate carboxylase.

$$\text{pyruvate} + CoA + NAD^+ \rightarrow \text{acetyl} - CoA + CO_2 + NADH + H^+ \quad (1.4)$$

$$\text{pyruvate} + NADH + H^+ \leftrightarrow \text{lactate} + NAD^+ \quad (1.5)$$

$$\text{pyruvate} + CO_2 + ATP + H_2O \rightarrow \text{oxaloacetate} + ADP + P_i + 2H^+ \quad (1.6)$$

The other metabolic route for glucose 6-phosphate is that of the pentose phosphate pathway (Fig. 1.3). This pathway consists of an oxidative phase in which glucose 6-phosphate is converted to ribulose 5-phosphate by several enzymes acting sequentially, glucose 6-phosphate dehydrogenase, lactonase and 6-phosphogluconate dehydrogenase (Reaction 1.7). This is an important reaction since it regenerates NADPH and associated reducing power. The second phase is a complex oxidative component consisting of a number of enzymes that yields ribose 5-phosphate (Reaction 1.8, catalyzed by phosphopentose isomerase), fructose 6-phosphate and glyceraldehyde 3-phosphate. Of the three pentose phosphates, ribose 5-phosphate (a 5-carbon sugar phosphate) is needed for the synthesis of nucleic acids, and the other two, fructose 6-phosphate and glyceraldehyde 3-phosphate, can enter as intermediates in the glycolytic pathway.

$$\text{Glucose } 6 - \text{phosphate} + 2NADP^+ + H_2O \rightarrow \text{ribulose } 5 - \text{phosphate} + 2NADPH \quad (1.7)$$
$$+ 2H^+ + CO_2$$

$$\text{Ribulose } 5 - \text{phosphate} \leftrightarrow \text{ribose } 5 - \text{phosphate} \quad (1.8)$$

From the point of entry of glucose into cells, the glycolytic pathway is composed of ten enzymatic reactions, all occurring in the cell cytoplasm, to give the 3-carbon product pyruvate. Pyruvate, in turn, can undergo one of several enzymatically catalyzed steps with its conversion to either of the following. (1) lactate: This reaction reduces pyruvate and occurs independent of the availability of oxygen; it is catalyzed

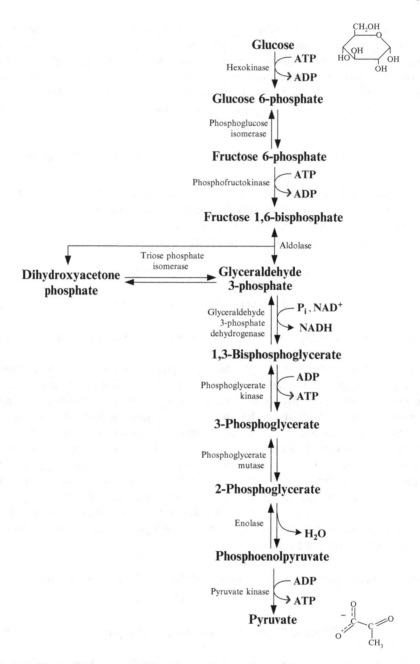

Fig. 1.2 The glycolytic pathway from the entry of glucose into a cell and the subsequent reactions that yield pyruvate. Note the conversion of the 6-carbon structure, glucose, into two molecules of pyruvate

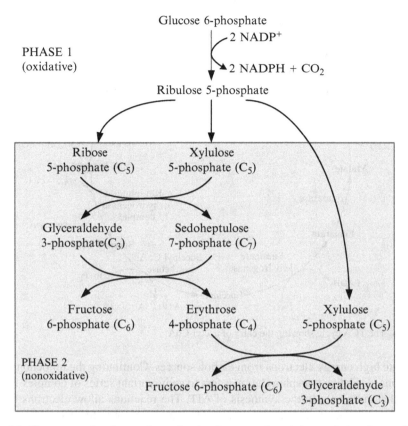

Fig. 1.3 The pentose phosphate pathway showing the conversion of glucose 6-phosphate to three pentose phosphates, ribose 5-phosphate, fructose 6-phosphate and glyceraldehyde 3-phosphate

by the enzyme lactate dehydrogenase. (2) oxaloacetate: This metabolite, obtained from the carboxylation of pyruvate by the enzyme pyruvate carboxylase, is an intermediate in the TCA cycle and also a precursor for the synthesis of glucose via a metabolic pathway, gluconeogenesis (the synthesis of glucose from non-glucose precursors). (3) acetyl-CoA: In the presence of oxygen the enzyme pyruvate dehydrogenase, in an oxidation reaction, catalyzes the conversion to acetyl-CoA, releasing CO_2, of which the body must rid itself, and reducing NAD^+ to NADH, thus the enzyme is catalyzing an overall oxidation-reduction reaction. Acetyl-CoA is an important intermediate for several pathways and serves as a convergent point for metabolism of carbohydrates, lipids and proteins.

Pertinent to our discussion, one major fate of acetyl-CoA is its entry into the TCA cycle, located in the mitochondrion and composed of eight enzymes (Fig. 1.4). This is an important component of metabolism, particularly when energy is needed and available to the cells. The entering acetyl group on acetyl-CoA is oxidized, i.e., loses electrons, and forms CO_2 (two molecules for each acetyl-CoA entering the pathway) in a series of oxidation-reduction reactions that

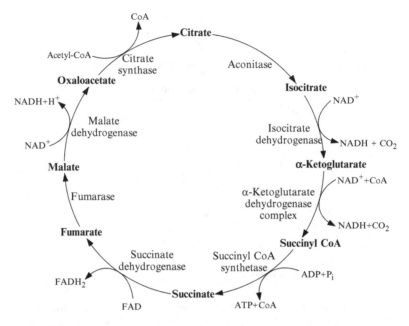

Fig. 1.4 The TCA cycle showing the entry of acetyl-CoA

generate high energy electrons from carbon sources. Continuing the mitochondrial reactions, oxidative phosphorylation refers to an important series of complex reactions that culminate in the synthesis of ATP. The reactions allow electrons from NADH and $FADH_2$ to transfer to oxygen (O_2) that is converted to water in the presence of hydrogen ions. This flow of electrons pumps protons into the region between the inner and outer mitochondrial membranes from the mitochondrial matrix (cf. the schematic in Fig. 1.5). The proton gradient is responsible for driving the synthesis of ATP from ADP and P_i, a reaction catalyzed by ATP synthase. Accounting for the ATP required to transport NADH into the organelle, a net production of 30 molecules of ATP for each molecule of glucose metabolized is realized. As mentioned earlier, even sedentary individuals require some 2,000 Cal per day, an amount that can increase significantly during vigorous exercise, and this requires some 80 kg of ATP biosynthesized per day, most of this from *de novo* synthesis by recycling ADP into ATP.

1.7.3 The Warburg Effect and Other Metabolic Alterations in Cancer

Working at the Kaiser Wilhelm Institute in Berlin, now the Max Planck Institute, Otto Warburg made the significant observation in the 1920s that cancer cells utilize more glucose than normal cells (Koppenol et al. 2011). Further, it was found that

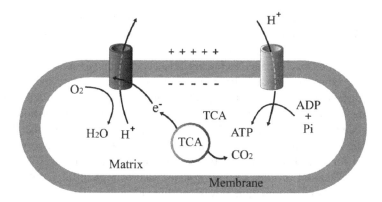

Fig. 1.5 A simplified and schematic representation of oxidative phosphorylation leading to the synthesis of ATP. The electron transport chain, shown as a *dark cylinder*, is responsible for transferring electrons from NADH and FADH$_2$ to oxygen and creating a proton gradient. The energy of the proton gradient that is used by the enzyme ATP synthase, indicated by a *gray cylinder*, to drive the synthesis of ATP from ADP and P$_i$. Adapted from (Tymoczko et al. 2013)

glucose was converted to lactate (lactic acid) via glycolysis (see Fig. 1.2 and Reaction 1.5). The confounding aspect of this finding, however, was that the increased level of glycolysis occurred even in the presence of oxygen. Under these conditions, i.e., ample oxygen, one expects aerobic respiration in which glucose is directed to pyruvate which is then converted to acetyl-CoA, not lactate.

Another surprising observation by Warburg was that aerobic respiration was like that of normal tissues, but it failed to prevent lactate formation. This is in contrast to aerobic metabolism in general since the well-known and accepted Pasteur effect leads to a reduction in lactate production in the presence of oxygen. In spite of these results, Warburg nonetheless believed that the pathway of aerobic respiration was damaged; it is now known however that it is the regulation of glycolysis that differs from normal cells in cancer cells. Warburg's experiments were initially conducted using thin slices of Flexner-Jobling rat liver carcinoma, and they were later confirmed with a number of human carcinomas.

The singular finding of enhanced anaerobic glycolysis in the presence of oxygen has led to numerous investigations in the subsequent years. Although a full explanation of the Warburg effect and its ramifications in cancer are still unfolding, a number of recent investigations have yielded many exciting and provocative observations that offer some rationale of why a less efficient energy-generating pathway, anaerobic metabolism, may be preferred over the more efficient aerobic respiration (Ferreira 2010; Cairns et al. 2011; Dang 2012; Bensinger and Christofk 2012; Icard and Lincet 2012; Oermann et al. 2012; Soga 2013). Moreover, elucidation of the distinctions between normal and cancer cell metabolism provides potentially new avenues to explore for therapeutic regimens (Jang et al. 2013). [*N.B. As a side note, the Warburg effect forms the basis of imaging by means of positron emission tomography (PET) in which patients receive 2-fluoro-2-deoxy-*d*-glucose (FDG), a radiolabeled (^{18}F) and non-metabolizable form of glucose that becomes concentrated in*

cancer cells at a higher level than in normal cells thus enabling imaging to occur.
Tumors in highly metabolically active tissues such as liver and brain, however, are
often difficult to detect because of the high background level.]

Through various mechanisms, lactate itself has been found to enhance angiogenesis, cell migration and escape from immune surveillance. Also, the increased lactate production reduces pericellular pH resulting in the activation of apoptosis in neighboring normal cells, the protection of the cancer cells by inhibition of the immune system and an elevation of a number of proteases, including metalloproteinases that can facilitate escape of tumor cells from their local environment, a requirement for metastasis to occur. In addition, the increased uptake of glucose, and hence the amount of glucose 6-phosphate, ensures a plentiful supply of substrate for the pentose phosphate pathway (cf. Fig. 1.3 and Reactions 1.7 and 1.8), products of which can be converted to nucleotides for nucleic acid synthesis or serve as intermediates to the glycolytic pathway.

Reinforcing the importance of metabolic changes in cancer, an exome sequencing study (175,471 exons from 20,661 genes) uncovered recurring mutations in the *IDH1* gene (Parsons et al. 2008). The protein encoded by *IDH1* is isocitrate dehydrogenase, an enzyme that converts isocitrate to α-ketoglutarate in the TCA cycle (cf. Fig. 1.4). Subsequent research by a number of investigators studying different cancers, reviewed by Garraway and Lander (Garraway and Lander 2013), showed that the mutations in *IDH1* led to gain-of-function in the enzyme and that, moreover, the enzyme product was an enantiomer of 2-hydroxyglutarate. This unexpected metabolite was found to inhibit α-ketoglutarate-dependent enzymes, including prolyl-4-hyroxylases that are important in regulating hypoxia inducible factor (*HIF*). Such *IDH1* mutations, surprisingly, correlated with the CpG island methylator phenotype; further, *IDH1* and *IDH2* (the mitochondrial homolog) mutations were found to be mutually exclusive with *TET2* mutations, the gene product being a methylcytosine dioxygenase that catalyzes methylcytosine to 5-hydroxymethylcytosine in DNA. Such unexpected observations and correlations reinforce the importance of metabolic alterations in cancer and emphasize the need for careful bioinformatic approaches when comparing large datasets; totally unexpected and potentially important new information can be forthcoming.

Another player that has emerged is the amino acid glutamine, and of interest is a role of the oncogene *MYC*, as well as other oncogenes and tumor suppressors, in regulating glutamine metabolism. Glutamine can function as a carbon source in the process of energy production; it can also regulate redox homeostasis, in large part through its role in the biosynthesis of the antioxidant glutathione. Lastly, glutamine can supply carbon and nitrogen to a number of cellular reactions. Regarding glutamine's role in energy production, the enzyme glutaminase is responsible for the conversion of glutamine to glutamate, the latter of which can be converted to α-ketogluterate that is an integral part of the TCA cycle (Fig. 1.3). This is particularly important in proliferating cells since citrate, another integral component of the TCA cycle (Fig. 1.3), is transported from the mitochondria to contribute to the synthesis of acetyl-CoA for lipid biosynthesis (Icard et al. 2012). Many other cellular proteins, including enzymes, oncogenes and tumor suppressors, are emerging as having important roles in cancer metabolism (Chen and Russo 2012; Oermann et al. 2012). While most of these are not discussed further, it is expected that they and others, as well as presently unknown regulators and processes, will materialize as important contributors to the altered metabolic status of cancer cells.

Hypoxia-inducible factor (HIF), notably *HIF1* (discussed in greater detail in Sect. 1.8), can escape its normal degradation under normoxic conditions due to mutations in certain enzyme-encoding genes, e.g., succinate dehydrogenase, fumarate hydratase, or prolyl hydroxylases or tumor suppressor proteins, e.g., von Hippel-Lindau (*VHL*), as well as higher cellular levels of metabolic intermediates, e.g., lactate, oxaloacetate and pyruvate (Cairns et al. 2011). The presence of *HIF1* under normal concentrations alters the expression of a number of genes participating in glycolysis, such as those for phosphofructokinase, hexokinase-II, pyruvate kinase M2, lactate dehydrogenase-A and glucose transporters, thus enhancing glycolysis; other genes are also affected that lead to reduced amounts of pyruvate entering the TCA cycle. *YC* and *HIF1* both activate the expression of the lactate dehydrogenase gene, *LDHA*, that favors the conversion of pyruvate to lactate; *MYC*, by suppressing two microRNAs (miR-23A and miR-23b), stimulates glutaminase gene expression resulting in a replenishment of intermediates in the TCA cycle (Oermann et al. 2012).

Another important metabolic component is that of the Ser/Thr kinase, AMP-activated protein kinase (*AMPK*), that serves to regulate metabolism and energy homeostasis. This regulatory kinase, depending upon the cellular conditions, can enhance or inhibit cancer cell growth (Faubert et al. 2014). In addition to these well documented changes, there are also other changes in cancer that impact on metabolism, but in the interest of brevity these will not be discussed.

Many years after the discovery of the Warburg effect, Warburg himself was still discussing the importance of mitochondrial alterations in giving a reduced ability of ATP synthesis via oxidative phosphorylation. Yet, more recent studies have shown unequivocally that cancer cells are not deficient in oxidative phosphorylation, at least for some cancers. On the other hand, some form of mitochondrial dysfunction or uncoupling has recently been noted. This involves elevated expression of certain uncoupling proteins (*UCPs*) that would lead to a reduction in effectiveness of the mitochondrial membrane potential. While not discussed herein this would result in a reduction in mitochondrial ATP synthesis, thus enhancing the cell's need for increased aerobic glycolysis.

Another aspect of the Warburg effect and mitochondrial function involves reactive oxygen species (ROS). An increase in oxidants such as ROS that are not countered by an increase in antioxidants, leads to oxidative stress in a cell. Since in mitochondrial respiration oxygen is the final acceptor for electrons in the formation of water, several ROS can arise: the superoxide radical ($\cdot O_2^-$), the hydroxyl radical (OH$\cdot$), and hydrogen peroxide (H_2O_2). These highly reactive species can damage all molecules, and proteins and DNA are particularly susceptible. There are enzymes to remove the free radicals, e.g., superoxide dismutase and catalase, but if ROS levels become too high, then cellular damage can occur. It is possible that the Warburg effect can reduce the level of ROS by increasing the amount of pyruvate produced since pyruvate can scavenge peroxides that result from the action of superoxide dismutase; moreover, the pentose phosphate pathway generates NADPH that is required for the conversion of glutathione disulfide to glutathione, important in the inactivation of hyperoxide. Lastly, the mitochondrial uncoupling discussed above may reduce oxidative stress.

An additional component of metabolism was recently found with regard to ROS (Anastasiou et al. 2011). Cancer cells, like normal cells, must protect themselves from high concentrations of ROS. Proliferation of the transformed cells requires reducing power from NADPH to support the biosynthesis of nucleotides and lipids. NADPH also acts to maintain glutathione in the reduced state, necessary for homeostasis of ROS. This increased demand for NADPH, supplied in large part through the pentose phosphate pathway (Fig. 1.3), was found to be facilitated by ROS-mediated oxidation of a particular cysteine on an alternatively spliced form of the glycolytic enzyme, pyruvate kinase that functions to convert phosphoenolpyruvate to pyruvate. This alternatively spliced form is designated pyruvate kinase M2 (*PKM2*) and is expressed in many cancer cells. Oxidation of the cysteine leads to enzyme inactivation, thus diverting glucose metabolism into the pentose phosphate pathway. This metabolic switch helps ensure synthesis of adequate amounts of NADPH to meet the needs for cell proliferation and protection from excess ROS.

The Warburg effect leads to interesting and, at first, paradoxical effects on the pH of cancer cells. It seems reasonable to expect the intracellular pH to decrease with the higher levels of lactate (lactic acid) and other acidic intermediates in the glycolytic pathway being produced. Yet, the opposite occurs with the intracellular pH increasing from its normal value of approximately 7.2 to about 7.4 or even greater. While this may appear to be but a minor alteration, it nonetheless represents a significant decrease in the concentration of hydrogen ions. Conversely the extracellular pH, normally some 7.3–7.4, becomes acidified. This unusual reversal of hydrogen ion fluxes can be attributed to increased expression of plasma membrane-associated acid transporters such as H^+-ATPase, the Na^+-H^+ exchanger NHE1, and the H^+-monocarboxylate transporter, all of which lead to increased efflux of hydrogen ions from the cell interior into the extracellular milieu (Webb et al. 2011). The latter also transports lactate out of the cells. Cell surface carbonic anhydrases increase as well, these being enzymes that catalyze the important reaction by which carbon dioxide (CO_2) from respiring cells interacts with water to form carbonic acid (H_2CO_3); this in turn, forms bicarbonate (HCO_3^-) and a hydrogen ion (H^+) as shown below (Reaction 1.9).

$$CO_2 + H_2O \leftrightarrow H_2CO_3 \leftrightarrow HCO_3^- + H^+ \qquad (1.9)$$

This simple reversible reaction can proceed non-enzymatically, but it is greatly accelerated by carbonic anhydrase. It shows how much of the carbon dioxide from respiring cells/tissues is converted to bicarbonate and how, in the lungs, carbon dioxide is formed that can be exhaled. In the vicinity of cancer cells overexpressing carbonic anhydrase there can be acidification from increased utilization of carbon dioxide, as well as the increase from hydrogen ions pumped from the cells, as further discussed in Chap. 8.

The small shift of the intracellular pH to one more alkaline can have profound effects on the cells. Numerous cellular pathways are altered by the pH change, in effect favoring cancer cell survival. For example, glycolysis, cell growth, and metastasis are enhanced while apoptosis is inhibited.

While this section has emphasized the alterations in carbohydrate metabolism, cancer cells also exhibit changes in other aspects of metabolism. For example, increased lipid biosynthesis often occurs in cancer (Yoshii et al. 2014), and lipids have been associated with maintaining redox potential in cancer cells, as well as enhancing tumor cell proliferation and survival (Santos and Schulze 2012). Altered amino acid metabolism and increased protein synthesis also accompany cancer development and growth. Recent studies have shown that *P53*, in addition to its known function as a tumor suppressor, is important in regulating glycolysis, oxidative phosphorylation, lipid metabolism, glutamine metabolism and ROS levels in non-transformed cells (Liang et al. 2013). Consequently, loss-of-function mutations in this gene can contribute significantly to the metabolic derangements in cancer,

In addition to the alterations in the cellular function of cancer cells, there are many other genes and pathways, some of which appear at first glance as being paradoxical, that can at least partially explain the Warburg effect. Of interest is the suggestion that epigenetics contribute to altered cell metabolism (Johnson et al. 2014). Importantly, what is emerging is a paradigm shift in our understanding and appreciation of the Warburg effect in that the metabolic perturbations may be important in driving tumor growth and survivability, not just the result of certain mutations that hinder carbohydrate metabolism. A comprehensive *omics* approach as discussed in this volume will contribute greatly to our understanding of this fundamental observation made many years ago that has withstood the test of time and countless studies, and along with genomic and proteomic investigations is surfacing again as a likely regulator, not a by-product, of cancer.

1.8   Emerging Roles of Hypoxia, Inflammation and Reactive Oxygen Species in Cancer

A general understanding now exists that hypoxia and inflammation are linked in cancer as well as in other pathological disorders. Hypoxia can lead to inflammation; in turn, inflammation can also lead to hypoxia, both of which can contribute to cancer formation and survival (Grivennikov and Karin 2010; Grivennikov et al. 2010; Eltzchig and Carmeliet 2011; Shay and Simon 2012; Ji 2014; Gorlach 2014). Adding to this pathophysiological interplay, ROS are associated with both hypoxia and inflammation, thus inextricably linking these three conditions and cellular components to cancer (Gorlach 2014; Costa et al. 2014). ROS, including the superoxide anion (O_2^-), hydroxyl radical (HO·) and hydrogen peroxide (H_2O_2), are highly regulated in cells through a combination of generation, e.g., mitochondrial metabolism, and elimination, e.g., via a variety of routes such as superoxide dismutases, catalase, glutathione peroxidase, thioredoxin and others. There are also reactive nitrogen species, but these are not discussed in this section. As briefly mentioned in Sect. 1.7 and discussed further in subsequent chapters, ROS is elevated in cancer and is believed to contribute to its initiation and subsequent cell growth (Waris and Ahsan 2006; Lu et al. 2007; Liou and Storz 2010; Catalano et al. 2013; Costa et al. 2014).

Hypoxia, or low oxygen tensions, is defined as cellular environments in the presence of 2 % or less oxygen. This is compared to normal, healthy cellular environments of oxygen in the range of 2–9 % (except at high altitudes the air we breathe is 21 % oxygen). Normal physiological responses to overcome hypoxia in the body include increased blood flow and respiration. Under more chronic conditions of hypoxia, two related heterodimers, *HIF1α/HIF1β* and *HIF2α/HIF2β*, respectively, are key players in regulating the myriad cellular responses to low oxygen (Wilson and Hay 2011; Shay and Simon 2012).

In normoxic conditions an enzyme, oxygen-sensitive prolyl hydroxylase, hydroxylates two prolines in *HIF1α*, which results in recognition by an E3 ubiquitin ligase, the von Hippel-Lindau tumor suppressor. The polyubiquinated *HIF* is then targeted for degradation by the 26S proteosome, thus rendering it inactive at normal oxygen tensions. Of interest, the degradation, even under normal oxygen concentrations, can be overcome by mutations in several proteins and by certain signaling pathways. As discussed earlier, *HIF* so stabilized is involved in enhancing glycolysis and inhibiting oxidative phosphorylation.

Another enzyme (factor-inhibiting *HIF*) is also oxygen-dependent and can inhibit *HIF* (via hydroxylation of asparagines on either of the two α subunits). It is the combined action of these two enzymes that monitor and respond to oxygen deprivation. At low oxygen concentrations prolyl oxidation is reduced and the *HIF1α* subunit accumulates and associates with *HIF1β*. This *HIF1* heterodimer is then translocated to the nucleus where it binds to a hypoxia-response element, thus transcriptionally activating several genes including those encoding nuclear factor κB (*NFκB*), toll-like receptors (*TLRs*), *VEGFA* and other growth factors, glucose transporters, most of the glycolytic enzymes (see Fig. 1.2), some enzymes in the pentose phosphate pathway (see Fig. 1.3) and others. These *HIF*-mediated gene activations lead to changes in metabolism, one such adjustment being that ATP production is shifted from oxidative respiration to glycolysis. This is a result of *HIF*'s role in stimulating gene expression of pyruvate dehydrogenase kinase 1, an enzyme that inhibits pyruvate dehydrogenase, the enzyme responsible for the reaction, pyruvate to acetyl-CoA (see Reaction 1.4).

Inflammation refers to a rather detailed and multifaceted process of vascular tissue in response to noxious or harmful stimuli, which can include hypoxia. The disorder, recognized some 2,000 years ago in the west by Celsus and Galen, is characterized by swelling, pain, redness, heat and loss of mobility (or function of a joint). Some of the normal responses of the body to overcome the harmful stimulus include vasodilation of the surrounding vessels to permit more blood flow and increased vessel permeability to permit leukocytes (mainly macrophages and other immune cells), antibodies, fibrin and other components to escape the blood and serve in a protective manner at the site of inflammation. Pertinent to our discussion is the observation that chronic inflammation can lead to cancer, for example hepatitis

B or C viruses give rise to liver cancer, *Helicobacter pylori* infections can result in gastric cancer and tobacco smoking can induce lung and other forms of cancer, to mention but a few.

It is now recognized that a number of mechanisms are involved in inflammatory-associated tumorigenesis (Grivennikov and Karin 2010; Grivennikov et al. 2010; Wu et al. 2014). Numerous signaling pathways lose their regulatory controls and result in pro-inflammatory gene expression related to cancer formation. Genes so activated include protein kinases, e.g., members of the *JAK* (Janus-activated kinase), *MAPK* (mitogen-activated protein kinase) and *PI3K/AKT* (phosphatidylinositol-3-kinase), thus impacting on cell proliferation. As discussed below, immune cells form an integral component of an inflammatory response. Moreover, as will be elaborated on later in the book, cancer cells develop an ability to escape immune destruction and instead use such cells, e.g., lymphocytes (T and B), macrophages, natural killer cells, neutrophils, and others, to produce cytokines that can function in a mitogenic or survival role for the developing, as well as established, cancer cells. For example, cytokines can activate transcription factors such as *STAT3* and *NFκB* that, in turn, can lead to the expression of many genes associated with tumorigenesis: angiogenic regulators, proliferation mediators and anti-apoptosis. Lastly, it has been shown that ultraviolet radiation to melanoma produces an inflammatory response that leads to metastasis (Bald et al. 2014), again documenting the important role of inflammation in cancer.

Recent studies have shown that hypoxia and inflammation are inextricably linked components of cancer. Solid tumors tend to be hypoxic and exhibit features of inflammation. For example, the presence of leukocytes in tumors was noted about 200 years ago. The main component of immune cells within solid tumors is now known to be macrophages, designated macrophage-associated tumors (TAMs). Hypoxia can give rise to inflammation; inflamed tissues are often hypoxic. Both hypoxia and inflammation trigger a series of biological responses that favor cancer growth. As described above, hypoxia of cancer cells, for example, leads to the transcriptional activation of *NFκB* and *TLRs*, as well as other genes encoding proteins involved in the endothelial-to-mesenchymal transition, metastasis, angiogenesis, cell proliferation (*HIF2α*, but not *HIF1α*, increases *c-MYC* activity) and activation of TAMs via secretion of chemokines and cytokines. In addition, hypoxia increases ROS and down-regulates DNA repair mechanisms. Similarly, leukocytes can be recruited and activated by the hypoxic cancer cells and, moreover, respond to hypoxia, also via *NFκB* and *TLRs*, by secreting chemokines and cytokines, as well as additional signals that enhance angiogenesis and other parameters favorable for cancer survival and growth. Necrotic cancer cells, acting through *TLRs*, also activate TAMs. Thus, rather than being benign or even negative aspects of cancer, hypoxia and inflammation participate in promoting cancer growth and metastasis.

Again the case is made for the need of incorporating *omics* approaches to aid in unraveling the many and often overlapping biological processes. The combination of experimental and computational biology is required to reduce the often confusing and, at times, paradoxical findings into a rational framework. Then and only then can the intricacies of cancer be fully appreciated and individual therapeutic regimens devised.

1.9 Overcoming Apoptosis

For survival all cancer cells must overcome apoptosis, i.e., programmed cell death (Elmore 2007). Apoptosis refers to a series of cellular events including plasma membrane breakage, reduction in cell volume, swelling of mitochondria, and chromatin fragmentation. There are two major pathways involved in apoptosis, namely an intrinsic and an extrinsic cascade; in addition, a third pathway, activated by natural killer (NK) cells and cytotoxic T lymphocytes, serves to lead to apoptosis of targeted cells.

The intrinsic pathway will be discussed first. This pathway is initiated by a variety of non-receptor-mediated factors that activate intracellular signaling pathways. These initiators can be quite diverse and include external and internal factors such as toxins, radiation, free radicals, viral infections and others. Also, certain proteins, e.g., cytokines, can initiate apoptosis simply by their absence, the presence of which inhibits apoptosis. The tumor suppressor protein *P53* is very much at the center of regulating this pathway, as are mitochondria. An important class of proteins is the *BCL2* family that contains both pro-apoptotic members (*BAX, BAK, BID, BOK* and others) and anti-apoptotic members (*BCL2, BCLXL, MCL1* and others). The link between *P53* and the *BCL* pro-apoptotic proteins is not well understood, but the proteins are known to act on the inner mitochondrial membrane and open the mitochondrial permeability transition (MPT) pore with a loss of the mitochondrial transmembrane potential and the initial release of cytochrome c, *SMAC/DIABLO* and a serine protease *HTRA2/OMI*. The heme-containing protein, cytochrome c, interacts with *APAF1* to make the apoptosome, and this structure activates procaspase-9, a member of the caspase (**c**ysteine **asp**artyl-specific prote**ase**s) family of proteases, converting it to the enzymatically active form, caspase-9. The activated caspase-9 then activates the first of the so-called executioner pro-caspases, pro-caspase-3; this in turn continues the proteolytic cascade via the activation of pro-caspases 6 and 7. These proteases serve to cleave a variety of proteins termed death substrates that contribute to the destruction of the cell. Two other mitochondrial proteins released in apoptosis are *SMAC/DIABLO* and *HTRA2/OMI*, which function to inhibit *IAP* (inhibitors of apoptosis), that otherwise would antagonize caspase-9. Later in apoptosis several additional proteins are released from the mitochondria, *AIF*, endonuclease *G* and *CAD*, three proteins that are responsible for fragmenting DNA and chromatin condensation.

A distinct pathway, the extrinsic pathway, can mediate apoptosis via transmembrane receptors (referred to as death receptors) that belong to the superfamily of *TNF* (tumor necrosis factor) receptors. Members of this family include *TNFR1, FASR, DR3, DR4* and *DR5*, and these bind, respectively, *TNFα, FASL, APO3L* and *APO2* (or *TRAIL*) that associates with both *DR4* and *DR5* (also termed *TRAILR1* and *TRAILR2*, respectively). The ligands bind to their cognate receptor ectodomain, thus triggering a conformational change transmitted through the membrane a cytoplasmic death domain, common to all death receptors. The death domain then binds and activates a *FAS*-associated death domain protein (*FADD*); the complex so formed is denoted as a death-inducing signaling complex (*DISC*). The role of *DISC* in apoptosis is to activate procaspase-8 (or in some cases procaspase-10) that, in turn, is responsible for activating pro-caspases 3, 6 and 7. This latter event represents a converging point for the intrinsic and extrinsic pathways. Moreover, a component of the intrinsic pathway can be recruited to enhance the extrinsic pathway. Here, *BID*, a member of the *BCL2* family, is activated by caspase 3 and functions to open mitochondrial channels, resulting in increased signaling for apoptosis.

The third pathway, also an extrinsic pathway but one requiring NK cells or cytotoxic T lymphocytes to initiate apoptosis, acts via two mechanisms. The most common one acts through the *FAS/FASR* interaction, and the other involves the proteases granzyme *A* and granzyme *B*. These enzymes enter the cell after perforin, a pore-forming protein that opens a channel into the targeted cell, and trigger apoptosis as follows. While both function, it appears that granzyme *B* is the more common of the two pathways. This protease exhibits specificity for cleaving proteins at Asp residues, and consequently serves to activate procaspase-3 and procaspase-10, as well as cleaving intracellular proteins. Granzyme *A*, acting independently of the caspase system, leads to DNA degradation by its actions on two proteins, DNAse *NM23H1* and *SET*, a nucleosome assembly protein.

These pathways, covered in greater depth elsewhere (Elmore 2007), represent challenges that cancer cells must overcome. The many mechanisms used by cancer cells to avoid apoptosis are discussed by Weinberg (Weinberg 2012) and reflect a multi-faceted approach to escape early destruction by the body. One of these responses by a variety of cancer cells is an inhibition, e.g., by overexpression of *MDM2* resulting in an inactivation of the *P53* pathway, thus diminishing its role in apoptosis, as well as permitting cells with damaged DNA to progress through the cell cycle. The same is true for the *RB* pathway, hence overcoming the negative regulation of the cell cycle exerted by this tumor suppressor protein.

Growth factors such as insulin-like growth factor *IGF1* are important in maintaining cell viability, and these may become overexpressed with a concomitant reduction in the expression or activity of the *IGF* binding proteins (*IGFBPs*) that otherwise would render them ineffective. Among the many intracellular signaling pathways activated by *IGF*, one important one for cancer cells is the activation of the *PI3K-AKT/PKB* pathway that results in anti-apoptotic signals. Another mechanism utilized by cancer cells is the overexpression of survivin, an inhibitor of caspases. An inhibitor of the extrinsic apoptotic pathway such as *FLIP* is often expressed to reduce apoptosis in cancer cells. These are but a few of the many changes that have been observed in cancer cells to overcome apoptosis. Indeed, cancer cells have devised multiple strategies for minimizing or even abolishing the three pathways used by normal cells for programmed death. Obviously there is much interest in the design of new drugs that act on these various steps found in cancer cells to aid their survival. This is another major area of interest where the use of *omics* can contribute significantly to new treatment options.

1.10   Contributions of the Extracellular Matrix and Stroma to Cancer

In addition to the changes in the cellular component associated with cancer, there is an important non-cellular component, the ECM, as well as the surrounding stromal cells, both of which have essential roles in the development and progression of cancer. Originally believed to be more of a static unit that maintains tissue integrity, it is now recognized that the ECM is vital to normal cellular function and has emerged as another key factor of cancer initiation and metastasis (Friedl and Alexander 2011; Jinka et al. 2010; Lu et al. 2012; van Dijk et al. 2013). Likewise, the neighboring stromal cells (e.g., fibroblasts), immune cells and endothelial cells (reflecting blood vessel formation), were originally believed to have no role in cancer, but there are now inconvertible data showing that these non-transformed cells contribute significantly to cancer progression (Bhowmick et al. 2004; Tripathi et al. 2012; Calona et al. 2014; Corteza et al. 2014; De Wevera et al. 2014; Escoté and Fajas 2014; Martinez-Outschoorna et al. 2014). Two key properties of ECM are particularly important in the current context: (1) a large number of growth factors tend to be stored with or linked to ECMs and (2) hyaluronic acid, a component of the ECM, has essential roles in all key transitions during carcinogenesis (see Chaps. 6 and 10).

ECM serves as a magnet and storage for a variety of growth factors released into the extracellular space, possibly as a way for their protection against degradation and to maintain them in close proximity to cells. ECM retains growth factors, e.g., bone morphogenetic protein (*BMP*), epidermal growth factor (*EGF*), fibroblast growth factor (*FGF*), hepatocyte growth factor (*HGF*), transforming growth factor β (*TGFβ*) and vascular endothelial growth factor (*VEGF*), by direct binding with ECM proteins such as fibronectin, collagens and proteoglycans (Schultz and Wysocki 2009). Biologically this seems logical since ECM serves as the basis for tissue cells; when a tissue is injured, the damaged ECM will lease its stored growth factors, thus facilitating tissue regeneration and repair.

Constituting a complex network, the ECM contains two main classes of extracellular macromolecules: proteoglycans and fibrous proteins. Several fibrous proteins constitute the non-proteoglycan portion, including the glycoproteins collagen, elastin, fibronectin and laminin. Proteoglycans are formed by the covalent attachment of glycosaminoglycans to proteins. The one exception is that of hyaluronic acid which is not attached to protein. Figure 1.6 shows a schematic illustration of the organization of the ECM.

Glycosaminoglycans refer to unbranched polysaccharide chains comprised of repeating disaccharide units, one of which is an amino sugar. The major amino sugars in glycosaminoglycans are N-acetyl-D-glucosamine and N-acetylgalactosamine, and the adjoining non-amino sugar is generally D-glucuronic acid or L-iduronic acid. Hyaluronic acid is a glycosaminoglycan like heparin, chondroitin-4-sulfate, chondroitin-6-sulfate, keratin sulfate and dermatan sulfate, and in general, is polydisperse and can contain up to some 250,000 units of the disaccharide D-glucuronic acid and N-acetyl-D-glucosamine connected in a β(1 → 3) linkage (Fig. 1.7).

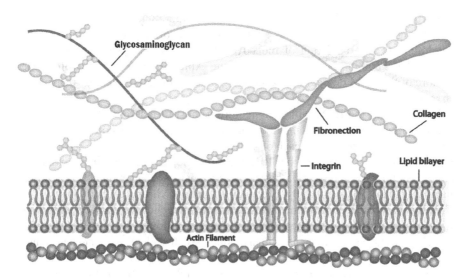

Fig. 1.6 A schematic of extracellular matrix

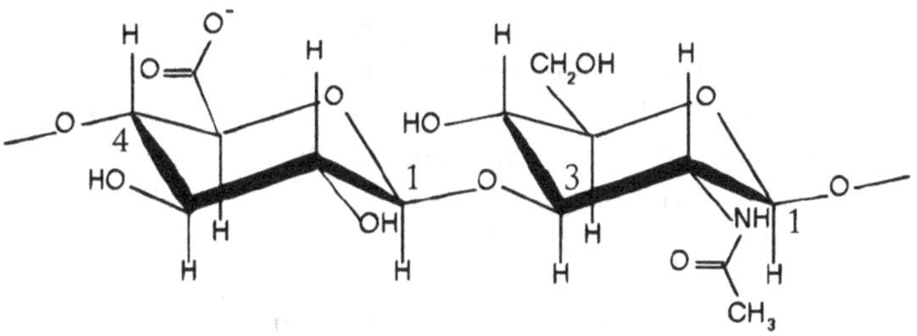

D-glucuronic acid   N-acetyl-D-glucosamine

Fig. 1.7 The structure of the repeating disaccharide, D-glucuronic acid and *N*-acetyl-D-glucosamine, that forms hyaluronic acid

The disaccharides are connected to each other in a $\beta(1 \rightarrow 4)$ linkage. Negatively charged due to the COO⁻ group on D-glucuronic acid, hyaluronic acid is a high molecular weight polyanion that binds several cations such as K⁺, Na⁺ and Ca²⁺. It forms a left-handed helix (single strand), one turn of which contains three disaccharides. Pertinent to our interest in this volume, fragments of hyaluronic acid, cleaved by hyaluronidase, have emerged as important structures in cancer, as will be discussed in Chap. 6.

The combination of these complex macromolecular structures yields special biochemical, biomechanical and biophysical properties to the ECM. Although highly complex in nature, the ECM is nonetheless highly regulated during development and tissue homeostasis. Such tight regulation implies well-controlled transcription, translation and post-translational modifications, and of course remodeling that, for example, may alter the synthesis of one or more components. In addition to the regulation of bioactive ECM macromolecules functioning as structural and cell interaction components, the expression and/or activation of one or more of the ECM degrading enzymes, e.g., matrix metalloproteinases, disintegrin and others, may be changed.

Interaction of the ECM with cells requires a host of macromolecular constituents. One interaction involves members of the class of proteins denoted as integrins. These are heterodimeric (one each of an α and a β subunit) cell surface receptors of which 24 are known, formed from 18 α and 8 β subunits. The integrin ectodomains exhibit binding specificity for a host of ECM macromolecular ligands, including members of the collagen family, fibronectin, laminin, vitronectin and elastin. The ECM ligand-integrin complex mediates its action intracellularly via focal adhesions in which the ligand-bound integrins form clusters, followed by interaction of the integrin cytoplasmic components with a number of cytoskeletal-associated proteins, actin, vinculin, talin and others. Such 'outside-in' signaling of ECM-cell interactions can lead to activation of a number of intracellular signaling pathways involving tyrosine kinases, e.g. *SRC*, focal adhesion kinase (*FAK*), integrin-linked kinase (*ILK*), extracellular-signal-regulated kinase (*ERK*) and others, and tyrosine phosphatases. A particular integrin, β1, is responsible for interacting with the ECM to regulate cell polarity, an important aspect of epithelial cells, particularly relevant to their division. Of interest, the complex just described in which integrin serves as a link between the ECM and the cell interior can function not only for the transmission of information from the extracellular milieu to the cell interior, but the various intracellular interactions with integrin can affect the type of ECM interaction, i.e., signaling from the cell interior to the cell exterior ('inside-out' signaling).

Throughout embryonic development and normal tissue differentiation and homeostasis, there is close interaction between the epithelial cells and the stromal cells. In addition to the important role of epithelial-stromal interaction in normal tissue function, such a cooperation functions in pathological states, e.g., wound healing and cancer, with elaboration on the latter below.

With this abbreviated introduction, the question arises as to how the ECM and stroma affect the initiation, development and metastasis of cancer. For one, the various ECM-cell interactions impact on processes critical to cancer such as proliferation, survival, invasion and migration. An important aspect of tumor initiation and progression involves a change in the integrin expression pattern (Jinka et al. 2012). Higher levels of expression of several integrins correlate with a host of cellular processes conducive to cancer growth and survival: cell proliferation, survival, tissue invasion, migration and new blood vessel formation (angiogenesis). The various integrins preferentially recognize different components of the ECM, e.g., collagen, laminin and fibronectin. These interactions, in turn, lead to activation of a variety of signaling cascades, including *RAS*, *SRC* and others. Several oncogenes, *MYC*, *SRC* and *RAS*, appear to be responsible for the transformation of anchorage-dependent cell growth (normal cells) to anchorage-independent cell growth (cancer cells), which is discussed in detail in Chap. 6.

Changing our focus to the stromal fibroblasts, it has been known for some time that cancer-associated fibroblasts differ from normal fibroblasts (Tripathi et al. 2012). For example, cancer-associated fibroblasts can respond to transformed epithelial cells with increased production of proteases, growth factors and collagen; moreover, the loss of transforming growth factor-β (*TGFβ*) on fibroblasts can serve as initiators of tumorigenesis. The stroma also responds to secretion of *VEGF* by cancer cells, a necessary event in the promotion of new blood vessels to provide blood-borne nutrients for a growing tumor and for colonization to distant sites.

In summary, the interactions of epithelial cells with the ECM and stroma contribute to the formation and growth of epithelial cell cancer, as further discussed in multiple chapters of the book. A better understanding of the various players and mechanisms could lead to new therapeutic modalities.

1.11 Challenging Questions in Classifying and Diagnosing Cancer

With the plethora of potential causes of cancer, ranging from metabolic alterations, hypoxia, inflammation, genomic changes and other changes, coupled with the known heterogeneity of this disease, it should be no surprise that attempts to consistently classify the extent and severity of cancer are challenging. In large part, the identification is based on the site of origin, the appearance of the cells, again compromised to some extent by the heterogeneity of cancer, and its spread to distant sites (often not known). This section provides a synopsis of the current methods in use for cancer diagnosis, grading and staging; a more detailed discussion and the introduction of emerging *omic* contributions are presented in Chap. 3.

Complementing the physical examination, there are a number of techniques in current use to aid in the identification of cancer. These include mammography, positron emission tomography (PET scanning), magnetic resonance (MR) imaging, and in some instances radiographic analysis and measurement of biomarkers, e.g. concentration of circulating prostate specific antigen (*PSA*). The final diagnosis is, however, based on pathological examination of tissue sections from biopsy or resection.

A specimen is judged to be benign or malignant and is then graded. The purpose of cancer grading is to provide an assessment of how abnormal the cells appear and indicate possible treatment modalities. In addition to visual inspection of the section, immunocytochemistry is often used to identify the presence of certain markers that impact on treatment and prognosis, e.g. estrogen receptor in breast cancer. Grading of most solid tumors is done using one of four possibilities, although prostate cancer grading is based on a different scale. Aside from GX which indicates that the grade cannot be assessed, grading will lead to one of the following, where high grade tumors require more aggressive treatment than low grade tumors: G1: well differentiated (low grade); G2: moderately differentiated (intermediate grade); G3: poorly differentiated (high grade); and G4: undifferentiated (high grade).

Cancer staging, on the other hand, is an assessment of the severity of an individual's cancer, and the stage assigned influences the choice of treatment and provides some information of the prognosis. The following components impact of the staging: tumor size and location, lymph node involvement, cell type and metastasis. The TNM staging system refers to the following three elements: T: extent of the tumor; N: whether or not the cancer cells are present in close proximity in lymph nodes; and M: whether metastasis has occurred.

The extent of the tumor, aside from TX (the primary tumor for whatever reason cannot be evaluated) or T0 (there is no evidence of a primary tumor), is given as Tis, referring to carcinoma *in situ* where the abnormal cells are localized and have not spread to other sites, or one of four designations: T1, T2, T3 and T4, reflecting the size and extent of the primary tumor. The regional lymph nodes, i.e., in close proximity to the primary tumor, can be designated as NX in which the neighboring lymph nodes cannot be evaluated; N0 which specifies that lymph nodes in the immediate vicinity are not involved; and N1, N2, N3, indicating the number of lymph nodes showing involvement. Distant metastasis is represented by MX, M0 or M1, referring to metastasis that cannot be evaluated, no metastasis, or the presence of metastasis, respectively.

It is mentioned only in passing that this staging method is not used for all cancers, but it covers the majority of solid tumors. Yet, the current grading and staging systems are quite subjective in many aspects and woefully inadequate in fully characterizing the important genetic changes leading to the particular molecular and cellular alterations in transformed cells; moreover, they lack discriminatory power when making choices for adjuvant treatment and for predicting likely outcomes with any degree of confidence.

The landscape of cancer characterization is rapidly changing with individual genome sequencing and the use of many of the *omics* techniques in this volume (Cowin et al. 2010). For example, a comprehensive study of breast cancer from 510 tumors obtained from 507 patients was conducted using a variety of methods: exome sequencing, microRNA sequencing, DNA methylation, genomic DNA copy number arrays, mRNA arrays and reverse-phase protein arrays (The Cancer Genome Atlas Network 2012c). Upon combining data from five platforms, they were able to classify four major classes of breast cancer in their starting population.

Studies such as this, now in the experimental stages, will surely emerge in time to offer a more meaningful and systematic classification of all cancers. Such detailed characterization should also prove very useful in deciding on treatment options and providing better prognoses for likely outcomes and disease recurrence. Detailed data of this type will also prove useful in distinguishing driver from passenger mutations and hopefully will provide specificity in seeking specific biomarkers in serum or urine.

1.12   Concluding Remarks

Cancer is a multi-faceted disease, a full understanding of which requires knowledge and information that span a number of scientific disciplines including biochemistry, genetics, and molecular, cellular and developmental biology. The material covered in this chapter on biochemistry and molecular and cellular biology provides the basic knowledge for the reader to follow the discussions in later chapters and to critically assess and utilize the material presented throughout the book for the reader's own research. It is worth emphasizing that cancer is a rapidly evolving system, so that the knowledge learned here, such as biochemical reactions or molecular interactions, is applicable to individual snapshots of an evolving system. Specifically, the environments where the biochemical reactions and molecular interactions take place continue to change. As the environment changes, the catalysts of these reactions and interactions will be altered according to the instructions encoded in the genome and the epigenome in response to the intra- and extracellular conditions, such as the oxygen level, the oxidative stress, the pH and a few others, which are determined by invading endogenous factors, immune responses, cellular metabolism, the genome and epigenomes of the relevant cells. Basically attention is drawn to the study of a dynamic biochemical reaction system. Superimposed upon this evolving cellular reaction system for individual cells, changes also occur at the cancer tissue level, which selects certain cells, and hence their reaction systems that best fit the current environment, and eliminate the others, i.e., Darwin's natural selection theory at work. Specifically, a cancer tissue is constantly changing its cell population by amplifying one sub-clone and inducing the demise of the other sub-clones as the disease evolves. The knowledge learned here is applicable to each snapshot as a static reaction system, and the information presented in Chaps. 3 through 13 will guide the reader to connect the snapshots along the possible evolution trajectories from multiple perspectives.

References

Alexandrov LB, Stratton MR (2014) Mutational signatures: The patterns of somatic mutations hidden in cancer genomes. Curr Opin Gen Devel 24: 52–60.

Alexandrov LB, Nik-Zainal S, Wedge DC et al. (2013) Signatures of mutational processes in human cancer. Nature 500: 415–421.

Anastasiou D, Poulogiannis G, Asara JM et al. (2011) Inhibition of pyruvate kinase M2 by reactive oxygen species contributes to cellular antioxidant responses. Science 334: 1278–1283.

Ashworth A, Lord, CJ, Reis-Filho JS (2011) Genetic interactions in cancer progression and treatment. Cell 145: 30–38.

Beck S et al. for the AACR Cancer Epigenome Task Force (2012) A blueprint for an international cancer epigenome consortium. A report from the AACR Cancer Epigenome Task Force. Cancer Res 72: 6319–6324.

Bald T, Quast T, Landsberg J, et al. (2014) Ultraviolet-radiation-induced inflammation promotes angiotropism and metastasis in melanoma. Nature 507: 109–113.

Bensinger SJ, Christofk HR (2012) New aspects of the Warburg effect in cancer cell biology. Sem Cell Devel Biol 23: 352–361.

Bhowmick NA, Neilson EG, Moses HL (2004) Stromal fibroblasts in cancer initiation and progression. Nature 432: 332–337.

Burrell RA, Swanton C (2014) The evolution of the unstable cancer genome. Curr Opin Gen Devel 24: 61–67.

Burrell RA, McGranahan N, Bartek J, Swanton C (2013) The causes and consequences of genetic heterogeneity in cancer evolution. Nature 501: 338–345.

Cairns RA, Harris IS, Mak TW (2011) Regulation of cancer cell metabolism. Nature Rev Cancer 11: 85–95.

Calona A, Taurielloa DVF, Batlle E (2014) TGF-beta in CAF-mediated tumor growth and metastasis. Sem Cancer Biol 25: 15–22.

Catalano V, Turdo A, Di Franco S, Dieli F, Todaro M, Stassi G (2013) Tumor and its microenvironment: A synergistic interplay. Sem Cancer Biol 23P: 522–532.

Chen J-Q, Russo J (2012) Dysregulation of glucose transport, glycolysis, TCA cycle and glutaminolysis by oncogenes and tumor suppressors in cancer cells. Biochim Biophys Acta 1826: 370–384.

Corteza E, Roswallb P, Pietras K (2014) Functional subsets of mesenchymal cell types in the tumor microenvironment. Sem Cancer Biol 25: 3–9.

Costa A, Scholer-Dahirel A, Mechta-Grigoriou F (2014) The role of reactive oxygen species and metabolism on cancer cells and their microenvironment. Sem Cancer Biol 25: 23–32.

Cowin PA, Anglesio M, Etemadmoghadam D, Bowtell DL (2010) Profiling the cancer genome. Ann Rev Genomics Human Gen 11: 133–159.

Dang CV (2012) Links between metabolism and cancer. Genes Devel 26: 877–890.

De Wevera O, Van Bockstalb M, Mareela M, Hendrixa A, Brackea M (2014) Carcinoma-associated fibroblasts provide operational flexibility in metastasis. Sem Cancer Biol 25: 33–46.

Duesberg PH (1987) Cancer genes: Rare recombinants instead of activated oncogenes. Proc Natl Acad Sci 84: 2117–2124.

Eifert C, Powers RS (2012) From cancer genomes to oncogenic drivers, tumour dependencies and therapeutic targets. Nature Rev Cancer 12: 572–578.

Elmore, S (2007) Apotosis: A review of programmed cell death. Toxic Path 35: 495–516.

Eltzchig HK, Carmeliet P (2011) Hypoxia and inflammation. New Engl J Med 364: 656–665.

Escoté X and Fajas L (2014) Metabolic adaptation to cancer growth: From the cell to the organism. Cancer Lett: in press. doi: http://dx.doi.org/10.1016/j.canlet.2014.03.034

Faubert B, Vincent EE, Poffenberger MC, Jones RG (2014) The AMP-activated protein kinase (AMPK) and cancer: Many faces of a metabolic regulator. Cancer Lett: in press. http://dx.doi.org/10.1016/j.canlet.2014.01.018

Fearson ER, Vogelstein B (1990) A genetic model for colorectal tumorigenesis. Cell 61: 759–767.

Ferreira LMR (2010) Cancer metabolism: The Warburg effect today. Exper Mol Path 89: 372–380.

Friedl P, Alexander S (2011) Cancer invasion and the microenvironment: Plasticity and reciprosity. Cell 147: 992–1009.

Fullgrabe J, Kavanagh E, Joseph B (2011) Histone oncomodifications. Oncogene 30: 3391–3403.

Garraway LA, Lander ES (2013) Lessons from the cancer genome. Cell 153: 17–37.

Gorlach A (2014) Hypoxia and reactive oxygen species. In: Melillo G (ed) Hypoxia and cancer, Humana Press/Springer, New York, pp. 65–90.

Grivennikov SI, Karin M (2010) Inflammation and oncogenesis: A vicious connection. Curr Opin Genet Dev. 20: 65. doi:10.1016/j.gde.2009.11.004

Grivennikov SI, Greten FR, Karin M (2010) Immunity, inflammation, and cancer. Cell 140: 883–899.

Hanahan D, Weinberg RA (2000) The hallmarks of cancer. Cell 100: 57–70.

Hanahan D, Weinberg RA (2011) Hallmarks of cancer: The next generation. Cell 144: 646–674.

Icard P, Lincet H (2012) A global view of the biochemical pathways involved in the regulation of the metabolism of cancer cells. Biochim Biophys Acta 1826: 423–433.

Icard P, Poulain L, Lincet H (2012) Understanding the central role of citrate in the metabolism of cancer cells. Biochim Biophys Acta 1825: 111–116.

Irmisch A, Huelsken J (2013) Metastasis: New insights into organ-specific extravasation and metastatic niches. Exp Cell Res 319: 1604–1610.

Jang M, Kim SS, Lee J (2013) Cancer cell metabolism: Implications for therapeutic targets. Exper Mol Med 45: e45. doi:10.1038/emm.2013.85

Ji R-C (2014) Hypoxia and lymphangiogenesis in tumor microenvironment and metastasis. Cancer Lett 346: 6–16.

Jinka R, Kapoor R, Pavuluri S, Raj AT, Kumar MJ, Rao L, Pande G (2010) Differential gene expression and clonal selection during cellular transformation induced by adhesion deprivation. BMC Cell Biol 11:93 doi: 10.1186/1471-2121-11-93

Jinka R, Kapoor R, Sistla PG, Raj TA, Pande G (2012) Alterations in cell-extracellular matrix interactions during progression of cancers. Internat J Cell Biology 2012: ID 219196. doi:10.1155/2012/219196

Johnson C, Warmoes MO, Shen X, Locasale JW (2014) Epigenetics and cancer metabolism. Cancer Lett: in press. doi.org/10.1016/j.canlet.2013.09.043.

Kandoth C, McLellan MD, Vandin F et al. (2013) Mutational landscapes and significance across 12 major cancer types. Nature 502: 333–339.

Klein CA (2013) Selection and adaptation during metastatic cancer progression. Nature 501: 365–372.

Koppenol WH, Bounds PL, Dang CV (2011) Otto Warburg's contributions to current concepts of cancer metabolism. Nature Rev Cancer 11: 325–337.

Lawrence MS, Stojanov P, Polak P et al. (2013) Mutational Heterogeneity in cancer and the search for new cancer-associated genes. Nature 499: 214–218.

Liang Y, Liu J, Feng Z (2013) The regulation of cellular metabolism by tumor suppressor p53. Cell Biosci 3: 9. doi:10.1186/2045-3701-3-9

Liou G-Y and Storz P (2010) Reactive oxygen species in cancer. Free Rad Res 44: 479–496.

Loeb LA (1989) Endogenous carcinogenesis: Molecular oncology into the twenty-first century-presidential address Cancer Res 49: 5489–5496.

Lu P, Weaver VM, Werb Z (2012) The extracellular matrix: A dynamic niche in cancer progression. J Cell Biol 196: 395–406.

Lu W, Ogasawara MA, Huang P (2007) Models of reactive oxygen species in cancer. Drug Disc Today Dis Models 4: 67–73.

Mack SC, Witt H, Piro RM et al. (2014)Epigenomic alterations define lethal CIMP-positive ependymomas of infancy. Nature 506: 445–550.

Martinez-Outschoorn UE, Lisanti MP, Sotgiab F (2014) Catabolic cancer-associated fibroblasts transfer energy and biomass to anabolic cancer cells, fueling tumor growth. Sem Cancer Biol 25: 47–60.

Meacham CE, Morrison SJ (2013) Tumour heterogeneity and cancer cell plasticity. Nature 501: 328–337.

Nakajima EC, Van Houten B (2013) Metabolic symbiosis in cancer: Refocusing the Warburg lens. Mol Carcinogen 52: 329–337.

Nowell PC (1976) The clonal evolution of tumor cell populations. Science 194: 23–28.

Oermann EK, Wu J, Guan K-L, Xiong Y (2012) Alterations in metabolic genes and metabolites in cancer. Sem Cell Devel Biol 23: 370–380.

Parker M, Mohankumar KM, Punchihewa C et al. (2014) C11orf95-RELA fusions drive oncogenic NF-κB signalling in ependymoma. Nature 506: 451–455.

Parsons DW, Jones S, Zhang X et al. (2008) An integrated genome analysis of human glioblastoma multiforme. Science 321: 1807–1812.

Pleasance ED, Cheetham RK, Stephens PJ et al. (2010) A comprehensive catalogue of somatic mutations from a human cancer genome. Nature 463: 191–197.

Santos CR, Schulze A (2012) Lipid metabolism in cancer. FEBS J 279: 2610–2623.
Schultz GS and Wysocki A (2009) Interactions between extracellular matrix and growth factors. Wound Repair Regen 17: 153–162.
Shay JES, Simon MC (2012) Hypoxia-inducible factors: Crosstalk between inflammation and metabolism. Sem Cell Devel Biol 23: 389–394.
Shen H and Laird PW (2013) Interplay between the cancer genome and epigenome. Cell 153: 38–55.
Soga T (2013) Cancer metabolism: Key players in metabolic reprogramming. Cancer Sci 104: 275–281.
Stratton MR, Campbell PJ, Futreal PA (2009) The cancer genome. Nature 458: 719–724.
Suva ML, Riggi N, Bernstein BE (2013) Epigenetic reprogramming in cancer. Science 339: 1567–1570.

The Cancer Genome Atlas Research Network [see cancergenome.nih.gov]. Nine of the multi-authored papers are listed below

McLendon R, Friedman A, Bigner D et al. (2011a) Comprehensive genomic characterization defines human glioblastoma genes and core pathways. Nature 455:1061–1068.
Bell D, Berchuck A, Birrer M et al. (2011b) Integrated genomic analyses of ovarian carcinoma. Nature 474: 609–615.
Muzny DM, Bainbridge MN, Chang K et al. (2012a) Comprehensive molecular characterization of human colon and rectal cancer. Nature 487: 330–337.
Hammerman PS, Lawrence MS, Voet D et al. (2012b) Comprehensive genomic characterization of squamous cell lung cancers. Nature 489: 519–525.
Koboldt DC, Fulton RS, McLellan MD et al. (2012c) Comprehensive molecular portraits of human breast tumors. Nature 490: 61–70.
Getz G, Gabriel SB, Cibulskis K et al. (2013a) Integrated genomic characterization of endometrial carcinoma. Nature 497: 67–73.
Creighton CJ, Morgan M, Gunaratne PH et al. (2013b) Comprehensive molecular characterization of clear cell renal cell carcinoma. Nature 499: 43–49.
Ley TJ, Miller C, Ding L et al. (2013c) Genomic and epigenomic landscapes of adult de novo acute myeloid leukemia. New Engl J Med 368: 2059–2074.
Weinstein JN, Akbani R, Broom BM et al. (2014) Comprehensive molecular characterization of urothelial bladder carcinoma. Nature 507: 315–322.
Timp W and Feinberg AP (2013) Cancer as a dysregulated epigenome allowing cellular growth advantage at the expense of the host. Nature Rev Cancer 13: 497–510.
Tomasetti C, Vogelstein B, Parmigiani G (2013) Half or more of the somatic mutations in cancers of self-renewing tissues originate prior to tumor initiation. Proc Natl Acad Sci 110: 1999–2004.
Tripathi M, Billet S, Bhowmick NA (2012) Understanding the role of stromal fibroblasts in cancer progression. Call Adhes Migr 6: 231–235.
Tymoczko JL, Berg JM, Stryer L (2013) Biochemistry: A short course, 2nd edn. W.H. Freeman and Company, New York.
Van Dijk M, Goransson SA, Stromblad S (2013) Cell to extracellular matrix interactions and their reciprocal nature in cancer. Exp Cell Res 319: 1663–1670.
Vogelstein B, Papadopoulos N, Velculescu VE, Zhou S, Diaz, Jr. LA, Kinzler KW (2013) Cancer genome landscapes. Nature 339: 1546–1558.
Vogt PK (2012) Retroviral oncogenes: A historical primer. Nature Rev Cancer 12: 639–648.
Waldman T and Schneider R (2013) Targeting histone modifications-Epigenetics in cancer. Curr Opin Cell Biol 25: 184–189.

Waris G and Ahsan H (2006) Reactive oxygen species: Role in the development of cancer and various chronic conditions. J Carcinogen 5: 14. doi:10.1186/1477-3163-5-14

Webb BA, Chimenti M, Jacobson MP, Barber DL (2011) Dysregulated pH: A perfect storm for cancer progression. Nature Rev Cancer 11: 671–677.

Weinberg RA (2012) The biology of cancer. Garland Science, New York.

Wilson WR, Hay MP (2011) Targeting hypoxia in cancer therapy. Nature Rev Cancer 11: 393–410.

Wu Y, Antony S, Meitzler JL, Doroshow JH (2014) Molecular mechanisms underlying chronic inflammation-associated cancers. Cancer Lett 345: 164–173.

Yoshii Y, Furukawa T, Saga T, Fujibayashi (2014) Acetate/acetyl-CoA metabolism associated with cancer fatty acid synthesis: Overview and application. Cancer Lett 347: 204–211.

Chapter 2
Omic Data, Information Derivable and Computational Needs

Cancer is probably the most complex class of human diseases. Its complexity lies in: (1) its rapidly evolving population of cells that drift away from their normal functional states at the molecular, epigenomic and genomic levels, (2) its growth and expansion to encroach and replace normal tissue cells; and (3) its abilities to resist both endogenous and exogenous measures for stopping or slowing down its growth. According to Hanahan and Weinberg, cancer cells, regardless of the type, tend to have eight hallmark characteristics (Hanahan and Weinberg 2011). As introduced in Chap. 1, these hallmarks are: (1) reprogrammed energy metabolism, (2) sustained cell-growth signaling, (3) evading growth suppressors, (4) resisting cell death, (5) enabling replicative immortality, (6) inducing angiogenesis, (7) avoiding immune destruction, and (8) activating cell invasion and metastasis. Other authors have suggested some additional hallmarks of cancer such as tumor-promoting inflammation (Colotta et al. 2009) and deregulated extracellular matrix dynamics (Lu et al. 2012). These recognized hallmarks have provided an effective framework for addressing cancer-related questions, having led to a deeper understanding of this disease. However, the reality is that our overall ability in curing cancer has not yet made substantive improvements, particularly in adult cancers that account for 99 % of all cancers since the start of the "War on Cancer" in 1971 (The-National-Cancer-Act 1971).

Major challenging issues that clinical oncologists have to deal with include not only considerable heterogeneity and different genetic backgrounds even within the same type of cancer, but also that effective medicines tend to lose their efficaciousness within a year, or often within a few months. A natural question to pose is: *What are the reasons for this loss of effectiveness?* Intuitively this is due to a cancer's ability to evolve rapidly, particularly in terms of generating drug-resistant sub-populations, which is facilitated by its abilities to proliferate and to accumulate genomic mutations rapidly. However, such an answer, plausible as it may be, has possibly missed the real root issue: *Why do these cells divide so rapidly in the first place?*

© Springer Science+Business Media New York 2014
Y. Xu et al., *Cancer Bioinformatics*, DOI 10.1007/978-1-4939-1381-7_2

The *Red Queen Hypothesis*, proposed by Leigh Van Valen in 1973, may provide a good framework for studying this and other cancer-related fundamental issues from an evolutionary perspective. The hypothesis states: <u>an adaptation in a population of one species may change the selection pressure on a population of another species, giving rise to an antagonistic coevolution</u> (Valen 1973). When in this frame of thinking, one may be inspired to ask: *What specific selection pressures must the evolving neoplastic cells overcome, pressures that may drive their rapid proliferation*? Currently we do not have an answer to this question yet. Among the many reasons that our knowledge is so sparse has been the lack of molecular-level data, full analyses and mining of which can potentially reveal the full complexity of an evolving cancer. While large quantities of *omic* data such as gen*omic*, epigen*omic*, transcript*omic*, metabol*omic* and prote*omic* data have been generated for a variety of cancer types, only a few cancer studies have been designed to take full advantage of all the information derivable from the available *omic* data (Cancer-Genome-Atlas-Research 2008, 2011, 2012a, b, c, 2013a, b; Kandoth et al. 2013). Integrative analyses of multiple data types may prove to be essential to gain a full and systems-level understanding about a cancer's evolution dynamics, including the elucidation of its true drivers as well as key facilitators at different developmental stages of a cancer. We anticipate that only when all of the key information hidden in *omic* data can be fully derived and utilized can we expect a meaningful breakthrough in our understanding of cancer.

2.1 Genomic Sequence Data

The Human Genome Project was initiated in 1986 by the US Department of Energy and the National Institutes of Health, which ultimately led to the generation of the first digital copies of two complete human genomes in 2001 (Lander et al. 2001; Venter et al. 2001), one by government agencies and one by a private organization. For the first time in history, the three billion base pairs (bps) of nucleotides comprising a complete human genome are represented in a digital form, directly readable by humans and computers, allowing researchers and clinicians to view and analyze the detailed genetic makeup of two healthy humans. This singular achievement has profoundly changed biological and medical sciences, clearly representing the most significant discovery since the finding of the double-strand helical structure of DNA in 1950s. Complementing and extending the invaluable genome sequence data, the major change that the Human Genome Project has brought about is that genetic science is now equipped with two powerful tools: rapid genome-sequence generation and computation-based information discovery from the genomic sequences. These tools along with the advances they have helped to make in the biological sciences, have fundamentally transformed the science of genetics, which is now data-rich and quantitative. This transition has attracted and continues to attract many mathematical and computational scientists to study problems related to genomes and other biomolecules represented in digital forms. The progress made has further transformed the general biological sciences and has substantially advanced our overall ability to study more complex biological problems than could be done before the *omic* era.

With the public availability of digitally represented human genomes in hand, scientists have computationally identified the vast majority of the ~20,000 protein-encoding genes in our genome, along with large numbers of single-nucleotide poly-morphisms (SNPs) and other types of genetic variations across individuals and different ethnic groups as well as various disease groups. Targeted sequencing of specific genomic regions deemed to be relevant to certain diseases has led to the identification of numerous genetic markers for multiple diseases. For example, Down syndrome is now understood to be caused by an extra copy of chromosome 21. A few additional examples include: (1) adrenoleukodystrophy, a progressive degen-erative myelin disorder caused by mutations in the *ABCD1* (ATP-binding cascade subfamily D) gene, which was made popular because of the movie "Lorenzo's Oil" in the early 1990s; (2) a class of hereditary breast cancers caused by mutations in the *BRCA* (breast cancer) genes; (3) familial hyperlipidemia attributable to mutations in the *APC* (adenomatous polyposis coli) gene; and (4) frontotemporal dementia, a form of inherited dementia, caused by mutations related to the splicing of exon 10 of the *Tau* gene (D'Souza et al. 1999). All these were detected through genome-scale or targeted gene sequencing and associated sequence analyses.

In addition to the Human Genome Project, a number of closely related genome sequencing projects have been established to provide a more comprehensive dataset for the human genome(s): (1) the Human Genome Diversity Project to document genomic differences across different ethnic groups (Cavalli-Sforza 2005); (2) the Human Variome Project to establish relationships between human genomic varia-tions and diseases (Cotton et al. 2008); (3) the International HapMap Project to develop a haplotype map of the human genome (International-HapMap 2003); (4) the 1000 Genome Project to establish a detailed catalog of all human genetic varia-tions (Service 2006); and (5) the Personal Genome Project to sequence the complete genomes and establish the matching medical records of 100,000 individuals (Church 2005). All these sequencing projects, along with other related ones, such as the Neanderthal Genome Project (Green et al. 2010) and the Chimpanzee Genome Project (Cheng et al. 2005; Green et al. 2010), could provide a comprehensive view of the genomes of healthy humans with normal polymorphisms as well as mutations associated with various diseases.

The Cancer Genome Atlas (TCGA) represents probably the most ambitious cancer-genome sequencing project, which aims to sequence up to 10,000 cancer genomes covering 25 major cancer types by 2014 and make the data publicly avail-able (Cancer-Genome-Atlas-Research et al. 2013). Such data will provide a sub-stantial amount of information about cancer-related genomic mutations. By comparing the genome sequences of a cancer and the matching normal tissue, one can identify all the genomic changes in the cancer genome, which generally fall into two categories: simple and complex mutations. Specifically, *simple* mutations refer to single base-pair mutations and DNA single or double-strand breaks; and *complex* mutations refer to duplications and deletions (together referred to as *copy-number changes*), translocations and inversions of genomic segments. Simple mutations can result from exogenous factors such as radiation, air-borne and food-related carcinogens in the environment, as well as from endogenous factors in the microenvironments inside our bodies, including ROS (reactive oxygen species) and other reactive metabolites plus random mutations. For example, ionizing radiation, including X-rays and gamma rays, can directly cause point mutations and DNA breaks. In addition, a variety of non-radioactive carcinogens have been identified that can damage DNA, including microbes, chemical compounds in the environment and reactive species inside our cells, as detailed in Chap. 5. Free radicals represent a large class of internal, potentially carcinogenic agents that are highly reactive molecules and can participate in undesired reactions, causing damages to cells and specifically to DNA. Infidelity of transcription and/or repair can also lead to simple mutations. While these carcinogens can produce simple DNA damages, it is the faulty or imprecise DNA replication and repair machineries that lead to the complex mutations, namely undesired duplications, deletions, inversions and translocations of large DNA segments.

There are multiple situations that can result in such complex genomic mutations. For example, under persistent hypoxic conditions, cells tend to use emergency mechanisms to repair simple mutations, but the inaccuracy of such mechanisms can lead to complex mutations as defined above (Scanlon and Glazer 2013). Here we outline one such mechanism, named *microhomology-mediated end joining* (MMEJ) for repairing double-strand DNA breaks, through which undesired DNA copy-number changes, inversions and translocations can result (Truong et al. 2013). Like the regular repair mechanism for double-strand breaks, MMEJ uses the sister chromosome as the template to replace the region with a break. The difference is that it uses a much shorter homologous region in the sister chromosome, typically 5–25 bps rather than the usual 200 bps required by the normal DNA repair mechanism, hence the designation microhomology-mediated. While the advantage is that this mechanism is substantially faster than the regular DNA repair machinery, which is needed under certain emergency conditions, it is error prone due to the less stringent requirement for finding the equivalent region in the sister chromosome, thus leading to various complex mutations (Bentley et al. 2004). This mechanism is used only under highly stressful conditions when the regular DNA repair mechanisms are functionally repressed (Bindra et al. 2007), and hence is often used in cancer-associated environments.

Knowing how different genomic mutations occur, one could possibly develop computational models to infer the evolutionary history of the mutations observed in a cancer genome from the matching reference genome. The idea is that one can first identify all the genomic differences between a cancer genome and the matching reference genome. For each identified complex mutation, one can apply a mechanistic model like the one outlined above (or from the literature) to predict how it occurs from the previous generation of the genome, while simple mutations can be assumed to take place randomly according to some stochastic models. It is worth noting that some of the evolutionary intermediates (mutations) may or may not be present in the cancer genome, due to the possibilities that some portions of the genome might have been deleted during evolution. In addition, it should be emphasized that such an approach (even when taking into consideration the other emergency DNA repair mechanisms) may not necessarily yield a unique evolutionary path from the reference to the cancer genome. One possible way to constrain this phylogenetic reconstruction problem to a solution space as small as possible is to find such a path under the parsimony assumption (Steel and Penny 2000), as often used in phylogenetic reconstruction algorithms. Specifically one can require that the predicted evolutionary path have either the smallest number of generations or the highest consistency with the occurrence probabilities of different types of mutations as documented in the literature. As of now, no such algorithms have been published for making evolutionary path predictions, but the need for such tools is clearly there in order to understand the evolution of a cancer genome.

Various other types of information may also be derivable from cancer genomes, such as: (1) oncogenes and tumor suppressor genes (see Chap. 1 for definition) that may be specific to a particular cancer type. Examples include gene fusions as in the case of the Philadelphia chromosome for chronic myelogenous leukemia (CML) (Nowell and Hungerford 1960); (2) potential integration of microbial genes into the cancer genomes as in the case of hepatitis B virus genes integrated into the host genome; (3) biological pathways that are enriched with genetic mutations in a particular cancer, leading to the loss of function at the pathway level; and (4) changes in mutation patterns as a cancer advances.

By systematically identifying mutations in the genomes of multiple patients of the same cancer type, one can identify biological pathways enriched with such mutations, using analysis tools like DAVID (Huang et al. 2009) against pathway databases such as KEGG (Kanehisa et al. 2010, 2012, 2014), BIOCARTA (Nishimura 2001) or cancer-related gene sets (Forbes et al. 2011; Chen et al. 2013; Zhao et al. 2013). For example, a study, published in 2007 on genomic mutations observed across 210 cancer types, discovered that the pathway having the highest enrichment with non-synonymous mutations is the *FGF* (fibroblast growth factor) signaling pathway, revealing one commonality among changes needed by cancer evolution across different cancer types (Greenman et al. 2007). With such information, one can further infer which cellular processes need to be terminated or become hyperactive in any specific order as a cancer evolves, hence possibly developing new insights about the evolutionary paths unique to particular cancer types or common among all cancer types.

2.2 Epigenomic Data

Epigenomic data provide information about all the chemical modifications on the genomic DNA and associated histone proteins in a cell, namely *DNA methylation* and *histone modification*, among a few other less studied epigenomic activity types. While epigenetic analyses are not new, it is the high-throughput array and sequencing techniques that have made such analyses at a genome scale possible and have clearly advanced our overall capabilities to study cancer.

DNA methylation is a process by which a methyl group is added to the carbon 5 position of cytosine residues (C) in CpG dinucleotides. This is accomplished through a group of enzymes known as *DNA methyl-transferases*, the reactions of which can be reversed by another group of enzymes termed *DNA demethylases*. When a CpG region is highly methylated, they attract a group of enzymes called *histone deacetylases* that will initiate chromatin remodeling to change the local structure of the DNA, hence changing its accessibility to large molecular structures such as the transcription machinery, RNA polymerase. Since long CpG regions (denoted as *CpG islands*) tend to be associated with the promoters of genes, methylation of such regions represses the expression of the genes.

Histones are proteins that bind with DNA to form the basic folding units, denoted as *nucleosomes*, of chromatin, as introduced in Chap. 1. The packing density of chromatin is closely related to the transcriptional state of a gene, i.e., lower packing density implying higher transcriptional activity. Cells change their chromatin structures through post-translational modifications on the relevant histones, including acetylation, deamination, methylation, phosphorylation, SUMOylation and ubiquitination. The understanding is that interactions between histones and DNA are formed by electrostatic attraction between the positive charges on the histone surface and the negative charges on DNA. Consequently, modifications on histones may change the charges of the surface residues, possibly changing the conformation and the transcriptional accessibility of a folded DNA and ultimately enhancing or repressing expression of the relevant genes (Strahl and Allis 2000; Kamakaka and Biggins 2005). Another mechanism is through recruiting and applying chromatin remodeling *ATPases*, where histone modifications can lead to disruptions of *ATPase* attraction to the chromatin, hence altering the DNA's physical accessibility to the RNA polymerase (Vignali et al. 2000).

Various techniques have been developed to reliably capture DNA methylations and histone modifications at a genome scale. Among the assays that have been used for detecting methylations is the *bisulfite* sequencing technique (Yang et al. 2004). By converting each methylated C to a T and removing the methylation, the bisulfite method utilizes the current sequencing techniques to produce the modified sequence and then recovers the methylation locations through comparisons between the sequenced Ts and Cs at the same locations in the original DNA and the modified DNA done as above.

Histone modification sites can be detected using the ChIP-chip array technique (Huebert et al. 2006), which has previously been used to identify the binding sites of transcriptional factors. The difference here is to detect the DNA binding sites with histones relevant to the packing of DNA. Comparisons between the identified DNA binding sites under different conditions can lead to the identification of modified chromatin structures. The advancement of sequencing techniques in the past few years has led to the development of the second generation ChIP technique, namely *ChIP-seq*, which can provide more quantitative and reliable data about histone modification sites.

From either of the two types of epigenomic data, one can infer genes that are primed to be repressed or enhanced transcriptionally at the epigenomic level.

These data, in conjunction with other *omic* data such as transcriptomic and genomic information, can be used to derive association relationships between epigenomic activities and the cellular as well as micro-environmental states. This can lead to identification of possible triggers and regulatory pathways of different epigenomic activities. Information of this type is clearly needed since, although numerous epigenomic effectors such the enzymes for DNA methylation and histone modifications have been identified, very little is known about the regulation of these effectors and under what conditions a specific set of genes will be methylated. As discussed in Chap. 9, epigenomic level changes can be considered as an intermediate step between (reversible) functional state changes of effector molecules and (permanent) genomic mutations. A detailed discussion regarding the possible relationships among these three types of changes needed by evolving cancer cells is given in Chap. 9.

A number of large-scale epigenomic sequencing projects have been initiated with similar ambitious goals to those of the genome sequencing projects outlined in Sect. 2.1. These projects include: (a) the NIH Roadmap Epigenomics Program, which started in 2008 with the aim of producing histone modification data for over 30 types of modifications in a variety of human cell types; (b) a component of the ENCODE (Encyclopedia of DNA Elements) project launched by the US National Human Genome Research Institute aiming as part of its goal the characterization of the epigenomic profiles of 50 different tissue types; (c) the International Human Epigenome Consortium having its goal to build on and expand the NIH Epigenomics Program to include nonhuman cells and tissues, and to make it a functional international program; and (d) some regional epigenomics projects such as the "Epigenetics, Environment and Health" project in Canada and the Australian Alliance for Epigenetics. A number of human epigenomic databases have been developed as the result of these and related projects (see Chap. 13 for details).

2.3 Transcriptomic Data

The advent of microarray technology in the mid-1990s has made it possible to measure in real time the expression levels of all the genes encoded in the human genome under defined cellular conditions. This methodology also applies to other genomes as long as their protein-encoding genes are known. This is one of the high-throughput techniques that has clearly fueled the revolution in biological sciences that we have been witnessing since the start of the Human Genome Project.

Comparative analyses of gene-expression data of cells collected under different controlled conditions or on disease *versus* control tissues can provide a large amount of information useful for studying human diseases at the molecular and the cellular levels. For example, by comparing gene-expression levels in a lung cancer tissue with those in the adjacent healthy tissue of the same patient, one can identify differentially expressed genes in the lung cancer *versus* the healthy lung. While not necessarily all the differentially expressed genes are directly relevant to cancer, this information provides a basis for further inference of genes that may be directly relevant to cancer. For example, one can compare such sets of differentially expressed genes across multiple patients of the same cancer type to eliminate those genes whose differential expressions are specific to a few individuals or cancers at a specific developmental stage. That is, one can identify genes that may be most essential to the development of a cancer type through the identification of genes that are commonly differentially expressed across all or the majority of the patients of the cancer type examined.

When applied in conjunction with pathway-enrichment analysis, particularly against the eight cancer-hallmark related pathways mentioned earlier, one can identify hallmark pathways enriched with up-regulated (or down-regulated) genes in a specific type of cancer. If the cancer data also have the stage information, one can further derive information about how each of the cancer hallmarks is executed at the molecular and cellular levels for this cancer type and in what order. By comparing such information across multiple cancer types, one can possibly detect which relative orders among the observed hallmark events are essential and which are coincidental. And by comparing such data between two subgroups of patients of the same cancer type, for example one with smoking histories and the other without, one can possibly derive how smoking may have contributed to the development of individual hallmark events. Similar analyses can be used to discover possible contributions by other lifestyle habits.

Actually, much more information can be derived through analyses of cancer transcriptomic data. For example, tiling array is a variation of the gene-expression technique used to detect DNA-binding sites of specific proteins through ChIP-chip experiments, hence making it possible to identify transcription regulators of specific genes under particular conditions (Ren et al. 2000; Iyer et al. 2001). RNA-seq is the new generation of techniques for transcriptomic data collection (Wang et al. 2009). It refers to the use of high-throughput technologies to sequence cDNAs that are reversely transcribed from the expressed RNA molecules. By doing deep sequencing, the dynamic range of RNA-seq can span five orders of magnitude, substantially larger than those of microarray-based techniques. This allows more accurate identification of differentially expressed genes, particularly those that tend to express at a relatively low or high level and where changes tend to be relatively small but statistically significant, such as those often observed with transcription factors. In addition, RNA-seq techniques are digital in nature, in comparison with the analog signals provided by microarrays. One advantage of digital signals is that the resulting measurements are more repeatable compared to analog signals and less prone to be affected by the experimental environments. The biggest advantage of RNA-seq data over microarray data is that it contains all the information about alternatively spliced variants since they do not rely on short sequence probes as in microarrays, instead producing the entire sequence for each transcript. Such information allows one to derive all splicing variants in specific cancers and cancer stages, thus enabling more detailed functional mechanism studies.

A few computer programs have been developed and made publicly available for inference of splicing variants based on RNA-seq data, such as Cufflinks, which requires a reliable reference genome for its inference of splicing isoforms (Trapnell et al. 2010). Another popular transcript-assembly program, Trinity, is a *de novo* method, i.e., no reference genome is required (Grabherr et al. 2011), but at the expense of less reliable assembly results compared to Cufflinks. The limitation of Cufflinks and similar programs is that they may not necessarily work well on cancer RNA-seq data when the underlying cancer genome is not available, which could be substantially different from the matching genome of healthy cells since cancer genomes tend to have large numbers of genomic reorganizations such as translocations, copy-number changes and inversions, as well as breaks as discussed in Sect. 2.2. Thus, more effective computational techniques are clearly needed for inference of splicing isoforms from cancer RNA-seq data.

Presently, a number of databases for microarray and RNA-seq gene-expression data have been developed and are publicly available. For example, GEO is a general-purpose gene-expression database consisting of both cancer and other tissue types (Edgar et al. 2002). A cancer-centric genome database that also contains epigenomic and transcriptomic data for numerous cancer types is TCGA (Cancer-Genome-Atlas-Research et al. 2013). Gene Expression for Pediatric Cancer Genome Project is a gene-expression database developed specifically for pediatric cancers (Downing et al. 2012). Overall these databases have genome-scale transcriptomic data for over 200 different types of cancer tissues and a substantially larger number of cancer cell lines. A tremendous amount of information could potentially be derived through comparative analyses of these data across different cancer types and cancers at varying stages or of distinct malignancy grades (see Chap. 3). For example, by simply plotting the average number of differentially expressed genes across cancer samples *versus* the 5-year survival rate for each of the following nine cancer types: melanoma, pancreatic, lung, stomach, colon, kidney, breast, prostate cancers and basal cell carcinoma, one can see that there is a close relationship between these two numbers (see Fig. 2.1).

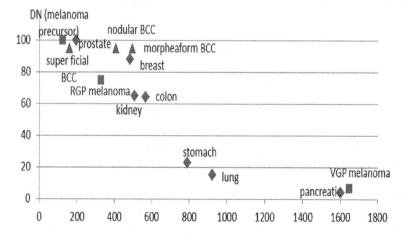

Fig. 2.1 The 5-year (y-axis) survival rate for each cancer type *versus* the average number of differentially expressed genes per cancer sample (x-axis) (adapted from (Xu et al. 2012))

By examining the average up- or down-regulation levels of genes in selected pathways of different cancer types, it is possible to derive information about activated energy metabolism in different cancer types, which vary from glucose- to lipid- to amino acid-based. As an example, Fig. 2.2 shows the activity levels of multiple energy producing metabolic pathways, covering glycolysis, the TCA cycle, oxidative phosphorylation and fatty acid metabolism in nine cancer types. One can see from the figure that pancreatic cancer has the highest up-regulation in glucose metabolism, followed by kidney and lung with breast cancer having the least changes in glucose metabolism when compared with their matching control tissues. One can also see that, while most of the seven cancer types on the left show down-regulation or no changes in oxidative phosphorylation, both skin cancer types, namely melanoma and basal cell carcinoma, show up-regulation in this pathway. [*N.B. Throughout this book, all the analyses of transcriptomic data across different patients samples are properly normalized, hence comparisons among fold-changes of genes across different samples are meaningful.*]

A variety of computational techniques have been developed for information derivation from gene-expression data, including: (1) identification of differentially expressed genes using simple statistical tests such as T-test or Fisher's exact test, (2) clustering analysis, (3) bi-clustering analysis and (4) pathway enrichment analysis for differentially expressed genes. The following discussion provides some basic ideas about these analysis techniques, followed with a list of novel techniques for more advanced analysis needs.

2.3.1   Data Clustering

Identification of co-expressed genes is a basic technique for gene-expression analysis, which has a wide range of applications in cancer studies. The idea is to identify all genes whose expression patterns exhibit statistical correlations over a time course (typically for cell line-based data) or among a collection of samples; such genes are called *co-expressed genes*. There are numerous online tools for identification of co-expressed genes such as DAVID, CoExpress (Nazarov et al. 2010) and GeneXPress (Segal et al. 2004). Co-expressed genes may suggest that the genes are transcriptionally co-regulated even though some co-expressed genes appear coincidentally, particularly when the number of conditions or the number of samples is small. One way to computationally "validate" such a prediction is through identification of conserved *cis* regulatory motifs within the promoter sequences of the co-expressed genes (Liu et al. 2009). The rationale is that if these genes are indeed co-regulated transcriptionally, they should share conserved *cis* regulatory elements for binding with their common transcription regulators. From the predicted co-expressed genes and *cis* regulatory motifs, one can predict with confidence that these genes are transcriptionally co-regulated, and even possibly predict their main transcription regulators using tools such as those by Essaghir et al. (2010) or by Qian et al. (2003).

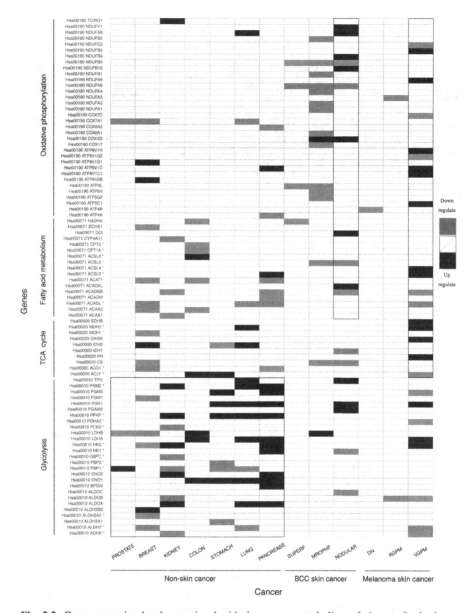

Fig. 2.2 Gene-expression levels associated with the energy metabolism of glucose (both glycolytic fermentation and oxidative phosphorylation) and fatty acids plus the TCA cycle in nine cancer types. The y-axis is a list of genes involved in four metabolic pathways: oxidative phosphorylation, fatty acid metabolism, TCA cycle and glycolysis; and the x-axis is a list of nine cancer types, including three stages of basal cell carcinoma (BCC) and melanoma, respectively. Each entry is the average log-ratio of expression levels between cancer samples and the matching control samples in different cancer types. The side-bar on the *right* shows the gray-level code for the expression level changes, with *"gray"* indicating down-regulation, *"white"* for no change and *"black"* denoting up-regulation. Adapted from (Xu et al. 2012)

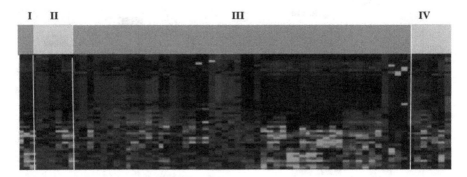

Fig. 2.3 A heat-map of gene-expression changes of 42 genes, with each row representing one gene and 80 gastric cancer samples *versus* the matching control samples, with each column representing one sample, which are grouped into four stages: I, II, III and IV, with *light gray*, *dark gray* and *black* representing up-, down-regulation and no changes, respectively. The figure is adapted from (Cui et al. 2011)

2.3.2   *Bi-clustering Analysis*

Bi-clustering is a generalized form of clustering analysis, which aims to identify co-expressed genes among some to-be-identified subgroups of samples, but not among all samples. Such a technique is particularly useful for discovering subtypes, stages or grades of a cancer (see Chap. 3 for details). Figure 2.3 shows one example of signature genes for gastric cancer stages identified through a bi-clustering analysis. Specifically, 42 genes are found to exhibit distinct patterns for a group of 80 gastric cancer samples (one sample from each patient) grouped according to their stages (Cui et al. 2011). Interestingly the samples assigned to stage III exhibit two distinct expression patterns, with samples on the left clearly showing different patterns from those on the right, suggesting that these patients may actually fall into five different stages such as stage I, II, IIa, III and IV, rather than four as proposed by the pathologists who analyzed these samples (Cui et al. 2011).

A bi-clustering problem is computationally much more difficult to solve than a clustering problem since it involves two variables, i.e., genes to be identified as co-expressed and samples to be found to have similar expression patterns, compared to only one variable, i.e., co-expressed genes in traditional clustering analyses. A few computer tools have been published for identification of bi-clusters in gene-expression datasets such as QUBIC (Li et al. 2009) and BicAT (Barkow et al. 2006). After bi-clusters are identified, similar analyses about regulatory relationships can also be carried out as above to predict the possible transcription regulators for each bi-cluster.

2.3.3 Pathway (or Gene Set) Enrichment Analysis

Pathway enrichment analysis is a way to map up- or down-regulated genes to higher level functional organizations such as biological pathways, networks or gene sets that are each known to be relevant to cancer or cancer-related. The basic idea is to homology-map the identified up-regulated (or down-regulated) genes to known pathways in pathway databases such as KEGG, REACTOME (Croft et al. 2011) or BIOCARTA, and then assess if a specific pathway has substantially more genes mapped to it than by chance, measured using statistical significance values. For example, DAVID is one popular tool for doing pathway enrichment analysis. Basically, it homology-maps a set of given genes to pathways in the above databases, then assesses the statistical significance of having K given genes in the given set mapped to a specific pathway using κ statistics, i.e., a chance-corrected measure of co-occurrence, and predicts that a pathway is enriched by the given gene set if its statistical significance is above some threshold (Huang et al. 2007). Figure 2.4 shows one enriched pathway by up-regulated genes in gastric cancer.

With the increasing needs for studying more complex analysis problems based on gene-expression data, there is clearly an urgent necessity for more powerful analysis techniques. A few are listed here, which could definitely benefit from the involvement of researchers equipped with advanced statistical analysis techniques.

1. _Inference of causal relationships_: Analyses discussed above, such as clustering or bi-clustering, can provide association relationships among activities of genes or pathways through detection of correlations among their expression patterns. Clearly cancer researchers could benefit even more if such analyses can be extended to infer causal relations among genes or pathways with altered expression patterns.

   Causality has been difficult to derive due to the nature of the problem (Pearl 2009). Many may remember the argument made by the tobacco industry when being presented with statistical data showing that smokers have higher probabilities of developing lung cancer than non-smokers. The industry officials argued that such data do not necessarily imply that smoking causes cancer, pointing out the following: there could be an unknown genetic factor that gives rise to a sub-population who enjoys smoking and has a higher propensity to develop lung cancer. Logically, this argument holds. Hence, in order to prove that smoking indeed causes lung cancer, one would need to demonstrate that individuals who are forced to smoke, regardless if they like it or not, are at higher risk of developing lung cancer than those who are forced not to smoke. This would then rule out a possible contribution from genetic factors as suggested by the defense lawyers of the tobacco industry. In general, inference of causality is fundamentally hard. Fortunately, there have been some interesting developments in theoretical studies on causal relationships. One example is the development of _causal calculus_ by Pearl (2009). Application of this or other causal theories to the information-rich gene-expression-based causality analyses would help to advance the field in a profound way.

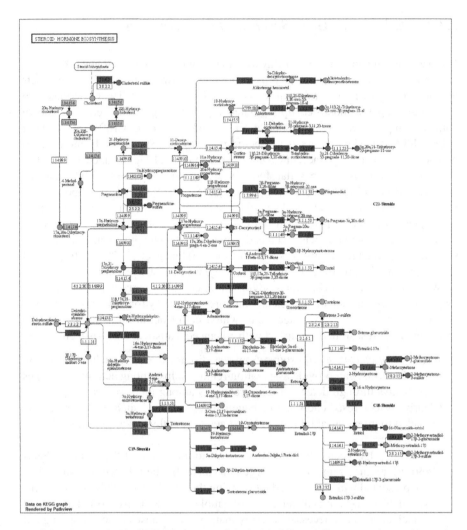

Fig. 2.4 An example of a KEGG pathway enriched with differentially expressed genes in gastric cancer *versus* matching controls. Each *rectangle* represents an enzyme-encoding gene and each *oval* represents a metabolite. An up-regulated gene is marked as *dark gray* and down-regulated gene marked in *light gray* while a *white rectangle* represents an enzyme whose gene is not identified yet. A metabolite with increased concentration is marked in *dark gray* and a decreased metabolite is marked in *light gray*

2. *De-convolution methods for expression data collected on cancer tissues*: One challenge in analyzing gene-expression data collected on cancer tissues is that the data are not from a homogeneous cell population, but instead a collection of different cell types with cancer cells as the dominating sub-population. It is well known that each sample of cancer tissue generally has other cell types such as macrophages and other immune cells, stromal cells, and blood vessel cells,

although there may have been attempts to make the cell population as homogeneous as possible, using techniques such as laser-directed micro-dissection (Emmert-Buck et al. 1996). The reality is that collecting highly homogeneous cell populations from cancer tissues is challenging and very time consuming.

Gene-expression data collected on a mixture of multiple cell types can easily lead to false conclusions if done without proper data processing. This issue has been reflected in a common complaint from gene-expression data analysts that tissue gene-expression data are not reliable and are difficult to compare across different samples. One key reason is that tissue samples collected by different labs may have been processed using different procedures so that the sub-populations of different cell types may be different from those *in situ*. Moreover, different sample-processing procedures may lead to systematic changes in the sub-populations but in different ways, thus making tissue gene-expression data not easily comparable.

It is our belief that techniques in statistical analysis, properly applied, can aid immensely in resolving the issue by de-convoluting the observed gene-expression data into expression levels contributed by different cell types. The basic idea of one such de-convolution technique is as follows. Each cell type has its unique functional characteristics. For example, cancer cells are the only cell type in the tissue that divides rapidly, while fibroblasts are the only cell type that synthesizes the components of the extracellular matrix. These unique functional characteristics of different cell types are reflected by their gene-expression data. Specifically, it is expected that each cell type can be represented (or approximated) by a set of expressed genes unique to the cell type, along with the cell type-specific correlations among the expression levels of different genes. Such cell type-specific (condition-invariant) correlations among their genes can possibly be represented in some generalized form of a covariance matrix, which can be considered as the *signature* of individual cell types. To derive such a signature, one needs unambiguous gene-expression data of specific cell types collected under multiple and different conditions, allowing the capture of the invariance among the correlations between expression patterns of individual genes.

With such a reliable de-convolution tool, one can decompose each gene-expression dataset collected on cancer tissues into gene-expression contributions from different cell types. Then, one can analyze the gene-expression data predicted to be solely associated with cancer cells or other cell types such as macrophages to understand the interplay between cancer and immune cells. Such decomposed datasets of cancer samples at different stages have the potential of enabling one to realistically study a range of important problems in elucidating the complex relationships among different cell populations in each cancer niche, which are not feasible with the current experimental techniques.

3. *Development of an infrastructure in support of the study of cancer systems biology*: Another area where computational statistical techniques can make a fundamental contribution is in characterization of cancer microenvironments and in linking micro-environmental factors to the evolutionary trajectories of specific

cancer tissues. While experimental studies of the evolving microenvironment of a cancer *in vivo* may not be feasible, computational analyses of gene-expression data could help to solve such a problem. The premise is as follows. When the microenvironment changes, such as changes in (1) the composition and physical properties of the pericellular matrix, (2) the level of hypoxia, (3) the ROS level, (4) the pH level and (5) the sub-population sizes of different cell types in the stromal compartments (see Chap. 10), some genes will respond by changing their expression levels. For example, when the cellular level of oxygen changes, the expression patterns of the *HIF1* (hypoxia-induced factor) and *HIF2* genes change, as discussed in Chap. 1. By carefully analyzing gene-expression data collected under specific conditions on relevant cancer cell lines, one should be able to train predictors for changes in each aspect of the microenvironment based on their relationships reflected by gene-expression data. Such prediction capabilities will enable cancer researchers to examine how micro-environmental factors change as a cancer evolves and to link such information to cancer phenotypes, hence possibly generating new understanding about how microenvironments affect cancer progression and cancer phenotypes.

2.4   Metabolomic Data

Our own experience has been that transcriptomic data represent probably the most information-rich data that are relatively straightforward to obtain for cancer studies. Such data are particularly useful for gaining a big-picture view and for the derivation of rough models for a specific mechanism, while genomic and epigenomic data can provide useful complementary information. Transcriptomic data, however, do not always portray an accurate picture regarding the activity of a pathway. This is because they measure only the intermediates for making the functional parts, the proteins, of the pathway; others, such as those constitutively expressed, will of course not appear as altered gene expressions. Clearly, it is highly desirable to have protein expression data. However, proteins are notoriously difficult to study, much more complex than, say, transcripts, as proteins may have different post-translational modifications and splicing variants, which are not amenable to the current high-throughput techniques. Consequently, proteomic data have not been as widely used as transcriptomic data in cancer studies. Metabolomic data can, however, assist in filling the void due to the lack of protein level information since they provide information on the substrates and products of proteins, specifically enzymes.

As of now, over 40,000 metabolites have been identified in human cells according to the Human Metabolome Database (HMDB) (Wishart et al. 2007, 2009, 2013). These metabolites can be intermediates or products of cellular metabolism, which include the basic metabolites such as amino acids, nucleotides, alcohols, organic acids and vitamins, and complex metabolites such as cholesterol and steroid hormones. By analyzing the quantitative data of metabolites associated with a specific metabolic pathway, it is possible to make a generally accurate estimate of the activity level of the pathway. For example, glucose-6-phosphate, fructose-6-phosphate, glyceraldehyde 3-phosphate, phosphoenol-pyruvate, pyruvate and lactate are the main metabolites of the glycolytic fermentation pathway (see Fig. 1.4), and their relative abundances provide accurate information about the activity level of the pathway. By carrying out metabolite flux analyses (Varma and Palsson 1994) based on the pathway information and the measured quantities of these metabolites, one can infer if some of these intermediates or end products may be directed towards other metabolic pathways, in addition to being part of the glycolytic pathway.

Metabolic flux analysis generally applies to any well-established biological pathway, such as those in central metabolism. That is, with all the relevant reactions and the encoding genes known, metabolic data can be used in conjunction with the matching transcriptomic data, to infer the flux of a specific molecular species such as carbon or nitrogen. In essence, this provides flux information of different elements across an entire network, which preserves balances between the total input and the output elements for each reaction, hence providing a systems-level representation of the flux distribution across all the branch points in the network. Identification of unbalanced reactions, i.e., the total number of carbons into a reaction is different from that out of the reaction, can help to detect previously unknown branches involved in the relevant reactions. This type of analysis can be used to identify possible relationships between two known metabolic pathways, such as detecting possible metabolic relationships between the glutaminolysis pathway (McKeehan 1982), which tends to be up-regulated in cancer cells, and other metabolic pathways, or detection of relationships between cholesterol metabolism and phospholipid metabolism in metastatic cancer (see Chap. 11). For example, an analysis like this has led us to detect that some metabolites of the glycolysis pathway become substrates of another metabolic pathway, the *hyaluronic acid synthesis* pathway (see Chap. 6). When reaction rate constants are available or can be estimated for all the relevant enzymes, one can identify the rate-limiting steps in a pathway, thus enabling one to undertake detailed mechanistic studies of a biological process.

Both high-resolution mass spectrometry (MS) and nuclear magnetic resonance (NMR) spectroscopy have been used to identify metabolites present in cells and in tissue samples, each having their own advantages and limitations. MS can provide quantitative measures for up to 1,000 different metabolite species, but it suffers from relatively low repeatability (Boshier et al. 2010). In comparison, NMR can provide highly accurate measurements of metabolites but is limited in the number of metabolite species in each experiment. With either type of instrument, one can obtain quantitative measures of numerous metabolite species.

When coupled with transcriptomic data and functional annotations of genes, metabolomic data can be used to infer the detailed metabolic pathway that may produce specific metabolites. Specifically, for each experimentally identified metabolite in a sample, one can search for enzymes among the expressed enzyme-encoding genes that may be responsible for its synthesis through comparisons against the Enzyme Classification (Bairoch 2000) or KEGG database. Both of these databases contain information about enzymes and substrates that can lead to the production of a specific metabolite. If there is more than one candidate, a selection can be inferred by finding the one that is most consistent with the available transcriptomic and functional annotation data, i.e., the enzyme-encoding gene is expressed and the substrate is among the identified metabolites. By repeating this process, one can create a pathway consisting of the identified enzymes along with the identified metabolites. Although one cannot expect to derive all the relevant enzymes along the pathway, it is generally possible to develop a crude model based on our own experience. In addition, it is possible to expand a pathway model through careful applications of the transcriptomic data and the metabolomic data collected to identify previously unknown or poorly studied branches of well-studied pathways. For example, by carefully analyzing the metabolites associated with glycolysis, one can possibly identify those that serve as intermediates between glycolytic metabolites and metabolites involved in the synthesis of hyaluronic acids as detailed in Chap. 6.

There are a number of databases for human metabolomic data in the public domain, including the HMDB, BiGG (metabolic reconstruction of human metabolism) (Schellenberger et al. 2010) and the Tumor Metabolism database (The-Tumor-Metabolome 2011). Another useful database is Brenda (Scheer et al. 2011), which provides the reaction parameters of various enzymes. All these databases provide useful information needed for reconstruction of specific metabolic processes in normal and cancer cells.

2.5 Patient Data

Knowledge of patient data is essential for the correct interpretation of their respective *omic* data. People of different gender, age and race, and with different histories of smoking, alcohol consumption and health problems, could have different baseline gene-expression levels. It was noted, from our previous studies, that some genes are sensitive to one aspect of a person's attributes, such as age or gender, while other genes may be more sensitive to other attributes. And some genes are attribute-independent. For example, based on our analysis on gene-expression data of 80 gastric cancer tissues and their matching tissues from 80 patients (see Appendix of Chap. 3 for details of the dataset), it was found that the expression levels of some genes are age-dependent, gender-dependent and smoking history-dependent, while other genes are, in large part, independent of any of these features (Cui et al. 2011). When working with these datasets, it was noted that the baseline expression levels of 143 genes were highly age-dependent, including *MUC1* (mucin 1), *UBFD1* (ubiquitin family domain 1) and *MDK* (neurite growth-promoting factor 2). In addition, 59 genes were gender-dependent; these included *WNT2* (wingless-type MMTV integration site family, member 2), *ARSE* (arylsulfatase E) and *KCNN2* (potassium intermediate/small conductance calcium-activated channel, subfamily N, member 2) (see (Cui et al. 2011) for details). Similar analyses can be carried out on dependence using various lifestyle habits such as smoking and medications.

Knowing such information, one then needs to make age or gender corrections on the observed gene-expression data before interpreting the data for functional inference. The detailed correction scheme depends on the actual relationship between a specific attribute and the gene-expression levels. Various normalization techniques and software tools are publicly available for this purpose.

2.6 A Case Study of Integrative *Omic* Data Analyses

We present an example here to show how integrative analyses of multiple *omic* and computational data types can lead to new insights about cancer mechanisms. The main question being addressed here is: *What makes metastatic cancers grow substantially faster than their primary cancer counterparts?* While a detailed model for this problem is given in Chap. 11, the current focus is on how this problem can be approached through transcriptomic data analyses coupled with limited metabolomic data analyses.

To address this question, all the transcriptomic data of metastatic cancers, along with their corresponding primary cancers, were collected on the Internet. Sixteen large sets of genome-scale transcriptomic data covering 11 types of metastatic and corresponding primary cancers were extracted from the GEO database, including breast to bone, breast to brain, breast to liver, breast to lung, colon to liver, colon to lung, kidney-to-lung, pancreas to liver and lung, and prostate to bone and liver. The detailed information of these datasets is given in Chap. 11.

The first question addressed is: *Which genes are consistently up-regulated in metastatic cancers in comparison with their corresponding primary cancers across all these datasets?* Simple statistical analyses led to the identification of about 100 such genes.

The second question asked is: *What do these genes do in terms of cellular function(s)?* Pathway enrichment analyses of these genes using DAVID against KEGG, REACTOME and BIOCARTA revealed that the most significantly enriched pathway was "cholesterol uptake and metabolism". Two questions were then asked: (a) *What does cholesterol do in metastatic cancer cells?* And (b) *Why do metastatic cancer cells need more cholesterol*, as suggested by the observation that at least one cholesterol-containing lipoprotein transporter gene, *SRB1* (scavenger receptor B), *LDLR* (low density lipoprotein receptor) or *VLDLR* (very low density lipoprotein receptor) was substantially up-regulated compared to the corresponding primary cancers except for some brain metastases. These metastases synthesize cholesterol *de novo* as cholesterol-containing lipoproteins probably could not enter brain tissue due to the blood-brain barrier (Bjorkhem and Meaney 2004).

Here, only the first question is considered. It was noted that multiple *CYP* (cytochrome P450) genes are up-regulated in each metastatic cancer type: these genes encode enzymes for oxidizing cholesterols to various oxysterols or bile acids. Some of these oxysterols are further metabolized to steroid hormones such as estrogens, androgens or steroidogenic derivatives by various enzymes whose genes show substantially increased expression levels in comparison with their corresponding primary cancers. A number of these steroid products can bind with and activate different nuclear receptors, such as *FXR* (farnesoid X receptor) and *ER* (estrogen receptor) (see Chap. 11). Various growth-factor receptors such as *FGFR* (fibroblast growth factor receptor) and *EGFR* (epidermal growth factor receptor) are up-regulated in different metastatic cancers, some of which can be directly activated by oxysterols and/or steroid hormones, whose abundances tend to be substantially elevated in metastatic cancers. For the other growth factor receptors, strong correlations between their gene-expressions and the expression patterns of the various nuclear receptors are observed across different metastases, thus suggesting the possibility of a functional relationship between the activation of the two sets of receptors. Based on more detailed analyses and validation, a mechanistic model for how metastatic cancers utilize oxidized cholesterols to accelerate their growth is presented in Chap. 11. Similar integrative analyses of multiple types of data can be carried out to derive the mechanistic models for a large variety of poorly understood cancer-related processes if one can ask the right questions that could be answered through analyses and mining of the relevant *omic* data.

2.7   Concluding Remarks

A substantial amount of information concerning the activities of individual biochemical pathways, their dynamics and the complex relationships among them, and with respect to various micro-environmental factors, is hidden in the very large pool of publicly available cancer *omic* data, including transcriptomic, genomic, metabolomic and epigenomic data. Powerful statistical analysis techniques can aid immensely in uncovering such information if one poses the right questions. Such focused questions create a framework for hypothesis-guided data analysis and mining to check for the validity of the formulated hypothesis, as well as for guiding the formulation of further questions, which may ultimately lead to the elucidation of specific pathways or even possibly causal relationships among the activities of different pathways. More powerful analysis tools for different *omic* data types are clearly needed in order to address more complex and deeper questions about the available data such as de-convolution of gene-expression data collected on tissue samples consisting of multiple cell types and inference of causal relationships. Integrative analyses of multiple types of *omic* and computational data will prove to be the key to effective data mining and information discovery. A large number of examples are presented throughout the following chapters regarding how best to address various cancer biology inquiries, including fundamental questions, through mining the available *omic* data.

References

Bairoch A (2000) The ENZYME database in 2000. Nucleic acids research 28: 304–305

Barkow S, Bleuler S, Prelic A et al. (2006) BicAT: a biclustering analysis toolbox. Bioinformatics 22: 1282–1283

Bentley J, Diggle CP, Harnden P et al. (2004) DNA double strand break repair in human bladder cancer is error prone and involves microhomology-associated end-joining. Nucleic acids research 32: 5249–5259

Bindra RS, Crosby ME, Glazer PM (2007) Regulation of DNA repair in hypoxic cancer cells. Cancer Metastasis Rev 26: 249–260

Bjorkhem I, Meaney S (2004) Brain cholesterol: long secret life behind a barrier. Arterioscler Thromb Vasc Biol 24: 806–815

Boshier PR, Marczin N, Hanna GB (2010) Repeatability of the measurement of exhaled volatile metabolites using selected ion flow tube mass spectrometry. J Am Soc Mass Spectrom 21: 1070–1074

Cancer-Genome-Atlas-Research (2008) Comprehensive genomic characterization defines human glioblastoma genes and core pathways. Nature 455: 1061–1068

Cancer-Genome-Atlas-Research (2011) Integrated genomic analyses of ovarian carcinoma. Nature 474: 609–615

Cancer-Genome-Atlas-Research (2012a) Comprehensive genomic characterization of squamous cell lung cancers. Nature 489: 519–525

Cancer-Genome-Atlas-Research (2012b) Comprehensive molecular characterization of human colon and rectal cancer. Nature 487: 330–337

Cancer-Genome-Atlas-Research (2012c) Comprehensive molecular portraits of human breast tumours. Nature 490: 61–70

Cancer-Genome-Atlas-Research (2013a) Comprehensive molecular characterization of clear cell renal cell carcinoma. Nature 499: 43–49

Cancer-Genome-Atlas-Research (2013b) Genomic and epigenomic landscapes of adult de novo acute myeloid leukemia. The New England journal of medicine 368: 2059–2074

Cancer-Genome-Atlas-Research, Weinstein JN, Collisson EA et al. (2013) The Cancer Genome Atlas Pan-Cancer analysis project. Nature genetics 45: 1113–1120

Cavalli-Sforza LL (2005) The Human Genome Diversity Project: past, present and future. Nat Rev Genet 6: 333–340

Chen JS, Hung WS, Chan HH et al. (2013) In silico identification of oncogenic potential of fyn-related kinase in hepatocellular carcinoma. Bioinformatics 29: 420–427

Cheng Z, Ventura M, She X et al. (2005) A genome-wide comparison of recent chimpanzee and human segmental duplications. Nature 437: 88–93

Church GM (2005) The personal genome project. Mol Syst Biol 1: 2005 0030

Colotta F, Allavena P, Sica A et al. (2009) Cancer-related inflammation, the seventh hallmark of cancer: links to genetic instability. Carcinogenesis 30: 1073–1081

Cotton RG, Auerbach AD, Axton M et al. (2008) GENETICS. The Human Variome Project. Science 322: 861–862

Croft D, O'Kelly G, Wu G et al. (2011) Reactome: a database of reactions, pathways and biological processes. Nucleic acids research 39: D691–697

Cui J, Chen Y, Chou WC et al. (2011) An integrated transcriptomic and computational analysis for biomarker identification in gastric cancer. Nucleic acids research 39: 1197–1207

D'Souza I, Poorkaj P, Hong M et al. (1999) Missense and silent tau gene mutations cause frontotemporal dementia with parkinsonism-chromosome 17 type, by affecting multiple alternative RNA splicing regulatory elements. Proceedings of the National Academy of Sciences of the United States of America 96: 5598–5603

Downing JR, Wilson RK, Zhang J et al. (2012) The Pediatric Cancer Genome Project. Nature genetics 44: 619–622

Edgar R, Domrachev M, Lash AE (2002) Gene Expression Omnibus: NCBI gene expression and hybridization array data repository. Nucleic acids research 30: 207–210

Emmert-Buck MR, Bonner RF, Smith PD et al. (1996) Laser capture microdissection. Science 274: 998–1001

Essaghir A, Toffalini F, Knoops L et al. (2010) Transcription factor regulation can be accurately predicted from the presence of target gene signatures in microarray gene expression data. Nucleic acids research 38: e120

Forbes SA, Bindal N, Bamford S et al. (2011) COSMIC: mining complete cancer genomes in the Catalogue of Somatic Mutations in Cancer. Nucleic acids research 39: D945–D950

Grabherr MG, Haas BJ, Yassour M et al. (2011) Full-length transcriptome assembly from RNA-Seq data without a reference genome. Nat Biotechnol 29: 644–652

Green RE, Krause J, Briggs AW et al. (2010) A draft sequence of the Neanderthal genome. Science 328: 710–722

Greenman C, Stephens P, Smith R et al. (2007) Patterns of somatic mutation in human cancer genomes. Nature 446: 153–158

Hanahan D, Weinberg RA (2011) Hallmarks of cancer: the next generation. Cell 144: 646–674

Huang D, Sherman BT, Tan Q et al. (2007) The DAVID Gene Functional Classification Tool: a novel biological module-centric algorithm to functionally analyze large gene lists. Genome Biol 8: R183

Huang DW, Sherman BT, Lempicki RA (2009) Systematic and integrative analysis of large gene lists using DAVID bioinformatics resources. Nature Protocols 4: 44–57

Huebert DJ, Kamal M, O'Donovan A et al. (2006) Genome-wide analysis of histone modifications by ChIP-on-chip. Methods 40: 365–369

International-HapMap (2003) The International HapMap Project. Nature 426: 789–796

Iyer VR, Horak CE, Scafe CS et al. (2001) Genomic binding sites of the yeast cell-cycle transcription factors SBF and MBF. Nature 409: 533–538

Kamakaka RT, Biggins S (2005) Histone variants: deviants? Genes & development 19: 295–310

Kandoth C, Schultz N, Cherniack AD et al. (2013) Integrated genomic characterization of endometrial carcinoma. Nature 497: 67–73

Kanehisa M, Goto S, Furumichi M et al. (2010) KEGG for representation and analysis of molecular networks involving diseases and drugs. Nucleic acids research 38: D355–360

Kanehisa M, Goto S, Sato Y et al. (2012) KEGG for integration and interpretation of large-scale molecular data sets. Nucleic acids research 40: D109–114

Kanehisa M, Goto S, Sato Y et al. (2014) Data, information, knowledge and principle: back to metabolism in KEGG. Nucleic acids research 42: D199–205

Lander ES, Linton LM, Birren B et al. (2001) Initial sequencing and analysis of the human genome. Nature 409: 860–921

Li G, Ma Q, Tang H et al. (2009) QUBIC: a qualitative biclustering algorithm for analyses of gene expression data. Nucleic acids research 37: e101

Liu R, Hannenhalli S, Bucan M (2009) Motifs and cis-regulatory modules mediating the expression of genes co-expressed in presynaptic neurons. Genome Biol 10: R72

Lu P, Weaver VM, Werb Z (2012) The extracellular matrix: a dynamic niche in cancer progression. The Journal of cell biology 196: 395–406

McKeehan W (1982) Glycolysis, glutaminolysis and cell proliferation. Cell Biology International Reports 6: 635–650

Nazarov PV, Muller A, Khutko V et al. Co-Expression Analysis of Large Microarray Data Sets using Coexpress Software Tool. In: Seventh International Workshop on Computational Systems Biology, WCSB 2010, 2010.

Nishimura D (2001) BioCarta. Biotech Software & Internet Report 2:

Nowell P, Hungerford D (1960) A minute chromosome in human chronic granulocytic leukemia. Science 132:

Pearl J (2009) Causal inference in statistics: An overview. Statistics Surveys 3: 96–146

Qian J, Lin J, Luscombe NM et al. (2003) Prediction of regulatory networks: genome-wide identification of transcription factor targets from gene expression data. Bioinformatics 19: 1917–1926

Ren B, Robert F, Wyrick JJ et al. (2000) Genome-wide location and function of DNA binding proteins. Science 290: 2306–2309

Scanlon S, Glazer P (2013) Genetic Instability Induced by Hypoxic Stress. In: Mittelman D (ed) Stress-Induced Mutagenesis. Springer New York, pp 151–181

Scheer M, Grote A, Chang A et al. (2011) BRENDA, the enzyme information system in 2011. Nucleic acids research 39: D670–676

Schellenberger J, Park JO, Conrad TM et al. (2010) BiGG: a Biochemical Genetic and Genomic knowledgebase of large scale metabolic reconstructions. BMC Bioinformatics 11: 213

Segal E, Yelensky R, Kaushal A et al. (2004) GeneXPress: A Visualization and Statistical Analysis Tool for Gene Expression and Sequence Data. Paper presented at the 11th International Conference on Intelligent Systems for Molecular Biology (ISMB),

Service RF (2006) Gene sequencing. The race for the $1000 genome. Science 311: 1544–1546

Steel M, Penny D (2000) Parsimony, likelihood, and the role of models in molecular phylogenetics. Molecular biology and evolution 17: 839–850

Strahl BD, Allis CD (2000) The language of covalent histone modifications. Nature 403: 41–45

The-National-Cancer-Act (1971) The National Cancer Act of 1971.

The-Tumor-Metabolome (2011) The tumor metabolome.

Trapnell C, Williams BA, Pertea G et al. (2010) Transcript assembly and quantification by RNA-Seq reveals unannotated transcripts and isoform switching during cell differentiation. Nat Biotechnol 28: 511–515

Truong LN, Li Y, Shi LZ et al. (2013) Microhomology-mediated End Joining and Homologous Recombination share the initial end resection step to repair DNA double-strand breaks in mammalian cells. Proceedings of the National Academy of Sciences of the United States of America 110: 7720–7725

Valen LV (1973) A new evolutionary law. Evolutionary Theory 1: 1–30

Varma A, Palsson BO (1994) Metabolic Flux Balancing: Basic Concepts, Scientific and Practical Use. Nature Biotechnology 12: 994–998

Venter JC, Adams MD, Myers EW et al. (2001) The sequence of the human genome. Science 291: 1304–1351

Vignali M, Hassan AH, Neely KE et al. (2000) ATP-dependent chromatin-remodeling complexes. Molecular and cellular biology 20: 1899–1910

Wang Z, Gerstein M, Snyder M (2009) RNA-Seq: a revolutionary tool for transcriptomics. Nat Rev Genet 10: 57–63

Wishart DS, Jewison T, Guo AC et al. (2013) HMDB 3.0–The Human Metabolome Database in 2013. Nucleic acids research 41: D801–807

Wishart DS, Knox C, Guo AC et al. (2009) HMDB: a knowledgebase for the human metabolome. Nucleic acids research 37: D603-610

Wishart DS, Tzur D, Knox C et al. (2007) HMDB: the Human Metabolome Database. Nucleic acids research 35: D521–526

Xu K, Mao X, Mehta M et al. (2012) A comparative study of gene-expression data of basal cell carcinoma and melanoma reveals new insights about the two cancers. PLoS One 7: e30750

Yang AS, Estecio MR, Doshi K et al. (2004) A simple method for estimating global DNA methylation using bisulfite PCR of repetitive DNA elements. Nucleic acids research 32: e38

Zhao M, Sun JC, Zhao ZM (2013) TSGene: a web resource for tumor suppressor genes. Nucleic acids research 41: D970–D976

Chapter 3
Cancer Classification and Molecular Signature Identification

Cancer is a family of diseases that share a common set of characteristics such as reprogrammed energy metabolism, uncontrolled cell growth, tumor angiogenesis and avoidance of immune destruction, referred to as cancer hallmarks, as introduced in Chap. 1. Based on their original cell types, cancers are classified into five classes: (1) *carcinoma*, which begins in epithelial cells and represents the majority of the human cancer cases; (2) *sarcoma*, derived from mesenchymal cells, e.g., connective tissue cells such as fibroblasts; (3) *lymphoma, leukemia* and *myeloma*, originating in hematopoietic or blood-forming cells; (4) *germ cell tumors*, developing, as the name implies, from germ cells; and (5) *neuroblastoma, glioma, glioblastoma* and others derived from cells of the central and peripheral nervous system and denoted as *neuroectodermal* tumors because of their beginning in the early embryo. Each class may consist of cancers of different types. For example, carcinoma comprises adenocarcinoma, basal-cell carcinoma, small-cell carcinoma and squamous cell carcinoma, independent of their underlying tissue types. Cancers of the same type and developing in the same tissue may have distinct properties in terms of their growth patterns, malignance levels, survival rates and possibly even different underlying mechanisms. They may respond differently to the same drug treatment and hence have different mortality rates. As of now, over 200 types of human cancers have been identified and characterized (Stewart and Kleihues 2003), the majority of which are determined based on the location, the originating cell type and cell morphology. It is now becoming evident that this type of classification, in large part subjective, is not adequate for developing personalized treatment plans, which are becoming increasingly desirable and clearly represents the future of cancer medicine.

With the rapid accumulation of high-throughput *omic* data for cancer, particularly transcriptomic and genomic data, it is now feasible to classify cancers based on their molecular level information. For example, this can be based on distinct expression patterns of certain genes or pathways shared only by samples of the same cancer

© Springer Science+Business Media New York 2014
Y. Xu et al., *Cancer Bioinformatics*, DOI 10.1007/978-1-4939-1381-7_3

type, or combinations of mutations that tend to co-occur (or be selected, to be more accurate) in certain cancer types. Such type-defining expression or mutation patterns of genes are referred to as the *signature* of a cancer type. This idea should be applicable to every kind of cancer as has been done for a few cancer types, such as Oncotype DX for one form of breast cancer (Albain et al. 2010), as long as transcriptomic or genomic mutation data are available for the cancer category. Similarly, it should also be possible to derive molecular signatures for cancer grades and cancer stages, with the former referring to the level of malignancy of a tumor and the latter representing the location of the cancer in its development towards the terminal stage, i.e., metastasis. Compared to the traditional definitions of cancer types, molecular signatures, as outlined here, can potentially provide more accurate characterization of a cancer and even reveal its underlying mechanisms, hence possibly having significant implications to cancer treatment and prognosis prediction. Here we use gene-expression data as an example to illustrate how cancer typing, staging and grading can be done using *omic* data, which could potentially lead to substantially more accurate characterization of cancers of different types, grades and stages. Similar ideas should be applicable to mutation-based cancer classification.

3.1 Cancer Types, Grades and Stages

The earliest description of cancer can be traced back to 2500 BC by Egyptian physician Imhotep (Mukherjee 2010). Evidence exists suggesting that Egyptian physicians at the time could distinguish between benign and malignant tumors. The study of cancer as a scientific discipline came in the nineteenth century when microscopes became widely available to physicians and surgeons. Microscopic pathology, pioneered by German doctor Rudolf Virchow, laid the foundation for the development of cancer surgery as practiced now. Since then, cancer tissues removed from patients are microscopically examined and classified based on their morphological characteristics. Scientific oncology was born out of the debate concerning a few competing hypotheses regarding the possible causes of cancer in the late 1800s through the early 1900s. It developed based on findings that linked microscopic observations made on cancer tissues to clinical data during the course of the disease development. The popular hypotheses included: (1) one proposed by Stahl and Hoffman, which suggested that cancer was caused by coagulated lymph; (2) a proposal by Johannes Muller who suggested that cancer cells arose from budding elements between normal tissues; and (3) the theory developed by Rudolph Virchow, which considered cancer as a disease of cells. The next major advance in attempts to elucidate the possible causes of a cancer came in the 1920s when the German biochemist Otto Warburg observed that cancer cells rely heavily on glycolytic fermentation rather than the more efficient oxidative phosphorylation for ATP generation, even when oxygen is available. This metabolic alteration is referred to as the *Warburg effect* (Warburg 1956) and remains under active investigation as discussed in depth in Chap. 5. Based on the accelerated glycolysis, some 10 to 20-fold over that of normal cells, Warburg attributed cancer to a malfunctioning mitochondria-induced metabolic disease. The discovery of oncogenes in 1970s by Bishop and Varmus, along with the discovery of tumor-suppressor genes by A. G. Knudson also in 1970s, represented the next key advancement, which started the era of classifying cancer as a genetic disease.

Early classification of cancers was based on a cancer's location, such as lung cancer, skin cancer or blood cancer (e.g., leukemia). Over time, oncologists began to realize that different types of cancers can develop from the same organ. The earliest classification of cancers from the same organ, in this case bone marrow which houses the hematopoietic stem cells, can be traced back to the early 1900s when it was found that there were at least four types of leukemia, namely ALL (acute lymphoblastic leukemia), AML (acute myelogenous leukemia), CLL (chronic lymphoblastic leukemia) and CML. This realization occurred about 50 years after the diagnosis of the first documented leukemia case (Beutler 2001). For other cancers, recognition of multiple cancer types originating from the same organ came rather late. For example, small-cell lung cancer was not considered as a separate type of lung cancer from the more prevalent and less aggressive non-small cell lung cancer until the 1960s. Gastric cancers were found to have at least two subtypes, intestinal and diffuse, in 1965 (Lauren 1965). It is worth noting that correct diagnosis of a cancer type has significant implications to designing the most effective treatment protocols and prognosis. For example, statistics show that the current 5-year survival rates for adult ALL, AML, CLL and CML patients are 50 %, 40 %, 75 % and 90 %, respectively, and the treatment plan for each of them is quite different. ALL is typically treated using chemotherapy followed by anti-metabolite drugs; AML is generally treated using chemotherapy; CLL, while incurable, is often being controlled with chemotherapy using a combination of fludarabine and alkylating agents; and CML is, in most cases, successfully treated using the so called "miracle" drug Gleevec, or else newer and improved drugs.

The multistage nature of a cancer was first discovered by Japanese researchers Yamagiwa and Ichikawa in the beginning of the twentieth century (Yamagiwa and Ichikawa 1918). Basically for most cancer types, the histological stage refers to the extent the cancer has spread, which is typically numbered from stage I through stage IV, with IV representing the most advanced stage. The stage of a cancer is an important predictor for survival, with the treatment plan often determined based on staging. Currently the stage of a cancer is generally determined by pathological analysis from a biopsied specimen of the cancer tissue, including lymph nodes, as well as analysis by imaging techniques with the results interpreted by radiologists; only limited molecular level information such as the expression levels of a few marker genes as determined by immune-detection.

In addition to type and stage, cancer grade is another important parameter that has been used by pathologists to represent the level of malignancy of a given cancer, determined based on surgical specimens. This parameter is largely independent of the type and the stage of a cancer. A popular grading system uses four grades: (1) G1 (highly differentiated), (2) G2 (moderately differentiated), (3) G3 (poorly differentiated) and (4) G4 (undifferentiated), with G4 representing the most malignant.

The level of differentiation refers to the maturity of a cell in developmental biology. In the current context, the more differentiated cancer cells resemble more of the normal mature cells, and they tend to grow and spread at slower rates than undifferentiated or poorly differentiated cancer cells. The grade of a cancer provides another key indicator for prognosis. While the term seems to be defined in terms of cellular differentiation, the actual determination of the cancer grade is often made based on a combination of the cellular appearance (degree of abnormality), the rate of growth and the degree of invasiveness.

The current availability of significant quantities of molecular level *omic* data on cancer, such as transcriptomic, genomic, epigenomic and metabolomic data, provides unprecedented opportunities for developing molecular-level signatures for each known cancer type, grade and stage, and, if needed, possibly reclassifying some of the previously determined cancer types, stages and/or grades. This has the potential to lead to more accurate classifications of a cancer for the purpose of improved treatment design and prognosis evaluation.

3.2 Computational Cancer Typing, Staging and Grading Through Data Classification

The main question addressed here is: For a given set of cancer samples, each marked with a specific type, stage or grade determined by pathologists, *is it possible to identify common characteristics, e.g., in terms of gene expression patterns among samples having the same class label*? If the answer is yes, such a capability could potentially be used to accurately define the type or subtype, stage or substage, grade or subgrade of a cancer. In the following sections, we demonstrate how this could be done to possibly provide a new way of classifying cancer based on molecular level data.

3.2.1 Cancer Typing

A basis for gene-expression data-based cancer typing is that cancers of various types have their distinct phenotypic characteristics such as differences in cellular shape, growth rates and responses to the same treatment regiments, and possibly distinct underlying mechanisms, while samples of the same type tend to share common characteristics. These phenotypic and mechanistic commonalities among cancer cases of the same type as well as differences across multiple cancer types are realized through molecular level activities and hence should be in general reflected by the expression patterns of some genes. A key in accomplishing cancer typing based on gene-expression data is to identify those genes whose expression patterns are shared by samples of the same type but not shared by samples of the other cancer types. This problem can be modeled computationally in various ways, depending on the specific purpose(s) of the cancer typing. For example, if the goal is to identify the defining characteristics of a cancer type, one may decide to identify a maximal gene set, whose expression patterns are similar across all the (available) cancer samples of the same type and different from those of other types. If, instead, the goal is to identify distinguishing characteristics between two (or more) types of cancers, one may want to find a minimal set of genes whose expression patterns can delineate among samples between the two (or more) cancer types, which may not necessarily contain any information about the distinct mechanisms of the different cancer types.

We now present one example to model the cancer typing problem and to illustrate how such a problem can be solved computationally. Consider two subtypes of gastric cancer, the intestinal (C_1) and diffuse (C_2) subtypes, each having genome-scale gene-expression data collected using the same platform on paired cancer and matching control tissue samples from the same patients. For each patient one can obtain the fold-change information for any gene between its expression in a cancer and its matching control, which is typically calculated as the logarithm of the ratio between the two expression levels, referred to as the *log-ratio* throughout this book. The present goal is to find a minimal subset of genes out of the total of ~20,000 human genes, whose expression patterns can unequivocally distinguish between the two subtypes, C_1 and C_2. Specifically, the aim is to identify a set G of genes and a discriminant function F() so that $F(G(x))>0$ for $x \in C_1$ and $F(G(x))<0$ for $x \in C_2$ for as many $x \in C_1 \cup C_2$ as possible, where $G(x)$ represents the list of fold-changes in expression levels of genes in G between cancer tissue x and its matching control. There are many classes of discriminant functions that can be used for solving this classification problem. Here a specific class of functions is used, the linear *support vector machine* (SVM) (Cortes and Vapnik 1995). The goal now becomes that of locating a minimal set G of genes and an SVM that achieve the best classification with the misclassification rate lower than a pre-defined threshold δ.

One method of solving this problem is by going through all combinations of K genes among all the human genes, searching from $K=1$ and up until an SVM-based classifier and a K-gene set G are found, which achieve the desired classification accuracy defined by δ. In practice, the search will not include all the human genes since the majority will not be expressed for any specific tissue type. For this problem, one only needs to consider genes that are differentially expressed between cancer samples and the matching controls. To get a sense of the amount of computing time that may be needed to exhaustively search through all K-gene combinations, consider the following typical scenario: the two gene-expression datasets with C_1 having 100 pairs of samples and C_2 consisting of 150 pairs of samples; and 500 genes showing differential expressions (see Chap. 2) between the two sets of samples. In this case, one would need to examine $\binom{500}{K}$ combinations to find a K-gene combination that achieves the optimal classification between the two datasets. For each K-gene combination, a linear SVM is trained to optimally classify the two datasets as discussed above; if a trained SVM achieves a classification accuracy better than δ, retain the SVM as a candidate classifier; then repeat this process until all

K-gene combinations are exhausted. The final classifier is the one with the lowest misclassification rate among all those retained. Our experience has been that K should be no larger than 8; otherwise the number $\begin{pmatrix} 500 \\ K \end{pmatrix}$ may be too large for a desktop workstation to handle. The following gives a detailed procedure of the search process:

Cancer classification algorithm

FOR $K = 1$ **TO** N **DO**

> **FOR** each K-gene combination from the pool of differentially expressed genes **DO**
>
>> a. **DO** the following **FOR** 1,000 times
>>
>>> 1. Randomly split C_1 and C_2 into C_1-training and C_1-testing, and C_2-training and C_2-testing, respectively, with C_x-training and C_x-testing having the same size, $x \in \{1, 2\}$;
>>> 2. Train a linear SVM based on the current K-gene combination on C_1-training and C_2-training, which achieves optimal classification between C_1-testing and C_2-testing;
>>> 3. **IF** the misclassification rate of the trained SVM is $< \delta$, **THEN** keep the SVM;
>>
>> b. **IF** at least one SVM for the K-gene combination has misclassification rates $< \delta$, **THEN** keep the K-gene combination with the lowest misclassification rate a candidate for the final classifier.

IF at least one final classifier candidate is found, **THEN OUTPUT** the one with the lowest misclassification rate, **ELSE OUTPUT** no classifier is found with at most N genes and misclassification rate $< \delta$.

where N is the upper bound (set by the user) for searching a satisfying K-gene discriminator, and 1,000 is the number of times used to find an optimal K-gene classifier over different partitions of the given datasets C_1 and C_2.

This simple procedure has been used to find an optimal SVM-based classifier between the two subtypes of gastric cancer based on gene-expression data collected on 80 pairs of gastric cancer and matching controls (Cui et al. 2011a). Figure 3.1 shows classification accuracies by the best K-gene classifiers for $K \leq 8$.

If one needs to search for a K-gene classifier with larger K's (> 8) for some application, a different search strategy may be needed to make it computationally feasible. One such strategy is called *recursive feature elimination*, a procedure often used in conjunction with an SVM application; together they are referred to as *RFE-SVM*. While the detailed information of an RFE-SVM procedure can be found in (Guyon et al. 2002; Inza et al. 2004), the basic idea is to start with a list of all genes, each having some discerning power in distinguishing between the two classes of samples, and to train a classifier, followed with the RFE procedure to repeatedly

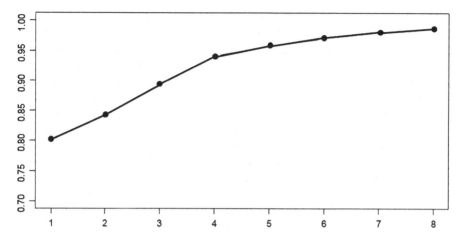

Fig. 3.1 SVM-based classification accuracy using the best K-gene combination, for $K = 1, 2, \ldots,$ 8, on 80 pairs of gastric cancer and control tissues

remove genes from the initial gene list as long as the classification accuracy is not affected until only K genes are left.

If desired, this idea for solving a 2-class classification problem can be generalized to M-class problems, for $M > 2$, so multi-type cancers originating from the same tissue, such as the different types of leukemia, can be classified based on identification and application of K-gene combinations as done above. One specific way to accomplish this is given as follow: a M-class classifier can be constructed by separately calculating M binary classifiers, each separating class i from the remaining classes, $i = 1, \ldots, M$. Then, an input sample is classified to class J if the sample has the highest classification significance by the Jth classifier. Such a method is regarded as one-*versus*-all multi-class SVM (Cui et al. 2011a). A detailed review on such classifiers can be found in (Duan and Keerthi 2005). Using this type of classification method, one can build classifiers for all the cancer types as long as they have gene-expression data available, along with labeled type information for each sample.

Numerous K-gene combinations, also referred to as *K-gene panels*, have been identified and used as signatures for various cancer types. For example, a panel of 104 genes has been identified for distinguishing cancer tissues (of multiple types) from healthy tissues (Starmans et al. 2008), aimed to detect if a tissue is cancerous or not. Other signature panels include: (1) a 70-gene panel for predicting the potential for developing breast cancer, built by MammaPrint (Slodkowska and Ross 2009); (2) a 21-gene panel, termed *Oncotype DX*, for a similar purpose; (3) a 71-gene panel for identification of cancers that are sensitive to *TRAIL*-induced apoptosis (Chen et al. 2012); (4) a 31-gene panel used to predict the metastasis potential of a breast cancer, developed by CompanDX (Cho et al. 2012); and (5) a 16-gene panel for testing for non-small-cell lung cancer against other lung cancer types (Shedden et al. 2008). Having a test kit for a specific cancer type, e.g., metastasis-prone or not, can enable surgeons to make a rapid and informed decision regarding the appropriate surgical procedure to adopt. Other test kits can assist oncologists in making an informed decision regarding the most appropriate treatment plan for a particular cancer case. For example, *TRAIL* (*TNF*-related apoptosis inducing ligand) is an anticancer-mediating protein that can induce apoptosis in cancer cells but not in normal cells. This makes *TRAIL* highly desirable; however, not all cancers are sensitive to *TRAIL*. Hence, having a test using such a kit can quickly determine if a cancer patient should be treated with *TRAIL* or not.

In order to ensure the general applicability of any identified signature genes, it is essential to carry out proper normalization of the to-be-used transcriptomic data that may be collected by different research labs, specifically to correct any systematic errors in the data caused by different sample-preparation and data-collection protocols. Batch-based normalization such as the model presented in (Johnson et al. 2007) may prove to be effective in removing so created systematic errors due to using different data-collection protocols.

Although a number of computational methods have been developed for defining cancer types using gene-expression data (Ramaswamy et al. 2001; Tibshirani et al. 2002; Weigelt et al. 2010; Reis-Filho and Pusztai 2011), none of them have achieved 100 % consistency with the typing results determined by cancer pathologists. There may be two key reasons for the less-than-perfect agreement. One is that some of the cancer typing decisions by pathologists may not necessarily be correct for various reasons: (a) a cancer identification protocol may use only limited molecular level and somewhat subjective visual information; and (b) there is always the possibility of human errors in executing a type-calling procedure, particularly when visual appearances may be borderline between different options. Another possibility could be due to limitations of the current classification techniques. For example, the above classification methods may be too simple to capture the complex relationships among the expression data of multiple genes, which are unique to a specific cancer type. Moreover, it may be due to something more fundamental, such as the gene expression data not necessarily having all the information needed to classify cancer types correctly, e.g., some of the needed information may be at the protein or the post-translational level. It is expected that answers to this question may emerge as more cancer *omic* data become available and/or when more advanced analysis techniques will be developed.

3.2.2 Cancer Staging

Cancer stages have been defined mainly in terms of the tumor size, cell morphology and the state of metastasis. Currently its determination involves some level of subjectivity by pathologists. Like cancer types, cancer stages can also be defined in terms of expression patterns of some subset of the human genes. A number of studies have been published on applications of computational techniques to predict the stage of a cancer based on gene-expression data (Eddy et al. 2010; Goodison et al. 2010; Liong et al. 2012). For example, a 7-gene panel (*ANPEP, ABL1, PSCA, EFNA1, HSPB1, INMT, TRIP13*) was used to measure the progression of prostate cancer and achieved high-80 % consistencies with pathologically-determined stages (Liong et al. 2012). Another example is a 4-gene panel (*IL1B, S100A8, S100A9, EGFR*) for assessing the progression of muscle invasive bladder cancer (Kim et al. 2011). Similar gene panels have been developed for a few other cancers, such as breast cancer (Rodenhiser et al. 2011; Arranz et al. 2012), colon cancer (Erten et al. 2012) and oral cancer (Mroz and Rocco 2012).

Potentially, one can develop such gene-panels for any cancer as long as transcriptomic data for cancer and control tissues, along with their stage information, are available. Here we use gastric cancer again as an example to illustrate how gene-expression data can be used to predict the developmental stage of a cancer.

The same set of gene-expression data collected on 80 pairs of gastric cancer and matching noncancerous gastric tissues used in Sect. 3.2.1 is again analyzed. Of the 80 cancer tissues, 4 were in stage I, 7 in stage II, 54 in stage III and 15 in stage IV. The detailed gene-expression data of these samples can be found in the Appendix. Note that these tissue samples are not evenly distributed across the four stages, but this may be a good representation of the actual stage distribution for gastric cancer patients presenting for resection, at least in China where the 80 samples were collected. The present goal is to identify a set of differentially expressed genes between cancer and the matching controls, where the expression patterns adequately reflect the stages of all the gastric cancer samples. On this data set of 80 pairs of samples, 715 genes were found consistently to be differentially expressed between the cancer and the matching controls (Cui et al. 2011a).

A simplified version of the staging problem is considered first, by merging stages I and II samples into one "early stage" group and stages III and IV samples into the "advanced stage" group, making this a 2-stage classification problem. From an analysis of all the differentially-expressed genes, four genes, *CHRM3* (cholinergic receptor), *PCDH7* (protocadherin), *SATB2* (special AT-rich sequence-binding protein) and *PPA1* (pyrophosphatase), were identified, each giving a consistency level with the two combined stages better than 80 % by using a simple fold-change cutoff. When using K-gene combinations for $K > 1$, the classification consistency (with pathologist-determined stages) continues to increase as K increases until it reaches 95 %, and then the improvement becomes asymptotic.

Using the generalized classification scheme outlined in Sect. 3.2.1, one can undertake the 4-stage classification problem. To ascertain if this problem is solvable, we have examined if there are genes whose (average) expression levels change monotonically with the progression of a cancer. Fortunately, numerous such genes are found, suggesting that the problem is solvable. Figure 3.2 shows three such genes, namely *LANCL3* (lanC lanti-biotic synthetase component c-like protein), *MFAP2* (microfibrillar-associated protein) and *PPA1* (pyrophosphatase).

While the average levels of these three genes each change monotonically with cancer progression, they may not necessarily represent the best genes whose

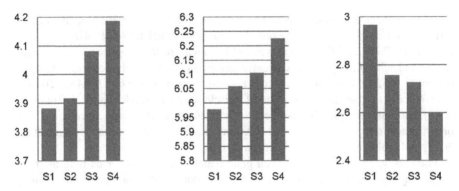

Fig. 3.2 The average gene-expression levels of three genes represented by three panels from *left* to *right*, *LANCL3*, *MFAP2* and *PPA1*, over all samples in each stage for stages S = 1, 2, 3 and 4. The y-axis is the average fold-change of gene-expression levels across all samples of a specific stage in cancer *versus* control samples, and the x-axis is the stage axis. The figure is adapted from Cui et al. (2011b)

expression levels are most informative in predicting cancer stages for individual tissue samples. To find out, an exhaustive search was made for the best *K*-gene discriminator, for $2 \leq K \leq 10$, for the 4-stage classification problem. The combination (*DPT, EIF1AX, FAM26D, IFITM2, LOC401498, OR2AE1, PRRG1, REEP3, RTKN2*) was found to be the best 9-gene signature for gastric cancer staging, and (*CPS1, DEFA5, DES, DMN, GFRA3, MUC17, OR9G1, REEP3, TMED6, TTN*) represents the best 10-gene marker, achieving 84.0 % and 90.0 % 4-stage classification consistencies with the pathologists who did the original staging, respectively (Cui et al. 2011b).

The following table lists the functions of these marker genes, which were retrieved from the GeneCards database (Rebhan et al. 1997), to give the reader a sense about what functional genes may serve as good markers for cancer staging. Interestingly, the two lists have very little in common with only one gene, *REEP3*, shared by the two lists plus a pair of homologous genes, *OR2AE1* and *OR9G1*, in the two lists as shown in the following table. Even by examining cellular level functions, the two sets of pathways enriched with the two gene lists have very little in common. This suggests that there is probably a sizeable set of genes whose expression patterns are informative for the determination of cancer stages, and it just happens that these two lists give rise to the two best discriminators (Table 3.1).

As in the case of cancer typing, the discrepancy between the pathologist-assigned stages and gene-expression-based staging could be due to various reasons as discussed in Sect. 3.2.1. One useful effort will be to refine both definitions through collaboration between cancer pathologists and cancer data analysts. Such a joint effort to identify reasons for staging discrepancies by the two approaches should lead to a refinement of the criteria used by both parties in an iterative fashion until there is convergence. Such an exercise could lead to improvement in cancer-staging based on gene-expression data in a systematic manner. Another important issue is that the current 4-stage classification scheme for measuring cancer progression is probably somewhat arbitrary. There is no strong evidence to support the operational premise

Table 3.1 Functional annotation of the signature genes

Gene name	Function
DPT (dermatopontin)	An extracellular matrix protein involved in cell-matrix interaction and matrix assembly
EIF1AX (ukaryotic translation initiation factor 1α)	An essential translation initiation factor
FAM26D (family with sequence similarity 26, member D)	A pore-forming subunit of a voltage gated ion channel
IFITM2 (interferon induced transmembrane protein 2)	An IFN-induced protein that inhibits the entry of viruses to the host cell cytoplasm
LOC401498 (a hypothetical protein)	No function has been identified
OR2AE1 (olfactory receptor 2AE1)	A hormone receptor responsible for recognition and G protein-mediated transduction of odorant signals
PRRG1 (proline-rich gamma-carboxyglutamic acid protein 1)	The protein containing two functional motifs generally found in signaling and cytoskeletal proteins
REEP3 (receptor accessory protein 3):	May enhance the cell-surface expression of odorant receptors
RTKN2 (rhotekin 2)	May have an important role in lymphopoiesis
CPS1 (carbamoyl-phosphate synthase):	Important in removing excess ammonia from the cell through the urea cycle
DEFA5 (defensin α5)	Has antimicrobial activity and kills microbes by permeabilizing their plasma membrane
DES (intermediate filament protein)	Forms a fibrous network connecting myofibrils to each other and to the plasma membrane
DMN (dystrophin)	A cohesive protein linking actin filaments to another support protein that resides on the inside surface of each muscle fiber's plasma membrane
GFRA3 (glial cell-derived neurotrophic factor family receptor)	Mediates the artemin-induced autophosphorylation and activation of the RET (rearranged during transfection) receptor tyrosine kinase
MUC17 (cell surface associated mucin 17)	Active in maintaining homeostasis on mucosal surfaces
OR9G1 (olfactory receptor, family 9)	May serve as a hormone receptor like OR2AE1 in the above
TMED6 (transmembrane emp24 protein transport domain)	A HNF1α (hepatic nuclear factor 1α) regulated transporter
TTN (connectin)	Contributes to the balance of forces between the two halves of the sarcomere by providing connections at the level of individual microfilaments

that the development of a cancer has four distinct phases, but not three or five or even a continuous progression without obvious phases and phase transitions, say, in terms of their probabilities to metastasize. To rigorously address this issue computationally, it will require not only transcriptomic data of cancer *versus* control tissues, but also data regarding metastases. This is clearly an area where computational approaches could assist in making fundamental and highly meaningful advances.

3.2.3   Cancer Grading

Cancer grading is a less developed area compared to cancer typing and staging. Only a handful of grading systems have been proposed for some cancer types since Bloom and Richardson developed the first grading system for breast cancer in 1957 (Bloom and Richardson 1957). Similar classifications include the Gleason system for prostate cancer (Gleason 1966; Gleason and Mellinger 1974), the Fuhrman method for kidney cancer (Fuhrman et al. 1982) and the approach proposed by Goseki et al. for gastric cancer (Goseki et al. 1992). As of now, only a few grading systems have been developed based on molecular information, such as the Nottingham grading system for breast cancer (Simpson et al. 2000) and the work by Cui et al. for gastric cancer (Cui et al. 2011b). The main challenge here is that, unlike cancer typing and staging, for which some molecular level information has already been used, cancer grading has been solely based on morphologic data of cancer cells and decided by cancer pathologists. Hence, there may be a large gap between pathologist-assigned grades and molecular-level commonalities among samples of the same grade. An example is given here to illustrate the possibility of using transcriptomic data to grade cancer tissues and point out possible issues with the existing grading procedures.

We continue to use the same gastric cancer dataset introduced in Sect. 3.2.1. Out of the 80 gastric cancer tissues, 54 have grades assigned by cancer pathologists (Cui et al. 2011b), so only these data are used for developing a computational method for grading a tumor based on its gene-expression data. Of the 54 tissues, 8 are well differentiated (WD), 9 moderately differentiated (MD), 35 poorly differentiated (PD) and 2 undifferentiated (UD), with the patients' data given in Table 3.2. The aim here is to identify a set of genes whose expression patterns can well distinguish among the four grades of gastric cancer.

As in cancer staging, one can determine if some genes have expression levels that change monotonically with change in cancer grades from highly differentiated to undifferentiated. Using this criterion, 99 such genes were found. For each of these genes, its average fold-change among samples of each grade exhibits a monotonic relationship with the grade list WD-MD-PD-UD from the least malignant to the most malignant, suggesting that the current grading scheme for gastric cancer does have some molecular basis. These genes include *POF1B* (premature ovarian failure 1β), *MET* (hepatocyte growth factor receptor), *CEACAM6* (carcinoembryonic antigen-related cell adhesion molecule), *ZNF367* (zinc finger protein involved in transcriptional activation of erythroid genes), *GKN1* (gastrokine-1 with strong anticancer activity), *LIPF* (gastric lipase with lipid binding and retinyl-palmitate esterase activity), *SLC5A5* (a glutamate transporter), *MUC13* (cell surface associated mucin), *CLDN1* (senescence-associated epithelial membrane protein), *MMP7* (matrix metalloproteinase) and *ATP4A* (ATPase, H+/K + transporting, α). Figure 3.3 shows four examples of these genes in terms of their averaged expression levels *versus* cancer grades across samples of each cancer grade.

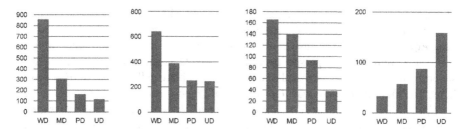

Fig. 3.3 The average gene-expression levels of four genes, *CEACAM6, MUC13, CLDN1* and *PGA4*, over gastric cancer samples of each grade for grades WD, MD, PD and UD. The definitions of the y- and x-axis are the same as in Fig. 3.2. Adapted from Cui et al. (2011b)

Intuitively one may expect that some combinations of the 99 genes should give a good classification among the four grades. However, this may not necessarily be the case for the same reason as discussed in Sect. 3.2.2. Instead, a 19-gene combination is identified, whose expression fold-changes gave a 79.2 % classification consistency with pathologist-assigned grades on two combined grades, namely "highly differentiated" covering the WD and MD samples and "poorly differentiated" for the PD and UD samples, using the algorithm of Sect. 3.2.1. It takes a minimum of 198 genes to give a 4-grade classification at a comparable classification consistency, specifically at 74.2 %.

There may be multiple reasons for the relatively low consistency levels between the pathologist-decided and gene-expression-based grading results, but one key reason, we suspect, may be that the morphological information-based grade arrived at by pathologists may not be as informative in terms of their prognostic values as it could be, at least not on this dataset, indicating the possible limitations of the current approaches and a need for improved techniques.

3.3   Discovering (Sub)Types, (Sub)Stages and (Sub)Grades Through Data Clustering

The analysis presented in Sect. 3.2 is based on the assumption that the pathologist-assigned cancer types, stages and grades are generally correct, i.e., they reflect, to a large extent, the true molecular level commonalities of cancer samples within each type (or stage, grade) and differences across cancer samples of different types (or stages, grades). A more general cancer typing (or staging, grading) problem is to identify cancer types (or stages, grades) when the information of human-designated types (stages and grades) is not available. The question addressed here is: *Can one possibly discover types or subtypes of a cancer based on the similarities among expression patterns of some (to-be-identified) genes* among a subset of cancer and matching control samples. To put it in a more specific context: when given a collection of gene-expression data collected on leukemia samples consisting of four types of leukemia, namely ALL, AML, CLL and CML, but without any labels, *is it possible to rediscover the four types of leukemia from the given samples based solely on their gene-expression data*? The answer is: Yes, but it may take a lot of computing time.

From a computational perspective, this represents a different type of data analysis problem from those discussed in Sect. 3.2, which are called *classification* problems. The main issue there was: *Given a set of objects, each labeled to belong to a specific class, can one identify "features" that can accurately predict the class label* (e.g., *stages or types) of each object based on the features*? For the current problem, the question is: *For the same set of objects, can one partition all the objects into a few classes so that objects in each class share some common features that are not shared by objects in other classes*? Using computer science terminology, this is a *clustering* problem.

Clustering techniques have long been used in gene-expression data analyses (Ben-Dor et al. 1999; Wu et al. 2004; D'haeseleer 2005). Through identification of sample groups sharing similar expression patterns of some genes, researchers have identified various previously unknown subclasses of human diseases. The earliest work in cancer class discovery based on gene-expression data was published by Golub et al., which showed that without prior knowledge, the algorithm "discovered" two subtypes of leukemia, namely, AML and ALL, based on the distinct gene-expression patterns among samples of the two subtypes (Golub et al. 1999). Other discoveries of cancer subtypes include: (1) the discovery of five subtypes of breast cancers based on gene-expression patterns, namely, luminal A, luminal B, basal-like, normal-like and *ERBB2*+ groups, which were found to have clinical implications (Livasy et al. 2006); (2) a recent study that classifies colon cancer into six subtypes based on distinct genomic mutation patterns in the samples, namely samples with or without *BRAF*, *KRAS* and *P53* mutations, CpG island methylation patterns, DNA mismatch repair status and the chromosomal instability level. The study also showed clinical relevance of the six subtypes (Marisa et al. 2013); and (3) a study that showed improvement in subtyping over the previously determined subtypes of leukemia using gene-expression data (Yeoh et al. 2002).

These examples signify the importance that the to-be-discovered new subtypes must have clinical relevance. Otherwise such an analysis may lead to clustering results that group cancer samples according to their growth rates, which may share similar expression patterns of some genes but not any common driving or facilitating mechanisms in cancer development, hence limiting their usefulness from a clinical perspective.

Recent studies have revealed one key inadequacy in the current clustering techniques in discovering subgroups having common or similar gene-expression patterns, which are distinct from other subgroups. Specifically, a major issue is that the clustering techniques require a pre-defined subset of genes, based on which tissue samples are grouped according to the similarities in expression patterns of these genes. This, however, is too restrictive for discovering novel subgroups that may have similar expression patterns of some genes that cannot be determined in advance. The computational difficulty in handling this more general clustering problem is that for a problem with m differentially expressed genes, 2^m combinations of genes need to be considered in order to identify a subset of the *m* genes sharing similar expression patterns among some samples. When *m* is relatively large, say even in the range of a few tens, this clustering problem becomes computationally intractable. A more powerful clustering strategy is needed to solve such problems, and *bi-clustering* is one such technique (Van Mechelen et al. 2004).

To understand the basic principles of a bi-clustering algorithm, one can represent a gene-expression dataset as a numeric matrix with each row representing a gene, each column representing a paired (cancer *versus* control) sample, and each entry in the matrix having the log-ratio value between the expression levels of the corresponding gene in the corresponding sample pair. Two genes are considered to have similar expression patterns for a subset of samples if the correlational coefficient between the two genes-corresponding rows across the samples-corresponding columns is above some defined threshold. A *bi-clustering* problem is defined as that locating all (maximal) sub-matrices, in each of which the correlational coefficient between each pair of rows across the samples defined by the sub-matrix is above the specified threshold. Each so defined sub-matrix is called a *bi-cluster*. Clearly, a bi-clustering problem is substantially more general than the traditional clustering problem, in that it enables one to discover previously unknown subclasses of a cancer class (e.g., type, stage or grade). The generality of a bi-clustering problem also makes it considerably more difficult to solve computationally.

A number of algorithms have been proposed to solve this challenging problem (Madeira and Oliveira 2004; Van Mechelen et al. 2004). To assess the effectiveness of the bi-clustering approach in subgroup discovery, we have applied QUBIC (Li et al. 2009), a bi-clustering method we previously developed, to gene-expression data of three leukemia types, ALL, MLL and AML, mixed together with their type information removed. The algorithm can accurately recover the three subtypes of leukemia as shown in Fig. 3.4, suggesting the general feasibility in discovering subtypes from gene-expression data of multiple samples of the same cancer type.

This technique has also been applied to the 80 pairs of gastric cancer expression data for the discovery of possible subgroups among the samples, which led to the identification of 20-plus bi-clusters. Some of these bi-clusters represent previously uncharacterized subtypes of gastric cancer. For example, Fig. 3.5 shows one bi-cluster defined by 42 genes, for which the 80 samples fall into two groups, each sharing common expression patterns of the 42 genes but different between the two groups, specifically the light-gray subset on the left and the dark-gray subset on the right in the figure. Further analyses suggest that the two subgroups may belong to two known subtypes of gastric cancer, namely intestinal and diffuse subtypes (Shah et al. 2011). This conclusion is based on the observation that six of the 42 genes, namely *CNN1, MYH11, LMOD1, MAOB, HSPB8* and *FHL1*, have previously been reported to be differentially expressed between the intestinal and the diffuse subtypes of gastric cancer, which all show similar expression patterns among samples in each subgroup in the figure.

Such a bi-clustering analysis can also be used for discovery of cancer stages and grades. The approach is to first identify genes whose expression patterns change with alterations in stage or grade and then conduct bi-clustering analyses using such genes as the gene set like the above analysis on cancer subtypes.

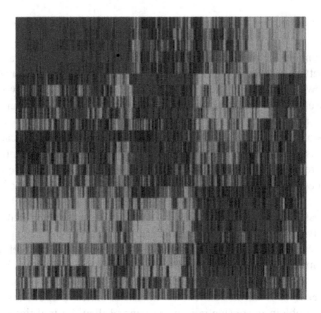

Fig. 3.4 An illustration of the identified three subtypes of leukemia based on gene-expression data using the bi-clustering method QUBIC without using *a prior* knowledge about the three subtypes. The *rows* and *columns* represent genes and samples, respectively, and *dark gray* and *light gray* represent up- and down-regulations, respectively

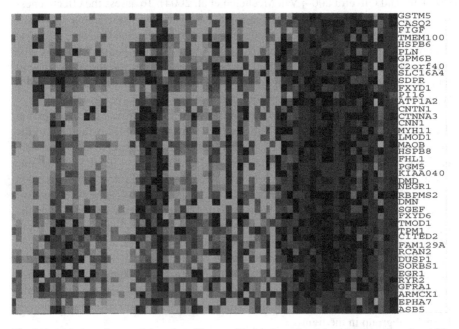

Fig. 3.5 A bi-clustering result based on 42 genes (*listed along the right side* of the figure) and 80 paired samples (*columns*). The patterns suggest that the 80 patients fall into two subtypes, intestinal and diffuse subtypes. Adapted from Cui et al. (2011a)

3.4 Challenging Issues

The availability of genome-scale transcriptomic data for a variety of cancer samples has enabled molecular information-based typing, staging and grading on more objective and scientific grounds. Along with this opportunity also comes a number of challenging technical issues in dealing with the complexity of the data and discovering samples sharing distinct gene-expression patterns with statistical significance. A few such challenges that must be addressed in order to make cancer typing, staging or grading analyses done in an informative and reliable manner are listed below.

3.4.1 Identification of Pathway-Level Versus Gene-Level Signatures

The basic premise for cancer typing (and similarly staging, grading) using classification or clustering techniques is that some genes exhibit similar expression patterns in cancer samples of the same type, which are not shared by cancers in other types. While this is probably true for some genes and cancers as shown in this chapter, there is no reason to believe that this has to be true universally. The reason is that cancers sharing certain phenotypic characteristics may tend to behave similarly at the biochemical pathway level rather than at the individual gene level. For example, the repression of the apoptosis system could be accomplished through functional state changes in numerous different ways such as the inhibition of *P53* transcription, *P53* gene mutations, over-expression of various survival pathways, the activation of anti-apoptotic members of the *BCL2* family, and over-expression of certain oncogenes. There are even multiple ways to repress apoptosis just through different ways of inhibiting the function of *P53*, such as repression of *P53*'s expression transcriptionally or epigenomically, over-expression of its inhibitory binding partner *MDM2*, prevention of the *P53* protein from entering the nucleus or inhibition of *P53*'s function through posttranslational modification (see Chap. 7 for details). Hence, an improved strategy for gene-expression-based cancer typing needs to take this fact into consideration. An improved strategy may need to first identify *equivalent* gene groups, each defined as genes whose expression changes may lead to the same effects at the pathway level. The challenge is how to identify such equivalent gene groups, which, we believe, requires novel ideas knowing that the current understanding of cancer-relevant pathways is far from complete.

3.4.2   Close Collaboration Between Data Analysts and Pathologists May Be Essential

Another challenge in using computational techniques for cancer typing (or staging, grading) lies in how to optimally integrate the experience of cancer pathologists in defining cancer types and the molecular information hidden in the *omic* data. A common practice, as shown above, has been to statistically link cancer samples, defined as the same type by pathologists, to a set of genes with common expression patterns, which are distinct from cancer samples of the other types. An issue encountered with such an approach is what to do next when the computational methods give rise to staging results different from those by pathologists, knowing that both approaches could have errors. An important message to convey here is that it is essential for cancer pathologists and *omic* data analysts to collaborate in order to resolve inconsistent results, and better yet to develop general protocols for mapping the knowledge of onco-pathologists to computer-based cancer typing, staging and grading procedures.

3.4.3   Capturing Complex Relationships Among Gene-Expression Patterns

Another challenging issue is to identify complex relationships among gene expression data. For example, some cellular regulation may be triggered when the difference between the concentrations of certain gene products exceed a certain range, rather than their actual expression levels increasing above some threshold. Oxidative stress, defined as the difference between the abundance of oxidant molecules (such as ROS) and that of antioxidants (see Chap. 8 for details), serves as a good example here. Specifically it is the difference between the abundances of ROS molecules and the antioxidant species, rather than the abundance of one individual molecular species like ROS, that triggers oxidative-stress responses when it is beyond some threshold. Basically more general models are needed for capturing the complex relationships among gene expression data than simply up-or-down expression levels. The problem here is to detect non-trivial mathematical relationships among some genes, which are shared by some subgroup of samples. Clearly this represents a substantially more complex problem in identifying genes similar expression patterns, which, if solvable, can help to solve substantially more complex clustering problems.

3.5   Concluding Remarks

The state of the art in cancer typing, staging and grading relies heavily on morphological information of cancer cells, along with limited molecular level data. The limitation of such approaches is obvious since they are not connected with the detailed molecular mechanism(s), raising an urgent need for improved cancer characterization using *omic* data. The importance in moving in this direction is clear, as knowing that typing, staging and grading have important implications to prognosis as well as selection of the optimum treatment plan(s). Large scale *omic* data, such as transcriptomic data, probably contain all or the majority of the information about the underlying cancer in terms of its driving force, growth mechanism and ability to invade and metastasize. By linking such information to typing, staging and grading, one can potentially develop more effective ways to assess the level of development and malignancy of a cancer. To render *omic* data-based cancer typing, staging and grading prediction impactful, collaboration between cancer pathologists and *omic* data analysts is the key.

There are two types of computational techniques that can assist in cancer typing, staging and grading. One relies on training datasets in which cancer samples are labeled with specific types, stages and grades by cancer pathologists; the problem is to extend this knowledge to enable computer programs to make the same calls by identifying genes whose expression patterns correlate well with the specified types, stages or grades. This is an example of what is termed a classification problem, or *supervised learning* as referred to in the field of data mining. The other does not require a training dataset; instead the problem is to determine if a given group of cancer samples can be partitioned into subgroups so that each shares common expression patterns among some to-be-identified genes, but distinct from other cancer samples. This approach is denoted as a clustering problem, or an *un-supervised learning* problem. Various challenging computational problems exist that await improved techniques, thus making computer-based decisions substantially more reliable than the state-of-the-art, including: (1) going beyond the simple similarity measures between gene expression to capture more complex relationships among gene-expression data of different cancer samples of the same type, stage or grade; and (2) more integrated approaches to cancer typing, staging and grading through a refinement of the existing classification schemes involving feedback from pathologists and computational prediction.

Appendix

Table 3.2 Patient data used in the analysis in Sect. 3.2

Patient ID	Age	Gender	Histologic type	Grade	Stage	Smoking	Alcohol	Weight
1	54	F	WMD	G2	III	0	0	70
2	62	F	WMD	G1	IIIA	0	0	60
3	53	M	WMD	G2	IIIB	0	0	60
4	51	M	WMD	G2	IIIB	1	0	–
5	73	M	WMD	–	IB	0	0	63
6	41	M	WMD	G2	II	–	–	–
7	59	M	WMD	G1	III	1	1	51

(continued)

Table 3.2 (continued)

Patient ID	Age	Gender	Histologic type	Grade	Stage	Smoking	Alcohol	Weight
8	68	M	WMD	G2	IV	0	0	48
9	56	F	WMD	G1	IIIA	0	0	45
10	43	F	WMD	G1	III	0	0	55
11	71	F	WMD	G2	III	0	0	42
12	65	M	WMD	G2	IIIA	0	0	70
13	55	M	WMD	G2	III	0	0	69
14	55	M	WMD	G2	IIIB	0	0	74
15	62	F	WMD	G1	IV	–	–	–
16	41	F	SRC	–	IV	0	0	43
17	42	M	SRC	–	III	0	0	60
18	68	M	SRC	–	III	0	0	50
19	50	M	SRC	–	III	0	0	62
20	55	M	SRC	–	III	0	0	50
21	34	M	SRC	–	III	0	0	90
22	63	M	PD	G3	IIIB	1	1	–
23	56	M	PD	G3	IIIB	1	1	–
24	71	M	PD	G3	IIIB	1	0	–
25	55	F	PD	G3	IIIB	0	0	63
26	64	M	PD	G3	IIIB	0	0	55
27	53	F	PD	G3	IIIB	0	0	77
28	56	M	PD	G3	IIIB	1	0	55
29	53	M	PD	G2–G3	III	0	0	62
30	71	M	PD	G3	III	0	0	60
31	58	M	PD	G2–G3	III	0	0	50
32	42	M	PD	G3	IB	0	0	52
33	65	F	PD	G3	IIIA	0	0	–
34	50	M	PD	G3	III	1	0	47
35	59	M	PD	G3	III	0	0	57
36	75	M	PD	G3	III	0	0	65
37	40	M	PD	G3	III	0	1	80
38	51	F	PD	G3	III	1	0	52
39	67	F	PD	G3	IV	0	0	48
40	65	F	PD	G3	IIIA	0	0	53
41	53	F	PD	G3	IIIA	1	0	60
42	60	F	PD	G3	IIIB	0	0	60
43	70	M	PD	G3	II	1	0	59
44	56	F	PD	G3	II	0	0	74
45	78	F	PD	G3	IIIB	0	0	39
46	65	M	PD	G3	III	0	1	70
47	68	M	PD	G3	III	1	1	69

(continued)

Table 3.2 (continued)

Patient ID	Age	Gender	Histologic type	Grade	Stage	Smoking	Alcohol	Weight
48	57	F	PD	G3	IIIA	0	0	61
49	68	F	PD	G3	III	–	–	–
50	61	M	PD	G2–G3	III	1	0	70
51	55	M	PD	G3	III	–	–	–
52	67	F	PD	G3	II	–	–	–
53	50	F	PD	G3	III	–	–	–
54	62	F	MC	–	III	0	0	70
55	55	M	MC	–	IIIB	0	0	60
56	57	M	MC	G2	IIIA		–	65
57	74	M	MC	–	IB	0	0	62
58	58	M	MC	G3	IV	0	0	66
59	76	M	MC	–	II	0	0	70
60	54	M	MC	–	III	1	1	49
61	47	M	(tublar)	–	IB	1	1	65
62	49	M	(tubular/papillary)	–	III	1	1	60
63	76	F	(undifferentiated)	G4	II	0	0	–
64	51	M	(undifferentiated)	G4	II	–	NA	70
65	69	F	(squamous cell)	–	III	0	0	50
66	65	M	(squamous cell)	G3	III	0	1	50
67	36	M	(ulcerative)	G3	IIIA	1	0	60
68	75	F	(ulcerative)	G2–G3	IV		–	40
69	69	M	(mucous cell type)	G3–G4	III	0	0	55
70	81	M	(adenosquamous)	–	III	1	0	56

References

Albain KS, Barlow WE, Shak S et al. (2010) Prognostic and predictive value of the 21-gene recurrence score assay in postmenopausal women with node-positive, oestrogen-receptor-positive breast cancer on chemotherapy: a retrospective analysis of a randomised trial. Lancet Oncol 11: 55-65

Arranz EE, Vara JA, Gamez-Pozo A et al. (2012) Gene signatures in breast cancer: current and future uses. Transl Oncol 5: 398-403

Ben-Dor A, Shamir R, Yakhini Z (1999) Clustering gene expression patterns. Journal of computational biology : a journal of computational molecular cell biology 6: 281-297

Beutler E (2001) The treatment of acute leukemia: past, present, and future. Leukemia 15: 658-661

Bloom HJ, Richardson WW (1957) Histological grading and prognosis in breast cancer; a study of 1409 cases of which 359 have been followed for 15 years. Br J Cancer 11: 359-377

Chen JJ, Knudsen S, Mazin W et al. (2012) A 71-gene signature of TRAIL sensitivity in cancer cells. Mol Cancer Ther 11: 34-44

Cho SH, Jeon J, Kim SI (2012) Personalized medicine in breast cancer: a systematic review. J Breast Cancer 15: 265-272

Cortes C, Vapnik V (1995) Support-vector networks. Mach Learn 20: 273-297

Cui J, Chen Y, Chou WC et al. (2011a) An integrated transcriptomic and computational analysis for biomarker identification in gastric cancer. Nucleic acids research 39: 1197-1207

Cui J, Li F, Wang G et al. (2011b) Gene-expression signatures can distinguish gastric cancer grades and stages. PLoS One 6: e17819

D'haeseleer P (2005) How does gene expression clustering work? Nature Biotechnology 23: 1499-1501

Duan KB, Keerthi SS (2005) Which is the best multiclass SVM method? An empirical study. Multiple Classifier Systems 3541: 278-285

Eddy JA, Sung J, Geman D et al. (2010) Relative expression analysis for molecular cancer diagnosis and prognosis. Technol Cancer Res Treat 9: 149-159

Erten S, Chowdhury SA, Guan X et al. (2012) Identifying stage-specific protein subnetworks for colorectal cancer. BMC Proc 6 Suppl 7: S1

Fuhrman SA, Lasky LC, Limas C (1982) Prognostic significance of morphologic parameters in renal cell carcinoma. Am J Surg Pathol 6: 655-663

Gleason DF (1966) Classification of prostatic carcinomas. Cancer Chemother Rep 50: 125-128

Gleason DF, Mellinger GT (1974) Prediction of prognosis for prostatic adenocarcinoma by combined histological grading and clinical staging. J Urol 111: 58-64

Golub TR, Slonim DK, Tamayo P et al. (1999) Molecular classification of cancer: class discovery and class prediction by gene expression monitoring. Science 286: 531-537

Goodison S, Sun Y, Urquidi V (2010) Derivation of cancer diagnostic and prognostic signatures from gene expression data. Bioanalysis 2: 855-862

Goseki N, Takizawa T, Koike M (1992) Differences in the mode of the extension of gastric cancer classified by histological type: new histological classification of gastric carcinoma. Gut 33: 606-612

Guyon I, Weston J, Barnhill S et al. (2002) Gene selection for cancer classification using support vector machines. Mach Learn 46: 389-422

Inza I, Larranaga P, Blanco R et al. (2004) Filter versus wrapper gene selection approaches in DNA microarray domains. Artif Intell Med 31: 91-103

Johnson WE, Li C, Rabinovic A (2007) Adjusting batch effects in microarray expression data using empirical Bayes methods. Biostatistics 8: 118-127

Kim WJ, Kim SK, Jeong P et al. (2011) A four-gene signature predicts disease progression in muscle invasive bladder cancer. Mol Med 17: 478-485

Lauren P (1965) The Two Histological Main Types of Gastric Carcinoma: Diffuse and So-Called Intestinal-Type Carcinoma. An Attempt at a Histo-Clinical Classification. Acta pathologica et microbiologica Scandinavica 64: 31-49

Li G, Ma Q, Tang H et al. (2009) QUBIC: a qualitative biclustering algorithm for analyses of gene expression data. Nucleic acids research 37: e101

Liong ML, Lim CR, Yang H et al. (2012) Blood-based biomarkers of aggressive prostate cancer. PLoS One 7: e45802

Livasy CA, Karaca G, Nanda R et al. (2006) Phenotypic evaluation of the basal-like subtype of invasive breast carcinoma. Modern pathology : an official journal of the United States and Canadian Academy of Pathology, Inc 19: 264-271

Madeira SC, Oliveira AL (2004) Biclustering algorithms for biological data analysis: a survey. IEEE/ACM Trans Comput Biol Bioinform 1: 24-45

Marisa L, de Reynies A, Duval A et al. (2013) Gene expression classification of colon cancer into molecular subtypes: characterization, validation, and prognostic value. PLoS Med 10: e1001453

Mroz EA, Rocco JW (2012) Gene expression analysis as a tool in early-stage oral cancer management. J Clin Oncol 30: 4053-4055

Mukherjee S (2010) The emperor of all maladies: a biography of cancer. Scribner,

Ramaswamy S, Tamayo P, Rifkin R et al. (2001) Multiclass cancer diagnosis using tumor gene expression signatures. Proceedings of the National Academy of Sciences of the United States of America 98: 15149-15154

Rebhan M, ChalifaCaspi V, Prilusky J et al. (1997) GeneCards: Integrating information about genes, proteins and diseases. Trends Genet 13: 163-163

Reis-Filho JS, Pusztai L (2011) Gene expression profiling in breast cancer: classification, prognostication, and prediction. Lancet 378: 1812-1823

Rodenhiser DI, Andrews JD, Vandenberg TA et al. (2011) Gene signatures of breast cancer progression and metastasis. Breast Cancer Res 13: 201

Shah MA, Khanin R, Tang L et al. (2011) Molecular classification of gastric cancer: a new paradigm. Clin Cancer Res 17: 2693-2701

Shedden K, Taylor JM, Enkemann SA et al. (2008) Gene expression-based survival prediction in lung adenocarcinoma: a multi-site, blinded validation study. Nat Med 14: 822-827

Simpson JF, Gray R, Dressler LG et al. (2000) Prognostic value of histologic grade and proliferative activity in axillary node-positive breast cancer: results from the Eastern Cooperative Oncology Group Companion Study, EST 4189. J Clin Oncol 18: 2059-2069

Slodkowska EA, Ross JS (2009) MammaPrint 70-gene signature: another milestone in personalized medical care for breast cancer patients. Expert Rev Mol Diagn 9: 417-422

Starmans MH, Krishnapuram B, Steck H et al. (2008) Robust prognostic value of a knowledge-based proliferation signature across large patient microarray studies spanning different cancer types. Br J Cancer 99: 1884-1890

Stewart BW, Kleihues P (2003) World cancer report. IARC Press,

Tibshirani R, Hastie T, Narasimhan B et al. (2002) Diagnosis of multiple cancer types by shrunken centroids of gene expression. Proceedings of the National Academy of Sciences of the United States of America 99: 6567-6572

Van Mechelen I, Bock HH, De Boeck P (2004) Two-mode clustering methods: a structured overview. Stat Methods Med Res 13: 363-394

Warburg O (1956) On the origin of cancer cells. Science 123: 309-314

Weigelt B, Baehner FL, Reis-Filho JS (2010) The contribution of gene expression profiling to breast cancer classification, prognostication and prediction: a retrospective of the last decade. The Journal of pathology 220: 263-280

Wu S, Liew AW, Yan H et al. (2004) Cluster analysis of gene expression data based on self-splitting and merging competitive learning. IEEE Trans Inf Technol Biomed 8: 5-15

Yamagiwa K, Ichikawa K (1918) Experimental study of the pathogenesis of carcinoma. The Journal of Cancer Research 3: 1-29

Yeoh EJ, Ross ME, Shurtleff SA et al. (2002) Classification, subtype discovery, and prediction of outcome in pediatric acute lymphoblastic leukemia by gene expression profiling. Cancer Cell 1: 133-143

Chapter 4
Understanding Cancer at the Genomic Level

According to mainstream thinking in the past three decades, <u>cancer is a disease of the genome</u>. That is, cancer evolves from benign to malignant lesions by accumulating a series of genetic mutations over time. This model was initially developed for colorectal cancers based on mutations in the *APC* gene (Fearon and Vogelstein 1990) and a few other recurring genomic mutations that have been observed in colorectal cancers. To drive the genetic basis of this and other cancers, extensive collaborative efforts have been established to sequence the genomes of numerous cancer types, predominantly solid tumors. This undertaking has led to the public availability of thousands of cancer genomes and the identification of myriad genomic mutations, including single-point mutations, copy-number changes and genomic rearrangements. Analyses of the sequenced genomes have observed that a cancer genome may harbor tens to a few tens of thousands of mutations across different cancer types. One somewhat surprising observation has been that cancer genomes tend to have a high degree of heterogeneity in terms of their mutation patterns among tissue samples of the same cancer type, even among different cells in the same cancer tissue (Xu et al. 2012). From this, an obvious question is: *Which of the observed mutations contribute to the initiation and development of a sporadic cancer, and how?* Or, from another perspective, *are any of these mutations responsible for tumor initiation, progression and metastasis?*

To address such and related questions, cancer genome analysts have cataloged all the genetic changes observed in cancer genomes (*versus* their healthy controls) and have identified numerous common changes across different genomes of the same as well as different cancer types. Interesting results have emerged. For example, ~50 % of the sequenced cancer genomes harbor mutations in the *P53* gene and ~90 % of colon cancer genomes have mutations in the *APC* gene. With all the cancer and their control genomes, one can start to address a variety of basic questions about cancer such as: (1) *Do genes of a specific pathway tend to have more mutations than other pathways?* (2) *Do mutations in certain pathways tend to take place before mutations in other pathways?* (3) *To which aspects of a cancer development do mutations tend to contribute during the process of tumorigenesis?* And even (4) *Do genomic*

© Springer Science+Business Media New York 2014
Y. Xu et al., *Cancer Bioinformatics*, DOI 10.1007/978-1-4939-1381-7_4

mutations really drive a cancer development as it has been widely believed? We fully expect that questions like these and possibly many beyond can be realistically addressed based on the available and emerging cancer genome data.

4.1 Basic Information Derived from Cancer Genomes

Since the first human genome was sequenced in 2001 (Lander et al. 2001; Venter et al. 2001), a number of large cancer-related genome sequencing projects have started as outlined in Chap. 2. These efforts have led to the generation of thousands of sequenced cancer genomes or exomes as of early 2014, all aimed to find the holy-grail: the driver mutations for each cancer. Clearly such data have provided ample opportunities for cancer researchers and genome analysts to characterize the genomic landscapes of mutations across different cancer types and to link such information to their clinical phenotypes. The rapid advancement in sequencing technology, along with the rapid price reduction, e.g., with a 100-fold price reduc-tion in sequencing one human genome from 2008 to 2014, has clearly fueled the competition for producing larger numbers of cancer genomes worldwide, thus increased opportunities to discover common genomic and pathway level character-istics across cancer samples of the same types. It has become common now for a published study to report sequencing and analysis results of dozens to hundreds of cancer genomes.

With the accumulation of the cancer mutation data, various statistics on mutation patterns have been compiled. For example, according to a recent review (Vogelstein et al. 2013), each solid-cancer genome for adults harbors a median of 25–80 genes having non-synonymous mutations with respect to the matching control. This con-clusion is based on data from multiple cancer types including colon, breast, brain and pancreatic cancers. More than 90 % of these mutations are single-base muta-tions (such as C to G substitutions), of which 90.7 % result in missense changes, 7.6 % result in nonsense changes and 1.7 % are alterations in splice sites or untrans-lated regions. The variation in the number of mutations per genome can be large across different samples of the same cancer type. For example, a median of ~9,600 mutations per genome was found in the gastric cancer genomes that our lab sequenced (Cui et al. 2014), and the genome with the highest number of mutations, namely ~50,000 mutations, in this set is from a patient with some 20 years of smok-ing history. The compiled statistics also revealed that different cancer types may have different ranges in the number of mutations per genome. For example, small cell lung cancer and melanoma genomes are at the high end in this spectrum, con-taining 23,000 and 30,000+ mutations per genome on average, respectively. In con-trast, pediatric cancers were found at the lowest end, harboring 9.6 point mutations per genome on average. Multiple explanations for cancer type-dependent mutation frequencies have been proposed, including histories of smoking and exposure to UV light, but few of them have been rigorously tested. We believe that cancer *omic* data analyses can provide useful insights in linking endogenous factors to the observed variations in mutation frequencies through association analyses between factors such as cellular ROS levels *versus* single-point mutation rates or hypoxic levels *versus* rates of complex mutations.

The public availability of sequenced cancer genomes has also made it possible to carry out in-depth genetic analyses regarding early mutations that may cause cancer, as well as statistical inference of the relative order of occurrences/selections of mutations associated with different pathways. Such information would be invaluable, for example, to inference of which mutation(s) may be the disease initiator(s) responsible for the onset of a cancer and which mutations subsequently contribute to progression and metastasis. The *APC*-based model for colon cancer represents the first such study, which predated the cancer genome sequencing projects (Fearon and Vogelstein 1990). According to this model, the *APC* mutation is probably the first or of the first few among other mutations involved in the formation of a colon adenoma. The normal function of the protein includes signal transduction in the *WNT*-signaling pathway, mediation of intercellular adhesion, stabilization of cytoskeleton and cell-cycle regulation (Fearnhead et al. 2001). The model predicts that the loss-of-function mutations in the gene may provide the host cells a growth advantage, hence allowing the cells with the mutations to outgrow the neighboring cells and become a microscopic clone, forming a small slow-growing adenoma. A significant tumor expansion will take place when a second mutation arises in a proto-oncogene gene, such as *KRAS*, which promotes the clonal growth. At this point, two cell types co-exist in the same colony, one with *APC* mutations only and the other with mutations in both genes. The latter may have a substantially larger cell population than the former because of the growth advantage provided by the *KRAS* mutation. As this clonal expansion continues, additional mutations in other genes, specifically *PIK3CA* (phosphatidylinositol 4,5-bisphosphate 3-kinase), *SMAD4* (deleted in pancreatic carcinoma locus 4) and *P53*, may occur and be selected, eventually leading to a malignant tumor (Vogelstein and Kinzler 2004). Here *PIK3CA* is a proto-oncogene and has key roles in cell proliferation, survival and migration (Murat et al. 2012); *SMAD4* is a tumor suppressor gene and a key gene in the *TGFβ* (transforming growth factor β) pathway (de Caestecker et al. 2000); and *P53* is a well-studied tumor-suppressor gene that has a variety of functional roles in cell cycle control, RNA repair and initiation of apoptosis (Lakin and Jackson 1999; Zilfou and Lowe 2009) (see Chap. 7 for detailed information about *P53*).

Substantially improved understanding has been gained about the biology of cancer formation since the initial proposal of the *APC* mutation-based cancer model in 1990. It is now, for example, well understood that this genetic model is too simplistic for explaining the actual formation of a human colon cancer (see Sect. 4.4 for details). Still this model has played a major role in driving the research on cancer genetics in the past two decades.

Mutations in the aforementioned genes are considered as *driver* mutations in colon cancer development, since each is believed to give the host cells a growth advantage. A recent study has quantified this advantage by estimating that one such loss-of-function mutation results in a 0.4 % advantage towards cell growth in the dynamic equilibrium between cell growth and death (Bozic et al. 2010). Hence, the compounding effect of these slight advantages over years, say 10–20 years (the typical duration needed for a cancer to fully develop in an adult), may lead to the formation of a large tumor. Compared to driver mutations, the vast majority of the somatic mutations observed in cancer genomes are considered as *passenger mutations*, i.e., they are believed not to give any growth advantage to the host cells. These passenger mutations may happen and be selected by chance, as the result of a faulty or imprecise DNA replication or repair machinery.

To date, over 300 candidate driver genes have been proposed in different cancers, the majority of which are tumor suppressor genes and only a few dozen are proto-oncogenes (Vogelstein et al. 2013). According to this study by Vogelstein and colleagues, a typical cancer may require 2–8 driver mutations for its full development. What has been surprising is that, aside from a few driver mutations such as *APC* mutations in colorectal cancer, *BRCA1-2* mutations in familial breast and ovarian cancers and the fused *ABL-BCR* gene (also known as the Philadelphia chromosome) in CML, the vast majority of the predicted driver mutations have very low recurrence rates among the genomes of the same cancer type. This observation, not surprisingly, has put the usefulness of the driver-mutation concept in question, as will be further discussed in Sect. 4.4.

4.2   General Information Learned from Cancer Genome Data

As of the end of 2013, complete genomes for over 20 cancer types have been sequenced on a few to a few hundred cancer and control tissue samples per sequencing project. We use the following three cancer types as examples to give the reader a sense about the type of information that has been learned, namely (a) lung cancer, the most common cancer in the US and worldwide; (b) colon cancer, the cancer with a genetic model; and (c) gastric cancer, sequenced and analyzed by our own team and representing the second leading cause of cancer-related mortality worldwide. In addition, two subtypes of leukemia are also included as representative non-solid tumors in the discussion.

4.2.1   Lung Cancer Genomes

Lung cancer is the deadliest cancer type among all cancers for both men and women in the US and worldwide, with an estimated 228,190 new cases and 159,480 deaths anticipated in 2013 in the US (ACS 2013). Lung cancer has two main subtypes: non-small cell lung cancer (NSCLC) and small cell lung cancer (SCLC). The latter is more aggressive and accounts for about 15 % of all lung cancer cases. Most SCLC cases are attributed to smoking, and the patients generally have poor prognosis. In 2010, one SCLC cell line, NCI-H209, was sequenced along with a control line, NCI-BL209, by the Sanger Institute (Pleasance et al. 2010b).

This sequencing project led to the identification of a large number of mutations in SCLC cells, such as 22,910 somatic mutations including 94 non-synonymous single-point mutations, 65 insertions and deletions, 58 genomic rearrangements and 334 copy-number changes. The $G \rightarrow T$ substitution was the most common substitution, accounting for over 1/3 of the observed single-point mutations; this may be partially related to the known chemical modification of purines (A/G) induced by tobacco mutagens. Tobacco mutagens are known to bind with and chemically modify genomic DNA, forming bulky adducts at the purine residues and leading to non-Watson-Crick pairing during DNA replication. Such mispairing may escape correction by a compromised DNA repair system that tends to be associated with cancer (Pleasance et al. 2010a).

In comparison, more sequencing studies have been carried out on NSCLC, the most common form of lung cancer. A number of driver mutations have been predicted in a few proto-oncogenes and tumor suppressor genes such as *AKT1, ALK, BRAF, EGFR, HER2, KRAS, MEK1, MET, NRAS, PIK3CA, RET* and *ROS1* (Serizawa et al. 2013). Mutations in these genes may lead to constitutive activation of a number of growth-signaling pathways and hence possibly drive tumorigenesis. Interestingly, it was found that these driver mutations rarely co-occur in the same cancer sample, suggesting the possibility that these mutations play similar roles in the evolution of different cancer samples.

178 lung squamous cell carcinomas (SCC), a subtype of NSCLC, were sequenced in 2012 by the NIH TCGA (The Cancer Genome Atlas) consortium. Analyses of the sequenced genomes have led to the identification of a mean of 360 mutations in protein-coding regions, 165 genomic rearrangements and 323 copy-number changes or variations (CNVs) per genome. A few recurrent mutations were found in multiple proto-oncogenes and tumor suppressor genes such as *CDKN2A, PTEN, PIK3CA, KEAP1, MLL2, HLAA, NFE2L2, NOTCH1, RB1* and *P53*, with *P53* being mutated in nearly all the 178 genomes. In addition, CNVs were found with a few genes such as *SOX2, PDGFRA, KIT, EGFR, FGFR1, WHSC1L1, CCND1* and *CDKN2A*, and significant amplifications were observed with the *NFE2L2, MYC, CDK6, MDM2, BCL2L1* and *EYS* genes, along with deletions of the *FOXP1, PTEN* and *NF1* genes (The-Cancer-Genome-Atlas 2012a). Unfortunately, no fundamentally new biology about the formation mechanism of this cancer type was revealed based on the discovery of these mutations, a common outcome in multiple other cancer-genome sequencing projects.

To assist the reader in understanding why each of these specific genes has a large number of duplications or are deleted in the SCC genomes, the following table gives a brief description about the function of each (Table 4.1).

From this table one may come to the conclusion that genes with significant amplifications tend to be related to cell growth, proliferation, inhibition of cell death and response to oxidative stress, while the deleted genes have antagonist functions to those that are amplified.

Table 4.1 A brief functional description of the amplified and deleted genes in SCC genomes

Gene symbol (gene name)	Function
NFE2L2 (nuclear factor, erythroid 2-like 2)	Is important for the coordinated up-regulation of genes in response to oxidative stress
MYC (avian myelocytomatosis viral oncogene)	Activates the transcription of growth-related genes
CDK6 (cyclin-dependent kinase)	Promotes G_1/S transition in cell cycle
MDM2 (E3 ubiquitin protein ligase)	Inhibits *P53*- and *P73*-mediated cell cycle arrest and apoptosis
BCL2L1 (Bcl-2-like protein 1)	A potent inhibitor of cell death
EYS (eyes shut homolog)	Contains multiple epidermal growth factor (*EGF*)-like domains
FOXP1 (Forkhead box P1)	An essential transcriptional regulator of B-cell development
PTEN (phosphatase and tensin homolog)	Antagonizes the *PI3K-AKT/PKB* survival signaling pathway
NF1 (neurofibromatosis)	Accelerates *RAS* inactivation

4.2.2   Colorectal Cancer Genomes

Colorectal cancer is the third leading cause of cancer-related deaths in the US. Analyses of cancer genome data revealed that more than 80 % of the human colorectal carcinomas (CRCs) have mutations in the *APC* gene (Kinzler and Vogelstein 1996), which is true for both sporadic and hereditary CRCs. The physiological role of *APC* as a tumor suppressor is to retain the β-catenin protein for phosphorylation, thus preventing it from entering the nucleus to function as a transcription factor for cell proliferation.

The first large-scale genome sequencing paper on CRCs was published by TCGA in 2012, which sequenced 276 CRC genomes (The-Cancer-Genome-Atlas 2012b). Twenty-four genes were found to be significantly mutated in CRC genomes. In addition to the five genes used in Fearon and Vogelstein's CRC model, *ARID1A, SOX9* and *FAM123B* have been found to harbor mutations most frequently in CRC genomes, suggesting their importance to the formation of a CRC. Among these genes, *ARID1A* is related to chromatin remodeling (Guan et al. 2011); *SOX9* is a developmental gene, involved in male sexual development (Kent et al. 1996); and *FAM123B* is a signaling protein, possibly involved in kidney development (Genetics-Home-Reference 2014).

A computational analysis of these mutations has provided new insights into the biology of CRC at the pathway level. For example, 16 genes in the *WNT*-signaling pathway were found to be mutated, suggesting the importance in altering the normal function of the pathway to CRC development. It is worth noting that a key function of this pathway is in coordinating cell proliferation, differentiation and migration activities, so multiple mutations in this pathway suggest that CRC may benefit from loss of coordination among these three essential cellular processes. Mutations in the *PI3K* and *RAS–MAPK* signaling pathways were also common in the sequenced samples, including mutually exclusive mutations in *PIK3R1* and *PIK3CA*, deletions in *PTEN*, and mutually exclusive mutations in *KRAS, NRAS* and *BRAF*, suggesting that simultaneous inhibition of the *RAS* and *PI3K* pathways may be required to achieve therapeutic benefit in treatment of a CRC (The-Cancer-Genome-Atlas 2012b). The *TGFβ* and *P53* signaling pathways were found to be frequently mutated in CRC. Amplifications of the *ERBB2* gene were found multiple times across different samples, and amplification of *IGF2* was also observed, both of which are growth factors. Lastly, fusions between the *NAV2* gene and the *WNT*-pathway member *TCF7L1* were identified in multiple samples.

Overall, a substantial amount of information has been derived from the mutation data, which should be highly informative for elucidating the specific evolutionary pressures that the CRC is under. Specifically, these mutations were selected to facilitate the CRC cells to evolve and survive the pressures to which they are exposed (see Chap. 5 for a detailed discussion). Advanced data analyses of these mutations, along with the known functions of the relevant genes under physiological conditions and the current knowledge about the cancer development, could lead to the establishment of logic models that can explain why these mutations are specifically selected by the CRC cells, which could result in new insights about the underlying mechanisms of the selected evolutionary trajectories by the CRC samples.

4.2.3   Gastric Cancer Genomes

Very little is currently known about the molecular basis of gastric cancer, although infection of bacterial *H. pylori* is believed to be a risk factor for its development. A number of large-scale genome sequencing analyses have been published on this cancer, including two exome sequencing projects and one genome-wide association study, which reported novel mutations in the chromatin remodeling gene *ARID1A* and two suspicious loci associated with non-cardia gastric cancers (Shi et al. 2011; Wang et al. 2011). Another genome analysis on two gastric adenocarcinomas revealed the architecture of a wild-type *KRAS* amplification along with three distinct mutational signatures in this cancer (Nagarajan et al. 2012). Further analyses of the observed mutations, in conjunction with genome data from 40 gastric cancer exomes and followed with a targeted screening of an additional 94 independent gastric tumors, uncovered recurring mutations in the *ACVR2A, RPL22, LMAN1* and *PAPPA* genes in multiple gastric cancer samples (Nagarajan et al. 2012). *ACVR2A* can activate *SMAD* transcription regulators, which are cofactors of signal transduction of *TGFβ*; *RPL22* is a ribosomal protein; *LMAN1* is mannose-specific lectin; and *PAPPA* is a metalloproteinase and related to the release of *IGF* (insulin like growth factor).

Our group has performed a whole-genome sequencing analysis on five pairs of gastric adenocarcinoma and matching control tissues (Cui et al. 2014). The goal was to elucidate not just which, but also how the genomic changes may have arisen and their possible roles in cancer progression. A particular focus was on associations between the identified genomic changes and likely cancer-causing factors relevant to the impaired DNA repair system, and potential integration of *H. pylori* DNA into the host genome. The analysis identified 407 non-synonymous point mutations, among which the most recurrent were in *MUC3A* and *MUC12* (mucins) and three transcription factors, *ZNF717* (zinc finger protein), *ZNF595* and *P53*, where both zinc finger proteins have been implicated in a number of cancers (Litman et al. 2008; Barbieri et al. 2012; Liu et al. 2012). 679 genomic rearrangements were detected, which disrupt 355 protein-coding genes; in addition, 76 genes were found to have copy-number changes. The most interesting finding of the analysis, however, was the observation suggesting potential integration of *H. pylori* DNA into the host genome. If proven to be true experimentally, this could potentially provide highly useful guiding information for effective treatment of the illness.

4.2.4 Leukemia Genomes

For non-solid tumors, ALL is the most common pediatric malignancy. Among different subtypes, early T-cell precursor ALL (ETP ALL) has high occurrences of copy-number variations and is known to have low success rates in treatment (Coustan-Smith et al. 2009). A whole-genome sequencing study on 12 ETP ALL samples was published in 2012 (Zhang et al. 2012). An average of 1,140 point mutations, including 154 non-synonymous ones, and 12 structural alterations per genome were detected, which overlap with a number of protein-coding regions such as those involved in cytokine-receptor regulation and *RAS* signaling (*NRAS, KRAS, FLT3, IL7R, JAK3, JAK1, SH2B3* and *BRAF*), hematopoietic development (*GATA3, ETV6, RUNX1, IKZF1* and *EP300*) and histone-modification (*EZH2, EED, SUZ12, SETD2* and *EP300*). Two genes, *DNM2* and *ECT2L*, were found to have recurring mutations across multiple samples, where *DNM2* (a cytoskeletal protein) is believed to be involved in endocytosis and cell motility, and *ECT2L* (epithelial cell transforming sequence 2) may be an oncogene that acts as a guanine nucleotide exchange factor. In addition, mutations in *JAK3* (leukocyte Janus kinase), *IL7R* (interleukin 7 receptor), *IFNR1* (interferon gamma receptor 1) and *BRAF* were considered as a possible common pathogenesis for the establishment of the ETP leukemic clone.

Among all leukemia types, CML is probably the best understood. In most cases the illness is believed to be caused by or closely associated with the formation of a Philadelphia chromosome (see Fig. 4.1 and Chap. 1). Specifically, the *ABL* gene (acquired from the *Ab*elson murine *l*eukemia virus) on chromosome 9 is fused with *BCL* (*b*reakpoint *cl*uster region) on chromosome 22, giving rise to constitutive activation of the *ABL-BCL* tyrosine kinase, which is considered to be the sole driver of the cancer.

A tyrosine kinase inhibitor drug, Gleevac, was developed and hailed to be extraordinarily effective in terminating the rapid malignant cell proliferation. However, in time, the drug begins to lose its effectiveness due to the ability of the

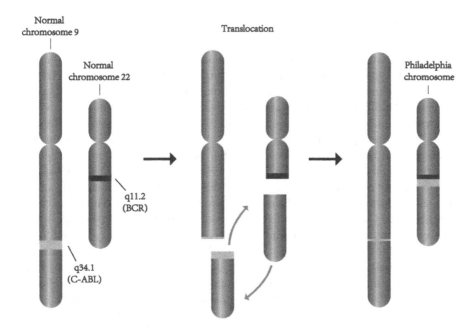

Fig. 4.1 A schematic for the formation of the Philadelphia chromosome

cancer to develop drug-resistant clones (Hochhaus 2006). Two general classes of resistance cases have been observed: one being *ABL-BCL*-dependent and the other *ABL-BCL*-independent. The *ABL-BCL*-dependent class tends to develop point mutations in the *ABL* gene that prevents Gleevac from binding (Deininger et al. 2005), while the *ABL-BCL*-independent class by-passes the drug-induced *ABL-BCL* inhibition by constitutively activating down-stream signaling proteins such as *SRC* kinases (Thomas et al. 2004). All these clearly raise an issue of <u>whether the Philadelphia chromosome is indeed the sole driver of the cancer</u>. If so, then what drives the cancer to return when the driver is inhibited?

4.3   Driver Mutations Considered at a Pathway Level: Case Studies

By mapping the observed driver mutations onto biological pathways in databases such as KEGG (Ogata et al. 1999), BIOCARTA (Nishimura 2001) and REACTOME (Croft et al. 2011), some pathways are found to be statistically enriched with such mutations. According to one report, all the known driver mutations enrich 12 signaling pathways (Vogelstein et al. 2013). The following discussion considers three pathways or biological processes enriched with driver mutations in various cancers: cell growth, cell survival and genome maintenance, to give the reader a sense of why these pathways tend to be highly mutated.

4.3.1   Cell Differentiation

Cell division and differentiation are two fundamental processes that are linked through cell-cycle controls, where a growth-to-differentiation transition (GDT) point exists in the G_2 phase of a cell cycle. A proliferating cell exits the cell cycle and enters the differentiation process only when the GDT check fails, e.g., when no adequate amount of nutrient is available in support of cell division. Thus, a normal proliferating cell proceeds to either of the two paths depending on the check result at the GDT. For cancer cells, the situation becomes quite complex. Specifically, cell division in cancer is a means for survival as detailed in Chap. 5. That is, without cell division, these cells will die. It is generally known that cancer cells tend to be low in ATP production compared to normal cells (Hirayama et al. 2009); also see Chap. 5 for a detailed discussion. When the ATP-deficient proliferating cells (another paradox of cancer) traverse the cell cycle, they should have been directed to cell differentiation at the GDT checkpoint. However, cells transitioned there will die, hence creating a pressure for them to select mutations that favor cell division over differentiation. Mutations found in the *APC, HH* and *NOTCH* genes are probably all relevant to this selection as these genes are known to be important to the GDT checkpoint (Jordan et al. 2006; Meza et al. 2008; Kahane et al. 2013). A specific mutation in *IDH* (isocitrate dehydrogenase), termed the 2HG-producing mutant *IDH*, has been found in glioma, acute myeloid leukemia and chondrosarcoma (Lu et al. 2012), which essentially serves the same purpose by blocking the cells from transitioning to differentiation through inhibition of a histone demethylation.

4.3.2   Cell Survival

As discussed above and detailed in Chap. 5, neoplastic cells proliferate as a way of survival. During this process, they gradually lose the normal functionalities of their cell cycle regulators. Survival has become the sole purpose of these cells, so consequently any genetic mutations that give these cells a survival advantage over death will be selected. Cancer genome analyses have shown that numerous mutations are related to cell survival such as mutations that can lead to: (1) prevention of apoptosis activation, (2) skipping checkpoints in cell cycles, (3) boosting cellular fitness levels as discussed in Chap. 8 and (4) enhanced proliferation that can also boost survival (Li et al. 1998). The following gives a few such genes: (1) tumor suppressor genes *PTEN, RB1, NF1, WT1, MYC, CDKN2A* and *VHL*, and (2) proto-oncogenes *MYC, EGFR, HER2, EGFR2, PDGFR, TGFβR2, MET, KIT, RAS, RAF, PIK3CA* and *BCL2*. Survival at any cost appears to be a key characteristic of all cancers.

4.3.3 Genome Maintenance

Normal human cells have a sophisticated machinery to maintain the fidelity of DNA replication to ensure that their genomes are faithfully copied from one generation to the next, which is required to execute their intended functions as encoded in the genome. An inaccurate copy of DNA may lead to elimination of the new cell under physiological conditions for the health of the whole tissue. It is known that neoplastic microenvironments tend to have a high level of oxidative stress due to their overproduction of the ROS and reactive nitrogen species (RNS), predominantly associated with increased metabolic activities. Under stressful conditions induced by ROS accumulation and DNA damage, DNA-integrity checks during the cell cycle may slow the cell-division process or even direct the cells to apoptosis, which cancer cells attempt to avoid. While healthy cells need such mechanisms for removing the malfunctioning cells, tumors cells do not; and losing such a capability may help them to gain efficiency for survival. Hence, numerous genes involved in genome-integrity maintenance, such as *P53* and *ATM* (ataxia telangiectasia mutated), are often mutated as widely observed across different cancers. For the same reason, genes responsible for DNA repair, such as *MLH1* (DNA mismatch repair protein 1), *MSH2* or *MSH6*, are also often mutated in cancers. This will lead to loss-of-function in DNA repair, which will, in turn, accelerate the accumulation of additional mutations.

Mutational analyses at the pathway level enables one to identify cellular processes that require inhibition or enhancement in cancer cells, hence providing a way to see the big picture of a forest (cellular pathways) rather than just individual trees (genes), and to understand the impact of individual mutations at a higher functional level. Further discussion along this line of thinking is carried out in Sect. 4.4, where all the mutations observed in cancer genomes of a specific type, including both driver or passenger mutations as defined in the current literature, are examined.

4.4 Information (Potentially) Derivable from Mutation Data of Cancer Genomes

A large number of cancer genomes have been sequenced and numerous genome-analysis papers have been published as outlined in the previous sections. These studies have uncovered mutations in genes encoding a wide range of cellular functions. However, what has been somewhat surprising and possibly disappointing is that not many breakthroughs in our understanding about the basic biology of cancer have resulted from these large-scale genome sequencing and analysis efforts. In the past few years, concerns have been raised by some, including leading cancer geneticists, questioning the true value of continual cancer-genome sequencing. A close examination of a presentation by Vogelstein in 2010 may give some hints about why this is the case (Kaiser 2010). The presentation predicts that the yet-to-be-identified driver genes will probably be part of the 12 pathways that are dominantly enriched with the 300+ predicted drivers, and does not expect that many new driver genes will be discovered from future cancer genome sequencing projects.

This is clearly surprising as this prediction seems to suggest that the genetic information of any cancer is dominantly encoded in these 300+ genes? To put it in a form of a question: *Have we really uncovered all the cancer-related genetic information from genomic mutation data, or have we been too limiting in our vision in information derivation from the sequenced cancer genomes?*

As a possible check on which one of these two possibilities is likely to be correct, one published set of 24 sequenced genomes of human colon samples have been chosen for further analyses, which cover different disease stages including colon polyps, precancerous colon adenoma (small), precancerous colon adenoma (large) and colon adenocarcinoma (Nikolaev et al. 2012). The selection of this particular dataset is mostly based on the consideration that it provides mutation information in precancerous colon tissues. Out of the 24 sequenced genomes, 4 were removed from our analysis since 3 are sessile serrated adenoma, which represent a type of colon tumor distinct from the other tumor samples, and 1 adenoma sample lacks a clear label for its developmental stage. The 20 samples used in our analysis consist of 1 polyp with 4 mutations, 8 mild and small adenoma samples harboring 272 mutations, 8 severe and large adenoma samples having 344 mutations and 3 adenocarcinoma samples with 198 mutations. Out of these 815 mutations, 9 are predicted to be driver mutations by the authors of the study, namely (*APC, KRAS, CTNNB1, P53, NRAS, GNAS, AKT1, ADRID1A, SOX9*), all of which are either tumor-suppressor genes or proto-oncogenes, while the remainder of the 815 mutations are considered as passenger mutations. Since this dataset has only three adenocarcinoma samples, too small to generate meaningful statistics, one published set of colon adenocarcinoma genomes (The-Cancer-Genome-Atlas 2012b) is also included in our analysis.

This combined dataset enables us to examine how the mutation-enriched pathways change as the disease progresses through different stages. The new dataset consists of complete genomes of 131 adenocarcinoma samples, 18 stage-1 samples with a total of 1,439 mutations, 47 stage-2 samples with 3,683 mutations, 43 stage-3 samples harboring 3,657 mutations and 23 stage-4 samples having 2,061 mutations. 32 of these mutations are predicted to be driver mutations by the authors of this dataset. The following analysis included **only** passenger mutations predicted by the original authors of these two datasets.

To check if the predicted passenger mutations may contain any interesting information, a simple pathway-enrichment analysis was carried out on these mutations using DAVID against three pathway databases, KEGG, BIOCARTA and REACTOME. Table 4.2 lists the pathways or gene groups that are enriched with genomic mutations of high statistical significance.

Each column of the table lists all the pathways enriched with (passenger) mutations having p-values <0.05 for the first dataset and <0.005 for the second set (using a more stringent cutoff to simplify the following analysis as an illustrative example since otherwise too many mutations need to be considered). From the table, one can

Table 4.2 Pathways/gene groups enriched with passenger mutations

Adenoma (small)	Adenoma (large)	Colon cancer of dataset 1	Colon cancer (stage 1)	Colon cancer (stage 2)	Colon cancer (stage 3)	Colon cancer (stage 4)
Cell adhesion (4.9E–8)	Glycoproteins (1.3E–9)	Cell adhesion (1.3E–10)	Glycoprotein (1.2E–23)	Glycoprotein (1.1E–27)	Glycoprotein (7.7E–25)	Glycoprotein (2.4E–24)
Fibronectins (5.0E–7)	Cell adhesion (7.1E–8)	Glycoproteins (3.0E–7)	Cell adhesion (8.7E–21)	Cell adhesion (3.7E–18)	Cell adhesion (6.5E–19) fibronectin (1.7E–15)	Cell adhesion (7.9E–21) EGF-like region (2.4E–13) fibronectin (2.3E–13) ATP-binding (6.7E–10)
Cell motion (6E–7)	Fibronectin (4.6E–5)	Extracellular matrix (3.9E–6)	Ion transport (1.9E–11)	Ionic channel (2.4E–16)	Immunoglobulin I-set (1.1E–13)	Plasma membrane (9.7E–9)
Morphogenesis (1.8E–5)	EGF-like genes (1.3E–4)	Immunoglobulin subtype (3.0E–4)	EGF-like region (2.7E–11)	Plasma membrane (6.4E–15) EGF-like region (7.7E–15) ion-binding (4.4E–13) ATP-binding (1.3E–11)	Plasma membrane (1.3E–12)	Immunoglobulin (3.8E–9)
Glycoproteins (1E–4) extracellular matrix (1.7E–4)	ABC transporters (2.0E–4)	Cell membrane (4.6E–4)	Plasma membrane (2.6E–10)	Extracellular matrix (3.0E–10) fibronectin (1.2E–10)	Ionic channel (9.4E–11) EGF-like (8.3E–11)	Extracellular matrix (4.0E–8) ion-transport (1.2E–8)
ECM-receptor interaction (2.2E–3)	Cadherin (2.3E–4)	EGF-like genes (1.1E–3)	Cell morphogenesis (2.5E–9)	Extracellular matrix (3.0E–9)	Ion-binding (5.6E–10)	Metal ion binding (3.0E–7)
Cell cycle (1.2E–2)	Extracellular matrix (8.5E–3)	Fibrinogen C-terminal (3.6E–3)	Fibronectin (4.3E–9)	Laminin (4.4E–8)	Extracellular matrix (6.7E–10)	Transmission of nerve impulse (2.4E–6)

(continued)

Table 4.2 (continued)

Adenoma (small)	Adenoma (large)	Colon cancer of dataset 1	Colon cancer (stage 1)	Colon cancer (stage 2)	Colon cancer (stage 3)	Colon cancer (stage 4)
	Actin-binding (2.2E–2)	Differentiation (5.4E–3)	Cytoskeleton (4.2E–9)	Synapse (2.3E–8) guanyl nucleotide exchange factor (1.3E–7)	Neuron differentiation (6.4E–10)	Neuron differentiation (3.1E–6)
		von Willebrand factor (9.2E–3)	Cadherin (1.5E–8)	Immunoglobulin subtype (3.9E–7)	Cell morphogenesis involved in	Cytoskeletal part (6.7E–6)
		Laminin G (9.4E–3)	Immunoglobulin (5.6E–8)	Motor protein (3.4E–7)	Differentiation (1.4E–9)	Sarcomere (6.1E–6)
		ECM-receptor interaction (1.2E–2)	Laminin (5.3E–8)	Cell morphogenesis (1.7E–6)	ATP-binding (1.2E–9)	Muscle cell differentiation (3.3E–5)
			Endometrial cancer (1.4E–7)	Ank-repeat (1.3E–6)	Protein kinase (5.4E–9)	Leucine-rich repeat (1.7E–5)
			Extracellular matrix (1.5E–6)	Cytoskeletal part (2.2E–6)	Transmission of nerve impulse (1.5E–7)	laminin G (2.9E–5)
			Collagen (5.0E–5)	Cell motion (3.8E–5)	Actin-binding (4.6E–7)	Cell motion (3.4E–5)
			Cytoskeleton organization (1.6E–4)	Actin cytoskeleton (1.3E–5)	Cytoskeleton (2.2E–7)	Triple helix and collagen (1.9E–5)

Morphogenesis (8.1E−4)	Embryonic development (1.7E−4)	Laminin (1.9E−7) microtubule (8.2E−7)	Dynein heavy chain, (8.1E−5)
Cell junction organization (1.9E−3)		Glutamate receptor activity (1.0E−6)	Calmodulin binding (2.3E−4)
complement control (1.5E−3)		Synapse (4.4E−6)	Dendrite (3.5E−4)
		Motor protein (2E−6)	Tyrosine protein kinase, active site (3.3E−4)
		Detection of abiotic stimulus (2.9E−5)	GTPase binding (6.1E−4)
		Calcium ion transport (2.7E−5)	
		Tyrosine-specific protein kinase (3.6E−4)	

Note: Mutations in the polyp sample do not enrich any pathway with statistical significance and hence are not included see that in the early adenoma stage, pathways (gene groups) enriched with mutations are related to cell adhesion, extracellular matrix composition and interaction (fibronectin, extracellular matrix (ECM), ECM-receptor interaction, glycoproteins), cell morphology, cell cycle and cell motility. These results strongly suggest that in early adenoma, changes are already made in: (a) the composition of the extracellular matrix as well as cell-ECM interactions, (b) cell-cell adhesion, (c) cell morphology and (d) cell cycle. It is noteworthy that all of these mutations are related to genes involved with tissue development!

As introduced in Chap. 1, the composition of the ECM plays an essential role in cancer (or any tissue) development. It has been well established that changes in the mechanical properties, via changes in the composition of the ECM, is a key step in the beginning of tissue development. The data in Table 4.2 suggest that altering the mechanical properties of the ECM by changing its composition is probably a first, or at least an early step that needs to be taken for a cancer (tissue) to develop. The changes observed here in ECM-cell interaction, cell-cell adhesion, cell morphology and cell cycle through mutations suggest that in colon cancer, and possibly other cancers as well, tissue development is not a **top-down** process as in a normal tissue-developmental process. In the latter, signals emanate to all the relevant processes (players) in a coordinated fashion to prepare and execute the actions needed for proper tissue development. Specifically the following changes must be in place for the cells to proliferate and the tissue to develop: (1) growth signaling; (2) material preparation for cell division; (3) cell cycle activation; (4) cell morphology changes induced by altered interactions with the extracellular matrix induced by (5); and (5) changes in the composition and hence the mechanical properties of the ECM, among a few other changes.

The observed mutation data suggest that the cancer tissue-development is a **bottom-up** process. Specifically cells are first pressured to divide (see Chap. 5 for the detailed driver information) but without the proper signals at the tissue level to inform all the relevant players. The initial cell-division signals, produced through the cells' altered metabolism (see Chap. 6) and specifically generation of hyaluronic acid fragments, may lead to the activation of some, but probably not all players involved in the tissue development machinery. At a minimum, this would not be at the same level of coordinated activities, which puts the relevant cells in a partially activated state for tissue development, waiting for the additional players to join. The selected mutations in the above categories may represent these awaited players, i.e., these mutations open all the doors needed for the cells to divide without full signals for tissue development.

This is not difficult to imagine since, although hyaluronic acid fragments can (theoretically) provide all the signals needed for the above (1)–(5), they are the result of random degradation of hyaluronic acid polymers by hyaluronidases and not designed to support tissue development in a well-coordinated fashion. In addition, knowing that cell growth signals will trigger cell death when no adequate macromolecules can be synthesized to support cell division (Vaux and Weissman 1993), we speculate that the un-coordinated hyaluronic acid fragment-based signaling for tissue development leads to cell-growth related stress, which may be the direct reason for the observed mutations selected by the neoplastic cells.

The mutation-enriched pathways in the large adenoma samples, as can be seen from the second column in Table 4.2, include: (1) *EGF*-like (epidermal growth factor like) domain, (2) *ABC* (ATP-binding cassette) transporters, (3) cadherin, and (4) actin binding in addition to those shared with small adenoma samples. It is worth noting that the *EGF*-like domain is part of the laminin protein, a key linker protein as an integral part of ECM. *ABC* transporters have long been known to be relevant to cancer, mostly because of their roles in the multi-drug resistance pathway (Szakacs et al. 2006). Recent studies have found that *ABC* transporters actually play active roles in cancer development based on the following observations: (a) the expression of *ABCB1* can delay the activation of apoptosis in leukemia; (b) *ABCC1* has been found to promote cell survival, and knock-down of this gene suppresses proliferation in neuroblastoma; and (c) cell proliferation was increased with reduced knock-down of *ABCG2*, all as reviewed in (Fletcher et al. 2010). Regarding (3), the loss of cadherin function reduces cell-cell adhesion, hence allowing cells to move and invade neighboring tissues. Regarding (4), actin-binding is the link between the ECM and the intracellular actin cytoskeleton, where cell division requires structural changes of the actin cytoskeleton, which is generally induced through interactions between actins and ECM-associated proteins as shown in Fig. 4.2.

Overall, it can be seen that, as an adenoma grows from a small to a larger size, it continues to alter the composition and hence the physical properties of the underlying ECM, possibly to enhance the effectiveness of growth signals and to induce

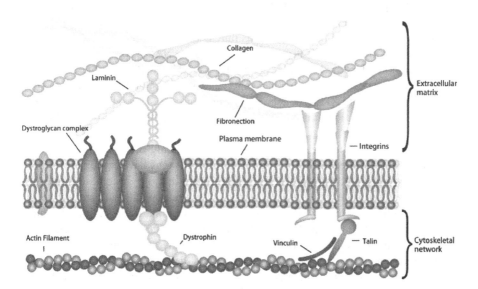

Fig. 4.2 A schematic of the actin cytoskeletal structure and interactions with the ECM-associated proteins. Cell division requires structural changes in the cytoskeleton, generally induced via interactions between actins and ECM-associated proteins such as integrins

changes in actin cytoskeletal structures. A previous study has shown, for example, that the effects of growth factors can increase 100-fold when the underlying ECM changes from very elastic to very stiff (Wells 2008). From the mutation data, we suggest that the increased change in cell morphology driven by mutations is possibly to stay abreast of the advances in other aspects of the poorly-coordinated tissue development, signaled by hyaluronic acid fragments (see Chap. 6 for details). In addition, cells seem to have initiated an effort to delay or repress apoptosis before other more permanent inhibitory measures are taken, like the loss of *P53* function.

For stage-1 adenocarcinoma, it can be seen from the fourth column of Table 4.2 that a number of pathways and gene groups enriched with mutations continue to be the same as in precancerous adenoma tissues, but with increased statistical significance and enhanced activities. These include cell adhesion, ECM composition, cytoskeletal structure and ATP binding, which tend to continue throughout the four stages of cancer development. These results clearly indicate that changing the functional states of these pathways through mutations are essential during the whole process of cancer (tissue) development, again possibly due to the lack of (sufficiently strong) signals for these aspects of (cancer) tissue development. In addition, mutations in a number of new pathways and gene groups start to emerge in stage-1 cancer, such as: (1) ion transporter, (2) plasma membrane, (3) immunoglobulins and (4) complement control.

Studies in the past decade have identified links of ion transporters (and channels) to the control of the timing of cell-cycle checkpoints. Specifically, it has been shown that ion channels mediate the calcium signals that punctuate the mitotic process (Becchetti 2011). Losing this capability has been found to promote neoplasia as one would expect. Regarding (2) above, changes in plasma membrane structures have long been known to be associated with cancer development (Weinstein 1976). One study has suggested that cancer cells may consume oxygen at the cell surface through plasma membrane electron transport to oxidize NADH to support glycolytic ATP production (Herst and Berridge 2006). Hence, it is possible that there is selection for glycolytic cancer cells that have altered plasma membrane structures to better facilitate glycolytic ATP production and cancer growth. Regarding the observed mutations in immunoglobulins in (3), a recent study found that the activities of these proteins may be relevant to the immune protection during carcinogenesis, making a possible link between the selected mutations in this family of proteins and cancer. Regarding (4), it was recently reported that cancer cells may have exploited the control of the complement pathway, as part of the innate immune system, to evade the immune attack on cancer cells through unwanted recognition of the altered self-cells, i.e., cancer cells (see Chap. 8) (Ferreira et al. 2010). Overall, these additional mutations in stage-1 cancer disrupt the normal timing control of cell cycle events, enhance glycolytic ATP production needed by cancer cells and disrupt the immune system.

The three sets of mutations in three consecutive developmental stages of colon tumor examined above provide a clear picture of the key events in the early developmental stage of a colon (pre) cancer. When such data are analyzed in conjunction with transcriptomic data of the same set of samples, one should be able to determine not only which functionalities must be inhibited, but also which functionalities must be activated (due to some mutations), hence providing a more complete picture of all the key events needed for the early development of a cancer.

To move on, one can see from columns 5–7 of Table 4.2 that, as the cancer progresses to stages 2, 3 and 4, new mutations occur in: (1) *Rho* guanyl-nucleotide exchange factor activity (stage 2), and the related microtubule (stage 3) and *GTPase* binding (stage 4); (2) cell motion (stages 2, 3 and 4); (3) differentiation (stages 3 and 4); (4) embryonic development; and (5) a tyrosine kinase (stages 3 and 4). Briefly, mutations in (1) are associated with the activation of cytoskeletal reorganization required by cell division, as discussed earlier. Mutations in (2) are clearly related to tumor invasion; both (3) and (4) are related to the so-called *de-differentiation* of cancer cells, a property that has been known to be associated with cancer cells (Medema 2013); and (5) is related to constitutive growth signaling, strongly suggesting that cell proliferation at this point is not driven by the early driver such as ridding cells of the accumulated glucose metabolites (see Chap. 5 for details) anymore; instead it is by something else, yet to be identified, that requires growth signaling in a more efficient manner than via hyaluronic acid-based signaling, specifically through mutation-facilitated constitutive activation of growth factors. Basically as a cancer progresses, cell motion, cell de-differentiation and tissue development become increasingly more important, as suggested by the mutation data.

It is worth noting that while mutations observed in precancerous and early stage cancer tissues are highly tissue-development related, mutations observed in the more advanced stages tend to be associated more to cell movement, cell de-differentiation and increased efficiency for growth signaling.

From this combined dataset on colon cancer genomes, it is ascertained that changes in eight major areas through genomic mutations are needed for cancer tissue development: (a) alteration in the composition and hence the mechanical properties of extracellular matrices; (b) ECM-cell interaction and cell-cell adhesions; (c) cell morphology, cytoskeletal structure and cell-cycle activation state; (d) ion channels and plasma membranes; (e) innate immune system; followed by (f) cell de-differentiation, (g) self-sufficiency in growth signaling and (h) tumor invasion. In addition, if one includes the "driver" mutations that were reported in the original studies and omitted in our data analysis, one should also see changes in (i) cell-cycle control; (j) evasion of apoptosis; and (k) growth signaling or activation of their receptors among a few other activities, which are clearly too narrowly focused on one aspect of cancer tissue development, i.e., cell proliferation and associated control.

It is expected that more careful analyses of such genomic mutation data could lead to a very detailed understanding about which cellular processes must be inhibited and which must be enhanced as a cancer develops, as well as the determination of the relative order of these changes. In addition, if analyzed together with gene-expression data, one may possibly derive which particular functional inhibition or enhancement must be accomplished through mutations and which can be done through either mutations or transcription or epigenomic level regulation. When comparative analyses of such data are carried out across different cancer types, it is expected to reveal which relative orders among mutated pathways are essential and which are probably accidental.

Overall, using the genome mutation data of one cancer type, it is demonstrated here that substantially more information, possibly orders of magnitude more, can be derived from the sequenced cancer genomes compared to the published studies on these two sets of cancer genomes. So the question is: *What has happened to the published studies*, i.e., *why has such information not been reported before*?

To answer this question, one needs to look carefully at the definitions of a few widely used terminologies such as oncogenes, tumor suppressor genes, and driver and passenger mutations. According to the widely accepted definition (Bozic et al. 2010), a driver mutation is a mutation that gives a selective advantage to a clone in its microenvironment, through either increasing its survival or reproduction. Based on this definition, all the mutations given in Table 4.2 should be driver mutations since they all helped the underlying cancer to develop, but none of them are included in the candidate driver genes discussed earlier. The reason is that, while they contribute to the growth of a cancer, they are not proto-oncogenes or tumor suppressor genes by the current definitions in the cancer literature.

Actually, according to the original definitions, oncogenes and tumor suppressor genes are both defined in terms of their relevance to cancer **tissue** development (see Chap. 1 for definitions). But their definitions in real practice have evolved. Basically they have been determined based on their **cellular** level functions. For example, the activation of the *MYC* gene can lead to cell proliferation in cell culture, an action that will not happen in a tissue environment since, as discussed above, many other conditions must be met in addition to an activated *MYC* before cell proliferation can take place. It can be checked that many of the proto-oncogenes reported in the literature are predicted to be oncogenic based on their observed functions in cell culture instead of tissue level studies! It is worth reemphasizing that the distinction between tissue and cellular level functions is vitally important since the latter definition has been clearly used in the published cancer genome analysis papers, which has led to the somewhat disappointing performance by these large-scale cancer genome sequencing and analysis studies. This vision is probably responsible for the rather limited information that has been derived from the cancer genomes.

For practical purposes, a valuable lesson learned here is that one should look at all the mutations, instead of only the putative "proto-oncogenes" and "tumor suppressor genes" identified based on their cellular functions in artificial environments. The pathway-enrichment analyses, without any pre-defined filtering, should inform us which genes are cancer-relevant with statistical confidence, and hence enable the identification of all the mutations that assist cancer in its development, the true driver mutations or genes by the original definition.

4.5   Limitations of Cell Line-Based Studies: A Prelude to Microenvironments Driving Carcinogenesis

It is important to point out that a substantial number of cancer-related studies have been carried out using cell lines or mouse xenograft models. Such experimental paradigms permit research to be conducted in a well-controlled and hopefully reproducible environment. This, in turn, enables elucidation of the specific signals that can activate specific pathways or those mutations that will change particular functional states. Virtually all the known molecular and cellular mechanisms for cancer have been derived in such *in vitro* cellular systems. However, one should not overlook the fact that cancer is not only a cellular level problem; instead, cancer is the result of very complex interactions between cells and their microenvironment(s) and that both evolve very rapidly. That is: the core issue of a cancer is a problem at the tissue level rather than at the cellular level (see Chaps. 5 and 6 for detailed discussion). Only with this understanding can one possibly find the right information encoded in the sequenced cancer genomes. The following example is used to illustrate why this is the case.

Oncogenes will lead to cell proliferation when activated, which is true only in cell lines or organ cultures lacking the actual tissue environment, while cancer *in vivo* is a tissue-developmental problem. It should be noted that cell proliferation and tissue development are fundamentally different problems. In a tissue environment, there are numerous constraints that define when a cell can start to divide. First, cells must be attached to their base, i.e., the ECM (also referred to as *basement membrane*, the portion in direct contact with the cells), to have a chance to grow. It also requires the underlying ECM to have certain mechanical properties to support cell growth. In addition, when proliferating cells come too close to each other, they will stop growing due to contact inhibition, an encoded mechanism in cells for the prevention of overgrowth. There are also a few intracellular conditions that the cells must meet before they can divide, even when an oncogene is activated, including: (1) the cell cycle must have been activated; (2) the proliferating cells must pass all the cell-cycle checkpoints; (3) the cells must have sufficient biomolecules to produce a new cell; (4) the cells must have a specific morphology; and (5) the cells must be relatively healthy (see Chap. 8). For cell division to occur, these conditions, and possibly additional ones, must be met or specific mutations must be selected to allow the checkpoints of some conditions to be by-passed. Basically cell division, possibly driven by oncogenes, is only part of a much larger machinery that controls tissue development. Cancer studies, including computational analyses such as prediction of oncogenes or analyses of genomic mutations, must be put into the context of tissue development.

This brings up a more general issue: genomic data of sporadic cancers may not necessarily contain much information on cancer drivers, as the driving forces for cancer initiation, progression and metastasis may predominately be attributable, at least in solid-tumor cancers, to the microenvironment(s) as is emphasized throughout this book. According to the current literature definitions, it is quite possible that the mutations in proto-oncogenes and tumor suppressor genes, selected by cancer evolution, may exert only late facilitator roles instead of early driver roles as repeatedly presented in the cancer literature.

4.6   Concluding Remarks

Thousands of cancer genomes have been sequenced and a large number of mutations have been identified in these sequenced genomes. However, questions are beginning to emerge from the cancer research community about the true value in further sequencing more cancer genomes. While interesting information has been forthcoming, with few exceptions the results have not led to groundbreaking discoveries about the biology of cancer or to new treatment modalities as many have expected. Yet, in this chapter we have demonstrated that considerable information is present in the cancer genomes that can be mined by using existing techniques. Indeed, it is estimated that perhaps an order of magnitude more information is available than that currently published in the genome sequencing and analysis papers. It is not that the needed methodologies are lacking, instead it is that we have limited our visions by concepts that were developed for other practical purposes. It is now time to cast aside those popular concepts and mine the data without any unnecessary constraints.

References

Barbieri CE, Baca SC, Lawrence MS et al. (2012) Exome sequencing identifies recurrent SPOP, FOXA1 and MED12 mutations in prostate cancer. Nature genetics 44: 685–689

Becchetti A (2011) Ion channels and transporters in cancer. 1. Ion channels and cell proliferation in cancer. American journal of physiology Cell physiology 301: C255–265

Bozic I, Antal T, Ohtsuki H et al. (2010) Accumulation of driver and passenger mutations during tumor progression. Proceedings of the National Academy of Sciences of the United States of America 107: 18545–18550

Coustan-Smith E, Mullighan CG, Onciu M et al. (2009) Early T-cell precursor leukaemia: a subtype of very high-risk acute lymphoblastic leukaemia. Lancet Oncol 10: 147–156

Croft D, O'Kelly G, Wu G et al. (2011) Reactome: a database of reactions, pathways and biological processes. Nucleic acids research 39: D691–697

Cui J, Yin Y, Ma Q et al. (2014) Towards Understanding the Genomic Alterations in Human Gastric Cancer. In review.

de Caestecker MP, Piek E, Roberts AB (2000) Role of transforming growth factor-beta signaling in cancer. Journal of the National Cancer Institute 92: 1388–1402

Deininger M, Buchdunger E, Druker BJ (2005) The development of imatinib as a therapeutic agent for chronic myeloid leukemia. Blood 105: 2640–2653

Fearnhead NS, Britton MP, Bodmer WF (2001) The ABC of APC. Human molecular genetics 10: 721–733

Fearon ER, Vogelstein B (1990) A genetic model for colorectal tumorigenesis. Cell 61: 759–767

Ferreira VP, Pangburn MK, Cortes C (2010) Complement control protein factor H: the good, the bad, and the inadequate. Molecular immunology 47: 2187–2197

Fletcher JI, Haber M, Henderson MJ et al. (2010) ABC transporters in cancer: more than just drug efflux pumps. Nature reviews Cancer 10: 147–156

Genetics-Home-Reference (2014) AMER1.

Guan B, Wang TL, Shih Ie M (2011) ARID1A, a factor that promotes formation of SWI/SNF-mediated chromatin remodeling, is a tumor suppressor in gynecologic cancers. Cancer research 71: 6718–6727

Herst PM, Berridge MV (2006) Plasma membrane electron transport: a new target for cancer drug development. Current molecular medicine 6: 895–904

Hirayama A, Kami K, Sugimoto M et al. (2009) Quantitative metabolome profiling of colon and stomach cancer microenvironment by capillary electrophoresis time-of-flight mass spectrometry. Cancer research 69: 4918–4925

Hochhaus A (2006) Chronic myelogenous leukemia (CML): resistance to tyrosine kinase inhibitors. Annals of Oncology 17: x274–x279

Jordan K, Schaeffer V, Fischer K et al. (2006) Notch signaling through Tramtrack bypasses the mitosis promoting activity of the JNK pathway in the mitotic-to-endocycle transition of Drosophila follicle cells. BMC Developmental Biology 6: 1–12

Kahane N, Ribes V, Kicheva A et al. (2013) The transition from differentiation to growth during dermomyotome-derived myogenesis depends on temporally restricted hedgehog signaling. Development 140: 1740–1750

Kaiser J (2010) UPDATED: A Skeptic Questions Cancer Genome Projects. Available at: http://news.sciencemag.org/2010/04/updated-skeptic-questions-cancer-genome-projects.

Kent J, Wheatley SC, Andrews JE et al. (1996) A male-specific role for SOX9 in vertebrate sex determination. Development 122: 2813–2822

Kinzler KW, Vogelstein B (1996) Lessons from hereditary colorectal cancer. Cell 87: 159–170

Lakin ND, Jackson SP (1999) Regulation of p53 in response to DNA damage. Oncogene 18: 7644–7655

Lander ES, Linton LM, Birren B et al. (2001) Initial sequencing and analysis of the human genome. Nature 409: 860–921

Li C, Bapat B, Alman BA (1998) Adenomatous polyposis coli gene mutation alters proliferation through its beta-catenin-regulatory function in aggressive fibromatosis (desmoid tumor). The American journal of pathology 153: 709–714

Litman T, Moeller S, Echwald SM et al. (2008) Novel human micrornas associated with cancer. Google Patents,

Liu P, Morrison C, Wang L et al. (2012) Identification of somatic mutations in non-small cell lung carcinomas using whole-exome sequencing. Carcinogenesis 33: 1270–1276

Lu C, Ward PS, Kapoor GS et al. (2012) IDH mutation impairs histone demethylation and results in a block to cell differentiation. Nature 483: 474–478

Medema JP (2013) Cancer stem cells: the challenges ahead. Nature cell biology 15: 338–344

Meza R, Jeon J, Moolgavkar SH et al. (2008) Age-specific incidence of cancer: Phases, transitions, and biological implications. Proceedings of the National Academy of Sciences of the United States of America 105: 16284–16289

Murat CB, Braga PB, Fortes MA et al. (2012) Mutation and genomic amplification of the PIK3CA proto-oncogene in pituitary adenomas. Brazilian journal of medical and biological research=Revista brasileira de pesquisas medicas e biologicas/Sociedade Brasileira de Biofisica [et al] 45: 851–855

Nagarajan N, Bertrand D, Hillmer AM et al. (2012) Whole-genome reconstruction and mutational signatures in gastric cancer. Genome Biol 13: R115

Nikolaev SI, Sotiriou SK, Pateras IS et al. (2012) A single-nucleotide substitution mutator phenotype revealed by exome sequencing of human colon adenomas. Cancer research 72: 6279–6289

Nishimura D (2001) BioCarta. Biotech Software & Internet Report 2:

Ogata H, Goto S, Sato K et al. (1999) KEGG: Kyoto Encyclopedia of Genes and Genomes. Nucleic acids research 27: 29–34

Pleasance ED, Cheetham RK, Stephens PJ et al. (2010a) A comprehensive catalogue of somatic mutations from a human cancer genome. Nature 463: 191–196

Pleasance ED, Stephens PJ, O'Meara S et al. (2010b) A small-cell lung cancer genome with complex signatures of tobacco exposure. Nature 463: 184–190

Serizawa M, Koh Y, Kenmotsu H et al. (2013) Multiplexed mutational profiling of Japanese lung adenocarcinoma patients for personalized cancer therapy. Cancer research 73: supplement 1

Shi Y, Hu Z, Wu C et al. (2011) A genome-wide association study identifies new susceptibility loci for non-cardia gastric cancer at 3q13.31 and 5p13.1. Nature genetics 43: 1215–1218

Szakacs G, Paterson JK, Ludwig JA et al. (2006) Targeting multidrug resistance in cancer. Nature reviews Drug discovery 5: 219–234

The-Cancer-Genome-Atlas (2012a) Comprehensive genomic characterization of squamous cell lung cancers. Nature 489: 519–525

The-Cancer-Genome-Atlas (2012b) Comprehensive molecular characterization of human colon and rectal cancer. Nature 487: 330–337

Thomas J, Wang LH, Clark RE et al. (2004) Active transport of imatinib into and out of cells: implications for drug resistance. Blood 104: 3739–3745

Vaux DL, Weissman IL (1993) Neither macromolecular synthesis nor myc is required for cell death via the mechanism that can be controlled by Bcl-2. Molecular and cellular biology 13: 7000–7005

Venter JC, Adams MD, Myers EW et al. (2001) The sequence of the human genome. Science 291: 1304–1351

Vogelstein B, Kinzler KW (2004) Cancer genes and the pathways they control. Nature medicine 10: 789–799

Vogelstein B, Papadopoulos N, Velculescu VE et al. (2013) Cancer genome landscapes. Science 339: 1546–1558

Wang K, Kan J, Yuen ST et al. (2011) Exome sequencing identifies frequent mutation of ARID1A in molecular subtypes of gastric cancer. Nature genetics 43: 1219–1223

Weinstein RS (1976) Changes in plasma membrane structure associated with malignant transformation in human urinary bladder epithelium. Cancer research 36: 2518–2524

Wells RG (2008) The role of matrix stiffness in regulating cell behavior. Hepatology 47: 1394–1400

Xu X, Hou Y, Yin X et al. (2012) Single-cell exome sequencing reveals single-nucleotide mutation characteristics of a kidney tumor. Cell 148: 886–895

Zhang J, Ding L, Holmfeldt L et al. (2012) The genetic basis of early T-cell precursor acute lymphoblastic leukaemia. Nature 481: 157–163

Zilfou JT, Lowe SW (2009) Tumor suppressive functions of p53. Cold Spring Harbor perspectives in biology 1: a001883

Chapter 5
Elucidation of Cancer Drivers Through Comparative *Omic* Data Analyses

Past statistics suggest that one out of every two men and every three women in developed countries will develop cancer during his or her life time. Cancer accounts for 12.5 % of all disease-induced deaths and ranks number three behind cardiovascular diseases and infectious and parasitic diseases worldwide. Its ranking moves up to number two when only developed countries are surveyed. A central question to be addressed here is: *What causes a cancer to initiate and develop*?

Clinically, different cancers seem to have different causes. For example, some cancers are known to be closely related to viral or bacterial infections. Cervical cancer results primarily from infection by human papilloma virus. Hepatitis viruses, such as HBV and HCV, can cause hepatic (liver) cancer. Similarly, *Helicobacter pylori* is believed to be responsible for some gastric (stomach) cancers. Skin cancer, such as basal cell carcinoma, is attributed to overexposure to UV light, particularly for individuals with fair skin. Other cancer-inducing factors include: (1) microbial products such as aflatoxin produced by flavus growing on stored grains, (2) industrial chemical compounds such as dioxins, benzene and asbestos, (3) tobacco products, and (4) nuclear radiation such as gamma rays and alpha particles, all of which are referred to as *carcinogens*. Other than the environment-induced cancers, there is a class of cancers that are considered as hereditary or familial, such as breast and ovarian cancers resulting from *BRCA* gene mutations or colon cancer attributable to *APC* gene mutations.

Although induced by different factors, various cancer types share certain common characteristics at the cellular and tissue levels such as: (1) uncontrolled cell proliferation, (2) reprogrammed energy metabolism, (3) development of angiogenesis, (4) evasion of apoptosis, (5) avoidance of immune destruction, and (6) cell invasion and metastasis; the so-called cancer hallmarks as discussed in Chap. 1. These common characteristics strongly suggest that different cancers may share something in common at the root level. This commonality could possibly be something intrinsic to

© Springer Science+Business Media New York 2014
Y. Xu et al., *Cancer Bioinformatics*, DOI 10.1007/978-1-4939-1381-7_5

our cellular systems, as well as similar characteristics in the abnormal conditions induced by the relevant exogenous or endogenous factors that may force the underlying cells to take similar evolutionary trajectories for their survival. The root-level commonalities that have been proposed across different cancer types are analyzed in this chapter, along with an analysis of what may be missing from the current thinking on this important issue. This will be followed by a proposal describing a new model about the possible root causes of cancers.

5.1   Two Distinct Schools of Thoughts About Cancer Drivers

5.1.1   Cancer as a Metabolic Disease Related to Reprogrammed Energy Metabolism

German biochemist Otto Warburg published one of the earliest papers in 1924 about the possible causes of cancer at the molecular and cellular level (Warburg et al. 1924). When studying cancer metabolism, Warburg noted that cancer cells utilize glycolysis followed by fermentation of pyruvate to lactic acid in cytosol as the main ATP producer. This mode of ATP production is in contrast with normal cells that use glycolysis followed by a more complete oxidation process, *oxidative phosphorylation* in mitochondria for ATP generation. A key difference is that the first process does not require oxygen (anaerobic) while the second uses oxygen (aerobic) as the terminal electron acceptor. In addition, the second process is about 18 times more efficient than the first in terms of the number of ATPs produced per mole of glucose oxidized, as introduced in Chap. 1. Warburg observed that cancer cells utilize the first process even in the presence of oxygen, the so called *Warburg effect* (Warburg et al. 1924). This metabolic alteration was considered by Warburg to be the main characteristic across all cancers and possibly the primary cause for a cancer to initiate and develop (Warburg 1956). In 1967 he explicitly stated: *"Cancer ... has countless secondary causes. But ... there is only one prime cause,[which] is the replacement of respiration of oxygen in normal body cells by a fermentation of sugar"*(Warburg 1967). He went further to state: *"... the de-differentiation of life takes place in cancer development. The highly differentiated cells are transformed into non-oxygen-breathing fermenting cells, which have lost all their body functions and retain only the now useless property of growth ... What remains are growing machines that destroy the body in which they grow"*, to offer his insights about the essence of tumorigenesis. While all of this was very thought-provoking, his theory about a metabolic cancer driver never became part of the mainstream thinking among cancer researchers during his active years.

Among the issues for which Warburg had difficulty in convincing his peers about his theory, he unfortunately gave a partially incorrect explanation about a "paradox" of cancer cells. As rapidly proliferating cells require more ATP than normal cells, one would expect *a priori* that cancer cells should use the more efficient oxidative phosphorylation, but instead they choose the less efficient glycolytic fermentation for ATP production. Warburg suggested that cancer cells must have damaged or dysfunctional mitochondria and hence have to use glycolytic fermentation even when oxygen is available. This proposal, however, was found not to be the case on cancer samples as revealed by later studies (Weinhouse et al. 1956; Pedersen et al. 1970). Warburg's explanation was probably correct on one class of cancer, namely hereditary cancers, as discussed in Sect. 5.5, but incorrect on some sporadic cancers, at least during some developmental stages. This discrepancy clearly did not help his case when trying to convince his colleagues of the validity of his proposal during his lifetime. Warburg's theory remained visible in the scientific literature for the next half century. It is worth mentioning that Warburg received the Nobel Prize in medicine in 1931, but the award was for his work on "Discovery of the nature and mode of action of the respiratory enzyme", which is unrelated to his cancer study. His theory was basically relegated to the sidelines by the time cancer came to be considered a genetic disease in the 1970s.

Interestingly, in the past few years Warburg's proposal has received renewed attention, reflected by the numerous publications in mainstream cancer journals, often containing words like "reexamination of the Warburg effect".

5.1.2 Cancer as a Genomic Disease

The discovery of the retrovirus oncogene, *SRC*, in the 1970s marked the beginning of a new era of cancer research (Stehelin et al. 1976). In 1976, Bishop and Varmus discovered that certain human genes, when multiple-copied, mutated or over-expressed, can become *oncogenes*, i.e., cancer-causing genes (Stehelin et al. 1976). These genes are referred to as *proto-oncogenes* in their normal functional states, as introduced in Chap. 1. For this discovery, Bishop and Varmus received the Nobel Prize in medicine in 1989, and their work has had an enormous impact on cancer research in the past three decades. The conventional thinking in cancer research subsequently has become greatly genome-centric up till now. A substantial effort has been invested into the study of oncogenes since then, with a major influence by governmental funding agencies. As of now, ~150 proto-oncogenes have been identified in the human genome, including the well-studied *RAS* (rat sarcoma protein), *WNT* (wingless-type MMTV integration site family, member 1) and *MYC* genes.

Another group of genes, referred to as *tumor suppressor genes* (introduced in Chap. 1), was also discovered to have essential roles in the initiation and development of cancers in the 1970s. *RB* was the first tumor suppressor gene discovered by A.G. Knudson when studying human retinoblastoma (Knudson 1971). Such genes can safeguard a human cell from developing into a cancerous cell, with the protection provided by such a gene being lost when both copies of the gene have loss-of-function mutations. Hence, having mutations in one copy of the gene will increase the risk of losing its tumor-suppression function. As of now, ~200 tumor suppressor genes have been identified (Zhao et al. 2013), including the well-known *P53*(tumor protein 53), *RB*(retinoblastoma protein) and *APC* (adenomatous polyposis coli) genes.

The concepts of proto-oncogenes and tumor suppressor genes have clearly provided an effective framework for the development of mechanistic models that link gene mutations to cancer initiation and progression. Specifically by identifying over-expression or amplification of specific proto-oncogenes (positive cell cycle regulators) and repression or mutations of certain tumor suppressor genes (negative cell cycle regulators) in a cancer, one can infer the main driver mutations of the cancer and the associated mechanistic models. As of now, a number of cancer models have been developed based on the identified proto-oncogenes and tumor suppressor genes, such as the widely cited *APC* mutation-based model for colorectal cancer by Fearon and Vogelstein (Fearon and Vogelstein 1990); the *BRCA* mutation-induced breast cancer model developed by Pollard and colleagues (Lin et al. 2003); and the *BCR-ABL* gene fusion (i.e., the Philadelphia chromosome) model for CML by Nowell and Hungerford (Nowell and Hungerford 1960). It is clear that this proto-oncogene/tumor suppressor gene framework has helped to accelerate the generation of new information and knowledge about cancer initiation and development at the molecular and cellular levels.

The advent of high-throughput sequencing techniques, e.g., next generation sequencing, has helped to further accelerate genome-based cancer research. As of the end of 2013, thousands of complete cancer genomes have been sequenced worldwide using public and private funds (Mwenifumbo and Marra 2013). Most of these cancer genomes have had their matching control genomes also sequenced, making the identification of genomic changes in cancer readily doable, which include point mutations, copy number changes, inversions and genomic translocations. A substantial amount of information about cancer-associated mutations has been derived in various cancer genomes, as detailed in Chap. 4.

Various oncogenes and tumor suppressor genes have been identified that are associated with specific cancer types. For example, the *APC* gene is considered as a tumor suppressor gene of colon cancer and *CDK8* (cyclin-dependent protein kinase 8) as an oncogene of the cancer (Firestein et al. 2008). *HER2* (human epidermal growth factor receptor 2, also known as *ERBB2*) and *MYC* are considered as oncogenes of breast cancer among a few other genes, while *BRCA1* and *BRCA2* are tumor suppressor genes for many breast cancers (Buchholz et al. 1999). The oncogenes of prostate cancer include *HER2* and *BCL2* (B-cell lymphoma 2) (Segal et al. 1994; Arai et al. 1997; Scholl et al. 2001), while the tumor suppressor genes of the cancer include *GADD45A* (growth arrest and DNA damage 45A), *GADD45B* and *IGFBP3* (insulin-like growth factor binding protein 3) (Isaacs and Kainu 2001; Ramachandran et al. 2009; Ibragimova et al. 2010; Mehta et al. 2011).

The discoveries of oncogenes and tumor suppressor genes, along with the large number of other mutations found in cancer genomes, have contributed to the now popular speculation: "cancer is the result of a sequence of genomic mutations" (Fearon and Vogelstein 1990; Budillon 1995), which has been widely publicized in both scientific and popular publications. In the past few years, active discussions have been on-going about driver *versus* passenger mutations (Greenman et al. 2007; Stratton et al. 2009; Bignell et al. 2010). The aim is to distinguish mutations that are positively selected by cancer evolution from those random mutations that are neutral to cancer, hence providing a tool to allow researchers to focus on genes that are essential to cancer initiation and/or development.

The view of "cancer as a genomic disease" has not only attracted many cancer researchers to study the disease from a highly genome-centric perspective, but also has profoundly influenced the priorities of federal funding agencies. A substantial level of funding has been invested into the sequencing of cancer genomes, with the aim of understanding the genomic level drivers and key mutations through consortia such as TCGA (The Cancer Genome Atlas) (The-Cancer-Genome-Atlas-Research-Network 2008) and ICGC (International Cancer Genome Consortium) (Hudson et al. 2010). As a considerable amount of sequence data of cancer genomes has been generated from these and other projects, there has been an increasing voice from the cancer research community in the past few years that questions the true value of the cancer-genome sequencing projects in terms of gaining a deeper understanding about cancer biology and in support of developing improved capabilities to fight against cancer. For example, after decades of popularizing the view of cancer being a genomic disease, very little has actually been established between the activation of oncogenes and cancer initiation in a real tissue environment (*versus* cell culture models in artificial environments). In a published study on the predictive power of whole genome sequencing for cancer, Vogelstein and colleagues concluded: *"[Their] research casts doubt on whether whole genome sequencing can reliably predict the majority of future medical problems"* (Roberts et al. 2012)!

As presented in Chap. 4, the vast majority of the mutations selected by cancer tissues are not associated with proto-oncogenes or tumor suppressor genes as defined in the current literature; instead they tend to be associated with genes related to tissue development such as changes in extracellular matrix (ECM) composition and cell morphology, and immune responses, among other biological functions. This analysis has clearly revealed limitations in the current proto-oncogene and tumor suppressor gene-centric views in studies of cancer genome mutations and their relevance to cancer initiation and development.

5.2   A Driver Model Based on APC-Gene Mutations in Colorectal Cancer

In their widely cited article published in 1990, Fearson and Vogelstein proposed a genetic-centric model for the initiation of non-hereditary colorectal cancer (Fearon and Vogelstein 1990), in which cancer initiation is attributed solely to mutations in both proto-oncogenes and tumor suppressor genes. Specifically the authors considered *RAS* gene mutations as the possible initiating event for some colorectal cancers; however, as discussed below, mutations in *APC* were eventually considered as a likely initiator. A later study speculated that the *RAS* mutations discussed in this work may lead to constitutive activation of the *RAS* protein (Vojtek and Der 1998).

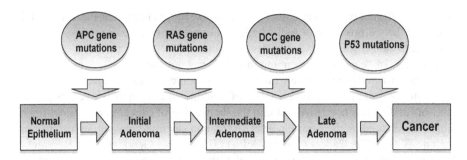

Fig. 5.1 A genetic model for colorectal cancer development (adapted from (Fearon and Vogelstein 1990; Martinez et al. 2006))

In addition, the model suggests that for the development of a colorectal cancer, the host cells need to lose part of their chromosome 5q, which was later found to be the region that encodes the *APC* (adenomatous polyposis coli) gene (Nishisho et al. 1991). Cancer-genome sequence data have confirmed that the vast majority of colon cancers have mutations in this gene (see Chap. 4). The model also suggests that the majority of colorectal cancers harbor mutations in *P53* as well as mutations in the *DCC* (deleted in colorectal cancer) gene. Overall, the model predicts that it takes mutations of at least these four genes for normal epithelial cells to develop to adenoma that progresses from the early to the advanced stage and then becomes adenocarcinoma when the cells lose the function of *P53* (as depicted in Fig. 5.1). The authors speculated that it is the accumulation of these mutations, rather than the relative order of the mutations, that really matters. When presenting the model, the authors made an important observation that neoplastic cells tend to have a small number of mutations initially, which continues to increase as the disease advances, a key point that will be further developed in our argument about cancer progression (see Chap. 9).

Since the publication of this work some 20 years ago, substantial progress has been made regarding the necessary conditions for cells to become malignant. Numerous other cellular and micro-environmental changes must take place before cells can become cancerous as discussed in Chap. 4, such as changes in energy metabolism, cell-cycle control, tumor angiogenesis, development of microenvironments with certain properties, and avoidance of immune destruction. In addition, the neoplastic cells must also develop a capability to enable their anchorage-independent proliferation and lose the contact-inhibition machinery encoded in their genomes, which has been widely observed in cancer tissues before they can start their neoplastic growth.

In the following sections, we consider the relationships between genetic mutations and cancer development in a larger and richer context, namely the overall micro- and intracellular environment needed for a cancer to initiate and develop.

5.3   Warburg's Thesis: Reprogrammed Energy Metabolism as a Cancer Driver

While the genetic-centric views remain a dominating school of thought concerning drivers of cancer initiation and development in the field of cancer research, increasingly more researchers have begun to reexamine Warburg's theory across a larger set of cancer types using the powerful *omic* techniques. The aim of these studies is to develop an improved understanding about how Warburg's observation relates to the fundamental biology of cancer, of which genetic mutations may be just a part. By going through the literature of cancer genetics and genomics studies, one surprising observation is made: the published studies seem to have ignored, for some reason, one basic issue, potentially a most important issue in cancer study: *What pressures do the cancer-forming cells evolve to overcome as they proliferate and select specific mutations?* In retrospect, it seems that this should have been an obvious question when studying the evolution of any cancer. In addition, no hypotheses or models have been proposed which aim to connect the numerous mutations observed in individual cancer genomes, essentially treating the observed mutations as independent events. This is clearly unsatisfying as discussed further in the following sections and later chapters.

Intuitively one would imagine that specific mutations are selected to better facilitate the evolution of the underlying cells to overcome some yet to-be-elucidated pressures cast on the cells by their microenvironment. An understanding of these "pressures" may provide functional links among the seemingly unrelated genetic mutations found in each cancer genome and possibly new understanding about the evolutionary trajectories selected by individual cancers. We suggest that the Red Queen Hypothesis (Valen 1973) outlined in Chap. 2 provides a useful framework of thinking about this issue, i.e., to guide one in elucidation of the evolutionary pressures that cancer cells need to overcome. The Hypothesis basically stated: *"the coevolution of interacting species drive molecular evolution through natural selection for adaptation and counter-adaptation"*. Here we use a more recent publication to further illustrate the essence of the Hypothesis and its possible relevance to cancer evolution, which reported an elegant study on coevolution and co-adaption between the bacterium *Pseudomonas fluorescens* and its viral parasite phage Φ2 that co-exist in equilibrium in the same environment (Paterson et al. 2010). The study demonstrated that the increased attacking ability, obtained through genetic engineering on the phage, hence shifting the equilibrium, will lead to accelerated evolutionary changes in the bacteria to regain the previously established equilibrium. The same result was also observed when the roles of the two organisms were switched, which is to enhance the defense ability in the bacteria, again through genetic engineering, leading to a shift in the dynamic equilibrium towards an increased bacterial population. This triggers accelerated evolutionary changes in the phage until the previous equilibrium is regained. A key point made by the authors of the study is that antagonistic coevolution is a cause of rapid and divergent evolution and likely to be a major driver of evolutionary changes within species.

Returning to the cancer evolutionary problem, the affected cells must be facing tremendous pressures, since they evolve rapidly. It is only natural to ask: *Can genomic mutations alone create such pressures that drive the affected cells to evolve*? Our answer is: **very unlikely** since: (1) changes in key functional states of cells, such as switch from non-dividing to dividing cells and then to a growing tissue, require substantial changes in the tissue environment, including changes in the functional states of cells and their ECM, along with various signaling molecules, as discussed in Chap. 4; (2) such changes, if only due to genomic mutations, will require a large number of mutations as discussed in Chap. 4 and shown in Table 4.2; and (3) the probability for so many co-occurring mutations in tissue development-related genes without being removed by the cellular, tissue or whole-body level surveillance systems is going to be extremely small if any!

One key new understanding about cancer development in the past decade came from the realization that the microenvironment of cells plays vital roles in cancer initiation and development (Witz and Levy-Nissenbaum 2006; Lorusso and Ruegg 2008; Sounni and Noel 2013), which is clearly consistent with the Red Hypothesis discussed above. It is now generally accepted that the following factors in the environment of cells contribute to tumorigenesis: (1) the physical properties of the underlying ECM (also see Chap. 8); (2) the level of intracellular hypoxia (Wilson and Hay 2011); (3) intra- and peri-cellular accumulation of ROS or reactive nitric species (RNS) (Wiseman and Halliwell 1996; Lu and Gabrilovich 2012); (4) population sizes of stromal and immune cells (Coussens and Werb 2002; Grivennikov et al. 2010; Chew et al. 2012); (5) the intra- and peri-cellular pH level (Estrella et al. 2013); and (6) certain signals from the local stromal cells. Here we focus on hypoxia and ROS, and discuss how these two factors may contribute to cancer initiation and development, leaving discussions on other micro-environmental factors to later chapters.

It has been well established that chronic inflammation can lead to hypoxia, and conversely hypoxia can also lead to inflammation (Eltzschig and Carmeliet 2011). In addition, chronic inflammation can lead to increased generation of ROS (Khansari et al. 2009), along with a number of other factors, including exogenous factors such as tobacco products and radiation, and endogenous factors such as oxidative phosphorylation, various chemical reactions and aging. One driver model for the early phase of carcinogenesis, presented in the next section and Chap. 6, is based on persistent hypoxia and the accumulation of ROS. This model provides one possible explanation of how Warburg's thesis, i.e., the primary cause of cancer is the replacement of respiration of oxygen in normal body cells by glucose fermentation, is related to cancer initiation and development.

First, to understand the generality of Warburg's observation, which was originally made on mouse ascites cancer, a larger-scale analysis of transcriptomic data of 18 types of cancers, namely bladder, brain, breast, cervical, colon, kidney, leukemia, liver, lung, melanoma, metastatic melanoma, metastatic prostate, ovarian, pancreatic, prostate, skin (basal cell), stomach and thyroid cancer, was carried out, focused on glucose metabolism including both glycolytic fermentation and oxidative phosphorylation (See Table 5.1). These cancers were selected because they represent a wide range of cancer types and each has a large number of genome-scale transcriptomic datasets in the public domain. Genes involved in the two forms of glucose-based energy metabolism, selected amino acid and fatty acid metabolism, are examined and their expression data compared in cancer *versus* adjacent normal tissues. The results of the analysis are shown in Fig. 5.2.

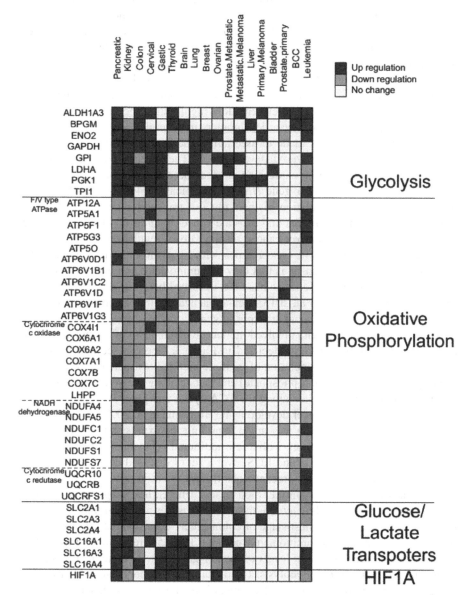

Fig. 5.2 Comparisons between gene-expression levels of glycolytic fermentation and oxidative phosphorylation pathways in cancer *versus* adjacent normal tissues for 18 cancer types. Multiple genes are used for each category. Each row represents a unique gene and each column represents a unique cancer type, with *dark gray*, *light gray* and *white* representing up-regulation, down-regulation and no change, respectively

One can see from the figure that in 16 of the 18 cancer types, the glycolytic fermentation pathway shows increased expression while the oxidative phosphorylation pathway shows decreased expression, consistent with Warburg's observation 90 years ago. Leukemia is a complex case as it shows increased glycolysis, but its oxidative-phosphorylation genes exhibit rather complex patterns of changes, having almost the same number of genes with increased expression as the number of genes with decreased expression. The only exception is bladder cancer, which has virtually no changes in gene expressions of either pathway, suggesting that the initiation mechanism of this cancer may be different from the other 17 cancer types. While these gene-expression patterns reveal that not all cancers show the same reprogrammed metabolism as observed by Warburg, 17 out of 18 cancer types have increased activities of glycolytic fermentation, highlighting the significance in the activation of this pathway in cancer development in general.

A further examination is made to ascertain if the expression of glucose transporter genes is altered as the hypoxia level changes. The lower part of Fig. 5.2 shows a high-level of consistency in expression changes between the glucose transporter genes and the hypoxia marker gene *HIF1α*. Specifically, 14 out of the 18 cancer types show overall up-regulation in their glucose transporter genes; two cancer types, skin basal cell carcinoma and prostate cancer, exhibit no changes in these genes; and one cancer type, liver cancer, displays down-regulation of one glucose-transporter gene. For the two cancer types with no changes in transporter-gene expressions, one possible reason is that they use amino acids and lipids as the main nutrient sources, but not glucose as has been observed before (Reitzer et al. 1979; Liu et al. 2010; Carracedo et al. 2013). Leukemia again shows a complex pattern of expression changes in its glucose-transporter genes.

In addition, metabolite data involved in glucose metabolism of colon and gastric cancers have been examined, as depicted in Fig. 5.3. One can see that there is a substantially increased accumulation of multiple metabolites such as G6P (glucose 6-phosphate), F6P (fructose 6-phosphate) and lactate along the glycolysis pathway, succinate, fumarate and malate along the TCA cycle, and glycerol, a substrate for gluconeogenesis, during cancer development (Hirayama et al. 2009).

An additional investigation was made regarding the expression patterns of the genes analyzed in Fig. 5.2 along with a few additional genes related to the cellular environment, namely hypoxia and ROS, over a set of diseased colon samples that range from different stages of precancerous tissues to colon cancer tissues. These include: (a) inflammatory sigmoid colon tissues (the earliest stage of adenoma), (b) inflammatory descending colon tissues, (c) tissues of inflammatory bowel disease, regarded as the earliest stage of colon cancer development, (d) colon adenoma and (e) colon adenocarcinoma tissues (see Table 5.1 for details of the data). On this dataset, the expression levels of marker genes for 10 biological processes are examined, namely (1) glycolysis, (2) oxidative phosphorylation, (3) hypoxia and glucose transporter genes, (4) cell cycle, (5) hyaluronic acid related genes, (6) apoptosis, (7) angiogenesis, (8) epithelial-mesenchymal transition (EMT), (9) inflammation and (10) immune response. The expression data of the diseased tissues are all normalized with respect to the matching normal colon tissues, as shown in Fig. 5.4.

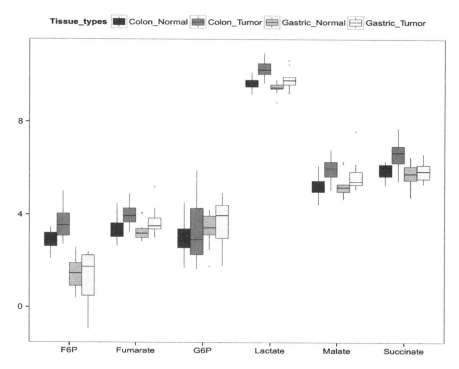

Fig. 5.3 An illustration of increased glucose metabolite accumulation in colon and gastric cancer tissues, adapted from (Hirayama et al. 2009)

One can see from the figure that: (1) inflammation marker genes tend to be up-regulated in early stage disease tissues but down-regulated in adenoma and adeno-carcinoma tissues; (2) hypoxia takes place in the early stage disease tissues and seems to be correlated with the expression levels of the inflammation marker genes; (3) glycolysis is generally up-regulated across all the diseased samples, (4) lactate exporter genes are generally up-regulated except for the adenoma tissues; (5) the majority of the oxidative phosphorylation genes are down-regulated across all disease stages; (6) hyaluronic acid synthesis and degradation genes are generally up-regulated except for *HAS2*, a hyaluronic acid synthase, across different disease stages; (7) cell cycle genes tend to be generally up-regulated in adenoma and adeno-carcinoma samples; (8) at least one of the anti-apoptotic genes among *BCL2*, *BAK1* and *BAX* is up-regulated across each of the disease stages; and (9) immune marker genes tend to be down-regulated in the adenoma and adenocarcinoma stages. These data, along with the ones in Fig. 5.3, set the stage for our model to be presented.

First, however, some background information is reviewed about published opinions regarding why cancer cells tend to have increased activities of glycolytic fermentation, a question that researchers have been groping with for some 90 years. A number of recent studies suggest that glycolytic fermentation is more beneficial than oxidative phosphorylation to cancer cells even when oxygen is available since, to maintain pace with rapid cell proliferation: (a) it generates ATP significantly

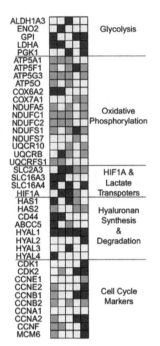
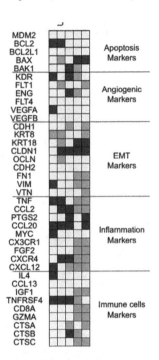

Fig. 5.4 Gene-expression level changes of *HIF1α* along with 10 sets of genes related to cancer hallmark pathways (Hanahan and Weinberg 2011) on precancerous and cancerous tissues of colon. In both heat-maps, each row represents a unique gene; and the five columns represent, from left to right, (1) inflammatory sigmoid colon tissues, (2) inflammatory descending colon tissues, (3) tissues of inflammatory bowel disease, (4) colon adenoma and (5) colon adenocarcinoma tissues. Each entry of the heat-map represents the log-ratio of the expressions of a gene in a diseased tissue *versus* the matching normal colon tissue, with *dark gray*, *light gray* and *white* representing up-regulation, down-regulation and no change, respectively

faster than oxidative phosphorylation because it has fewer reactions (Pfeiffer et al. 2001); and (b) it produces building blocks for DNA synthesis via the pentose phosphate pathway, that can be up-regulated by increased anaerobic fermentation (Lunt and Vander Heiden 2011).

Based on the above analyses of genome-scale transcriptomic data of pre-cancer and cancer tissues presented here, and the analyses in Chap. 6, we propose that: (1) it is the to-be-identified micro-environmental pressures, not the mutations, that derive the evolution of the underlying cells, possibly different pressures at different developmental stages; (2) cell proliferation is a feasible and sustained way for the cells to reduce these pressures; and (3) selection of particular mutations is probably dictated by the need to up-regulate or inhibit specific functions constitutively as demanded by the evolution, which are probably already being accomplished through other means, e.g., functional regulation.

As one will see in the next section, the key pressure that the evolving cells need to overcome is to remove the accumulation of glucose metabolites resulting from energy metabolism reprogramming. Within our proposed model, DNA synthesis from the accrued glucose metabolites is a way to dispose of these products, rather than needed solely in support of cell proliferation; moreover, cell proliferation is driven by the need for survival as it provides a way to remove the accumulated metabolites by consuming them towards macromolecular synthesis for the new cells. Consequently, our view regarding the reason for utilization of glycolysis fermentation in cancer cells, at least in the early stage, is fundamentally different from, actually opposite to the aforementioned view in the literature, in terms of its cause-and-effect relationship with proliferation.

5.4   Cell Proliferation as a Way of Survival: Our Driver Model

When asking: *What drives a cancer to grow*, a common answer that one will likely receive is: oncogenes! But if one carefully examines some examples of oncogenes, the answer may be not as simple as that. Consider the Philadelphia chromosome (Nowell and Hungerford 1960) as an example, which is believed to be the sole oncogene for CML. Specifically, the fusion of the *BCR* gene and the *ABL* gene gives rise to a new constitutively expressed tyrosine kinase *BCR-ABL* gene, which inter- acts with the *IL3β(c)* (interleukin 3β) receptor and continuously activates the cell cycle. It is worth reemphasizing that the activation of this fused gene alone is not adequate to drive cancer tissue development since this process requires numerous coordinated signals relevant to cell survival, cell-ECM interaction, by-passing anchorage-dependent growth requirement and a few others as detailed in Chap. 4. Gleevec was once considered a miracle drug for stopping CML through inhibiting the activation of this fused tyrosine kinase (Sawyers et al. 2002). However, long- term studies indicate that drug resistance becomes a common issue for CML patients due to additional genetic mutations (Roche-Lestienne et al. 2002). This raises an issue: *Are there driving forces at a deeper level than the fused BCR-ABL gene, or does this fused oncogene serve a similar role to those of mutated tyrosine kinase proteins in solid tumors such as platelet-derived growth factor receptors (PDGFRs) or the KIT gene?*

Genomic sequence data of CML, particularly bone marrow tissues where pre- CML cells are generated, should help to answer this question. However, unfortu- nately only one CML genome has been sequenced as of early 2014, and the only relevant information released is a one-page abstract (Sloma et al. 2013). From the very limited information given in the abstract, the authors report that the genome harbors 845,175 point mutations and 68,817 short indels, in addition to the Philadelphia chromosome. Virtually nothing is known currently about the molecular mechanisms of the CML development prior to the formation of the Philadelphia chromosome as an oncogene. Among all the mutations, eight genes are revealed: *JAK2,ASXL1, CTNNA1, AIDA, RAS, ULK1, GSR* and *NUP160*. Interestingly these mutated genes resemble mutated genes in some solid tumors: (1) *CTNNA1* is

involved in the association between catenins and cadherins that link to actin filament, hence relevant to cell morphological changes; (2) *JAK2, RAS* and *ULK1* are
involved in cell growth and development; (3) *AIDA* is related to embryogenesis,
hence possibly related to cell de-differentiation; (4) *GSR* is an antioxidant, suggesting high oxidative stress level; (5) *ASXL1* is a member of the Polycomb group that
may be the main regulators of epigenomic responses; and (6) *NUP160,* a nucleopo-
rin, is involved in RNA transport. This information from eight out of some 800,000
point mutations raises the possibility that CML may share similar or common
mechanisms with solid tumors, and further raises an issue: *Is the Philadelphia oncogene the root driver, or is it the result of some other events and serves as the main
facilitator, like oncogenes in solid tumors, for the host cells to escape certain pressures through cell proliferation?*

5.4.1   ATP Demand Versus Supply

Returning to the issue of root causes of a cancer, recent medical research has established that chronic inflammation is at the origin of many human diseases, including
cancer (Khansari et al. 2009), diabetes (Donath and Shoelson 2011) and dementia
(Blasko et al. 2004). The current understanding is that chronic inflammation leads
to hypoxia and an increased ROS production as discussed in Sect. 5.3. Transcriptomic
data analyses across a large number of tissue samples of different cancer types
reveal that hypoxia occurs in cancer-forming cells before any of the cancer hallmark
events (see Fig. 5.4 as an example). This has led us to approach the cancer driver
issue from the perspective of ATP supply *versus* demand in human cells under
chronic hypoxic conditions, inspired by Warburg's observation 90 years ago.

As mentioned earlier in this chapter, ATP production decreases in human cells
under hypoxic conditions due to the lower ATP-generation efficiency per mole of
glucose metabolized by glycolytic fermentation as compared to oxidative phosphorylation. This is also true for other vertebrates in general. The main question to be
addressed here is: *How does the ATP demand change under hypoxia versus normoxia?* This is addressed through a comparative analysis over eight organisms:
human, mouse, rat, hypoxia-tolerant rat, naked mole rat, blind mole rat, turtle and
frog. These organisms were selected because they are known to either develop cancer as in the case of human, mouse and rat, or rarely do as in the case of the other
five organisms. Moreover, their ATP-consumption data under normoxia *versus*
hypoxia are publicly available or can be reliably estimated based on the available
transcriptomic data. The specific question is: *What percentage of the ATP-consuming
proteins is substantially repressed during hypoxia versus normoxia?*

It has been established that the following six classes of enzymes and pathways
consume on average 84 % of the ATP in vertebrate cells: translation, Na^+/K^+ ATPase,
Ca^{2+} ATPase, gluconeogenesis, urea synthesis and actin ATPase (Rolfe and Brown
1997). (The list of genes is given in Table 5.3.) So we address the above question by
examining only these six classes of proteins, including all those associated with the
relevant pathways.

ATP consumption data by these proteins in hypoxia-tolerant rat, naked mole rat, frog and turtle during hypoxia (1–5 % oxygen in the experimental environments) *versus* normoxia (21 % oxygen) are available in the public domain (Buttgereit and Brand 1995; Hochachka et al. 1996; St-Pierre et al. 2000; Larson et al. 2012; Nathaniel et al. 2012). In addition, matching gene-expression data under the two conditions are also available for hypoxia-tolerant rat and naked mole rat (See Table 5.4). No ATP-consumption data are publicly available for human, mouse, rat and blind mole rat, but they each have gene-expression data collected under conditions of hypoxia *versus* normoxia. We will thus predict their reduced ATP consumption based on their reduced gene-expression data under hypoxia *versus* normoxia.

5.4.2   A Regression Model of ATP-Consumption Reduction Versus Reduced Expression Levels of Relevant Genes

A linear regression model is derived between the reduced ATP consumption and the reduced gene expressions for naked mole rats and hypoxia-tolerant rats, using publicly available data, namely the reduced ATP consumption ΔE and the averaged reduced expression level ΔELS of the relevant genes in each of the six groups of proteins. The validity of using gene-expression level to approximate protein-expression level is assured by a recent study on the detailed relationship between gene and protein expression levels (Evans et al. 2012). The parameters a and b in a linear model: $\Delta E = a * \Delta ELS + b$ are estimated for the two organisms using linear regression based on the ΔE and ΔELS values for each of the six groups of proteins collected from the published literature, along with the assumption, without loss of generality, that $\Delta E = 0$ when $\Delta ELS = 0$, which is introduced for the mathematical rigor and should not affect the data as shown in Fig. 5.5.

ΔE	a	b
Translation	−5.056	1.009
Sodium-potassium exchange ATPase activity	−0.6026	1.110
Calcium transporting ATPase activity	−0.8246	0.997
Gluconeogenesis	−1.919	1.002
Urea genesis	−1.012	0.995
Actin-activated ATPase activity	−0.979	1.004

This model is first applied to the ΔELS values for mouse, rat and blind mole rat based on their gene-expression data under hypoxia *versus* normoxia. The validity of the predicted results is assessed as follows. From Fig. 5.5, one can see: (1) the blind mole rat is predicted to have a larger reduction in its ATP consumption than that of the naked mole rat, which is consistent with the general knowledge that blind mole rats can tolerate more hypoxic conditions than naked mole rats, 3 % *versus* 5 ~ 10 % in terms of the minimal level of environmental oxygen needed by the two organisms, respectively (Edrey et al. 2011; Manov et al. 2013); (2) the blind mole rat is predicted

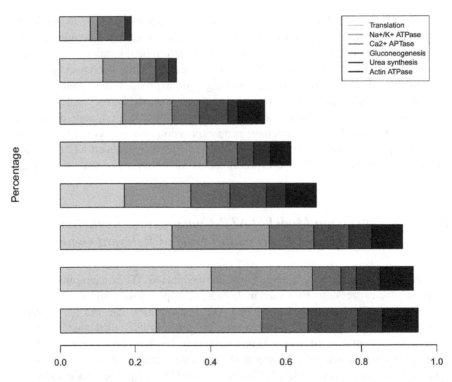

Fig. 5.5 For human, rat, mouse, naked mole rat, hypoxia-tolerant rat, blind mole rat, frog and turtle (from *bottom* to *top*), 1.0 along the x-axis represents the total ATP demand by the six groups of proteins during normoxia. The length of each bar represents the percentage of ATP consumption under matching hypoxia, hence the difference between 1.0 and the percentage representing the percentage in reduction. Each bar is divided into six gray-level coded sections, each representing one group of relevant proteins under consideration

to have a smaller reduction in its ATP consumption than those by frog and turtle, which is also consistent with the fact that frogs and turtles can live virtually without oxygen for extended periods of time; and (3) both mouse and rat are predicted to have substantially smaller ATP-consumption reductions than those by naked, blind mole rats and hypoxia-tolerant rats, which is clearly consistent with our understanding about the basic oxygen needs by mouse and rat. Based on this qualitative validation, we posit that the model is meaningful; hence we have applied it to predict the reduction in ATP consumption by human cells under hypoxia *versus* normoxia. Again the prediction result is consistent with our general knowledge about these organisms in terms of their relative abilities to deal with hypoxia.

A literature search revealed that organisms, shown in the figure, with larger energy reductions tend to have lower chances for developing cancer, suggesting a possible causal relationship between an organism's ability to adequately minimize certain parts of their ATP-consuming metabolism to keep the ATP demand consistent with the ATP supply under hypoxia *versus* the organism's potential for cancer development.

From an evolutionary perspective, this is not surprising as animals like blind mole rats have been using both (deep) underground and ground as their living habitats. Their metabolic systems have adapted to living in two different environments with substantially different levels of oxygen, and their cells have been trained to switch on and off certain parts of their house-keeping system when different levels of oxygen are available to keep their ATP demand within the ATP supply. This capability may have already been encoded in their genomes through adaption and natural selection during their evolution. In contrast, humans, as a population, never lived under hypoxic conditions for extended periods of time throughout evolution to the current stage. As a result, our systems have not been trained (or selected) to adequately reduce portions of our metabolism during hypoxia to keep ATP demands within the ability to supply, hence leaving a gap, seemingly a large gap, between energy demand and supply under persistent hypoxic conditions as shown in Fig. 5.5. Mice and rats seem to behave in the same way as humans.

5.4.3 A Driver Model

Because of the energy gap, human cells will need to substantially increase their glucose uptake during persistent hypoxia to meet the ATP demands of the cells, which has been widely observed in the majority of cancers as increased glucose metabolism has been the basis for PET/CT scans for cancer detection. In contrast, organisms like blind mole rats or turtles, do not increase glucose uptake under hypoxic conditions. Another widely observed phenomenon is that these cells accumulate glucose metabolites, as shown in Fig. 5.4. Hence one may speculate that the fundamental reason for the accumulation is the result of a mismatch between the influx rate of glucose, which is regulated by the ATP deficiency, and the maximum flux rate capable by the glycolytic fermentation pathway, which is determined by evolution. Knowing that humans have not lived under hypoxic conditions for extended periods of time during evolution, one can infer that their glycolytic fermentation pathway has been used only as a supplement to the aerobic respiration system for ATP production and for only short periods of time when humans are in oxygen demand. One can thus posit that the maximum flux rate of this system intrinsically cannot meet the need for dealing with the substantially increased influx of glucose under hypoxia. In addition, one can further speculate that this accumulation does not have a feedback mechanism developed to regulate the glucose transporters to cease functioning when glucose metabolite accumulation takes place, which may also be due to the lack of training during evolution for adapting to such a situation. Clearly all these speculations need to be experimentally validated in order to validate the hypothesis.

The continual accumulation of the glucose metabolites will lead to cell death if not removed (Kubasiak et al. 2002; Schaffer 2003), and this may continue as long as the hypoxic condition persists. Thus, we have the main hypothesis of the model: the need for removing the accumulated glucose metabolites casts strong initial pressure

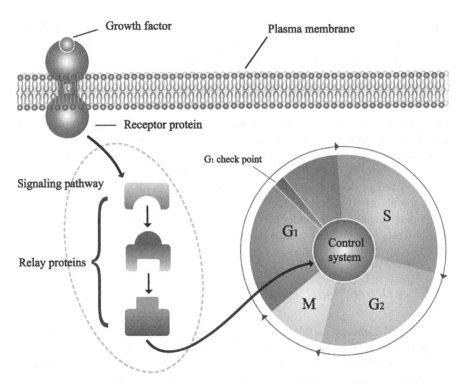

Fig. 5.6 A schematic of the cell cycle and growth factors from stromal cells

for the underlying cells to evolve, and cell division may represent a feasible and sustained way for the affected cells to rid themselves of the glucose metabolites. This defines the direction for the needed evolution for survival, i.e., cell proliferation is dictated by the need for survival and probably not by oncogenic mutations.

While the accumulated glucose derivatives may exert the initial pressure for the cells to evolve to eliminate the accumulation, the cells require signals to change their cellular state from the non-dividing G_0 phase to the dividing phase in the cell cycle to start division (see Fig. 5.6) and to enable them to overcome multiple tightly controlled conditions designed for preventing uncontrolled growth, which involves at least three sets of signals: (1) cell biomass growth, (2) cell division, and (3) cell survival with the ECM in certain states (see Chap. 6). When searching for links between accumulated glucose metabolites and cell proliferation, an association connecting these two is found, namely hyaluronic acid. While a detailed discussion regarding this connection is given in Chap. 6, some basic information is briefly provided here in order to complete the current development.

Hyaluronic acid is a long polysaccharide chain and serves as a key component of the ECM along with collagen fibrils and a few linker proteins such as fibronectins, elastins and laminins. Under normal conditions, hyaluronic acid is synthesized (see Fig. 5.7) in order to accommodate tissue development, remodeling and repair

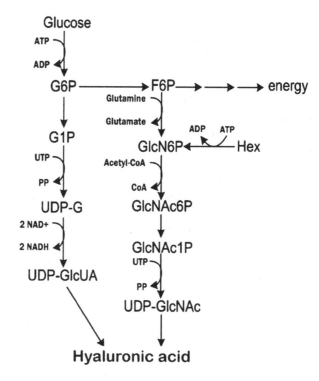

Fig. 5.7 The synthesis pathway for hyaluronic acid from glycolytic metabolites (adapted from (Vigetti et al. 2010))

(Chen and Abatangelo 1999; Noble 2002; Stern et al. 2006; Jiang et al. 2007), and is generally integrated into the ECM. However, under inflammation-induced hypoxic conditions with a plentiful supply of glucose metabolites, the biosynthetic pathway of hyaluronic acid will be activated. The newly synthesized hyaluronic acid is then exported to the extracellular space and degraded into fragments (see Chap. 6).

Coincidentally, studies on tissue injury and repair have discovered that when a tissue is injured, it releases the hyaluronic acid fragments from its ECM. These fragments of different sizes serve as signals for various purposes related to tissue repair, including signals for inflammation, anti-apoptosis, cell survival, cell proliferation and angiogenesis (Stern et al. 2006), as well as signals that allow anchorage-independent growth (Kosaki et al. 1999; Toole 2002) and loss of contact inhibition (Itano et al. 2002), basically all the signals needed for cancer development.

As detailed in Chap. 6, the released hyaluronic acid fragments, arising from glucose metabolite accumulation, are treated as signals for tissue injury, leading to the continuous process of "tissue repair" as long as the hypoxic condition persists. This progression of events probably serve as the initial driver and facilitator of cell proliferation, with strong supporting evidence as shown in Chap. 6.

One fundamental difference between this model and those in the literature is that DNA synthesis is initiated by the accumulation of glucose metabolites that in turn lead to the synthesis of hyaluronic acid and its subsequent fragmentation which promotes cell proliferation. Clearly, this is in contrast with a popular view that increased DNA synthesis is necessary for the rapid cell proliferation. The cause-effect relationship of these two processes is exactly opposite between the model presented here and much of the current thinking.

A recent study on the modulation of the architecture of fibroblasts by hypoxia strongly suggests that hypoxia may exert a more direct role in mediating cell division. Specifically, the study demonstrated that hypoxia can substantially change the organization of the actin cytoskeleton (Vogler et al. 2013), leading to morphological changes of the cells, a key step towards cell division. It has been previously established that the state of actin filament organization directly controls cell-cycle progression (Assoian and Zhu 1997; Thery and Bornens 2006). This observation raises the possibility that hypoxia may directly mediate cell division, at least by increasing the possibility for cell hyperplasia. Actually the same study also showed that hypoxia leads to increased cell volume, i.e., hypertrophy, which could be related to the glucose metabolite accumulation discussed above. If this proves to be true through experimental validation, it is likely that hypoxia has a double role in early tumorigenesis: creating the pressure for cells to evolve to remove the accumulated glucose metabolites and facilitating the removal of the accumulation through cell division.

Hypoxia is also known to mediate a number of other events that may further facilitate sustained cell division and cancer initiation, such as up-regulation of telomerase (*TERT*) (Nishi et al. 2004), genomic instability (Huang et al. 2007), angiogenesis (Moeller et al. 2004; Liao and Johnson 2007) and cell migration (Fujiwara et al. 2007). As discussed in Sect. 5.3, hypoxia takes place prior to other cancer hallmark events. Based on this and later discussions throughout the book, it should be pointed out that hypoxia can lead to most of the cancer hallmark events.

5.5   Roles Played by Genetic Mutations in Tumorigenesis

Previous studies in this area have mostly focused on the impact of mutations in proto-oncogenes and tumor suppressor genes, specifically in terms of the driving or inhibitory roles in cell division by these two classes of genes. Data presented in Chap. 4 clearly show that genetic mutations selected by cancer have much broader roles than just driving or inhibiting cell division, as cell division is only a part, albeit an important part, of cancer tissue development. Here we continue the discussion about genetic mutations and their roles in two areas: (a) ROS accumulation and implications to cancer development, and (b) replacement of persistent and abnormal functions to provide sustainability and energy efficiency.

5.5.1   Genetic Mutations Related to Hereditary Cancers

While the above model applies to sporadic cancers, it is natural to ask if this or a similar model may apply to familial cancers. To address this question, seven types of the most common familial cancers with known genetic mutations were examined: (1) breast cancer due to the *BRCA* mutations (Lin et al. 2003); (2) kidney cancer resulting from mutations in fumarate hydratase (*FH*) (Toro et al. 2003); (3) *APC* mutation-induced colon cancer (Morin et al. 1997); (4) retinoblastoma induced by *RB1* mutations (Murphree and Benedict 1984); (5) Li-Fraumeni syndrome due to *P53* mutations (Srivastava et al. 1990); (6) syndrome caused by *PTEN* mutations (Liaw et al. 1997); and (7) syndrome because of *VHL* mutations (Lonser et al. 2003). Each of these genes has multiple functional roles in normal cells such as *P53* (see Chap. 7 under "*P53* network"). All these genes are classified as tumor suppressor genes in the literature, because of their roles in cell cycle regulation and apoptotic activation (i.e., as gatekeepers). However as we will see below, these genes actually play driver roles in familial cancer development.

A literature survey revealed that these mutational effects are similar in one respect: they all lead to ROS accumulation in mitochondria. This event, in and of itself, will ultimately lead to the repression of mitochondrial function, including oxidative phosphorylation. Hence, it will ultimately force the activation of glycolytic fermentation to compensate for the reduced ATP production in mitochondria. We suspect that this may be the basis of Warburg's observation 90 years ago, which consists of normoxic cells with repressed mitochondrial functions due to ROS accumulation. Potentially, this common functional role by the loss-of-function mutations in the seven genes (see below) may prove to be the most essential role in the tumorigenesis of the relevant cancers, where they exert a "driving" instead of a gate keeping role as generally believed. Details follow.

Recent studies have shown that *BRCA* mutations in normal breast cells can lead to generation of hydrogen peroxide, as one of the normal functions of *BRCA* is to neutralize this ROS (Martinez-Outschoorn et al. 2012). The same study also observed increased glycolysis and decreased oxidative phosphorylation, revealing the repression of the mitochondrial activities, which forces cells to increase their activity of glycolytic fermentation regardless of being cancer or non-cancer cells.

Regarding fumarate hydratase, it has been shown that loss-of-function mutations in *FH* leads to the constitutive state of pseudo-hypoxia (e.g., increased expression levels of the *HIF* genes) and of increased ROS. These, in turn, further lead to increased glycolysis and decreased oxidative phosphorylation due to repressed mitochondrial function, hence in time leading to induction of the glycolytic fermentation pathway (Sudarshan et al. 2009).

Loss-of-function mutations in the *APC* gene have been found to lead to constitutive activation of the *WNT*-signaling pathway (Sunaga et al. 2001) since *APC* is a negative regulator of the pathway. This pathway activates a downstream gene, *RAC1* (a GTPase), the activation of which leads to ROS production (Sundaresan et al. 1996). From these observations, one can speculate that the gradual production and accumulation of ROS will progressively result in a reduction of mitochondrial function, including the repression of oxidative phosphorylation, and hence the activation of glycolytic fermentation pathway and the likelihood of cancer development at some point during the lifetime of the patient.

P53 gene mutations have long been linked to ROS production (Polyak et al. 1997). A recent study suggests the following mechanism for this observed activity (Kalo et al. 2012). Loss-of-function mutations in *P53* can interfere with the normal response of human cells to oxidative stress by attenuating the activation and function of *NFE2*-related factor 2, a transcription factor that induces antioxidant responses. This effect is manifested by decreased expressions of phase 2 detoxifying enzymes *NQO1* (NAD(P)H dehydrogenase, quinone 1) and *HMOX1*(heme oxygenase (decycling) 1) and an increased ROS level, ultimately leading to the repression of mitochondrial function, activation of the glycolytic fermentation pathway and possibly the development of cancer.

The relationship between *PTEN* (phosphatase and tensin homolog) mutations and ROS production is interesting. A recent study reported that loss-of-function mutations in the ATP-binding domain of *PTEN* lead to disruption of the correct subcellular localization of the protein, resulting in a significantly decreased nuclear *P53* protein level and its transcriptional activity, and hence increased production of ROS (He et al. 2011). Ultimately, this will lead to the activation of the glycolytic fermentation pathway and possibly cancer.

VHL-deficiency was recently found to constitutively activate *NOX* oxidases to maintain the protein expression of *HIF2α* (hypoxia inducible factor-2α), while NADPH oxidases of the *NOX* family are the major sources of ROS (Murdoch et al. 2006; Nauseef 2008; Frey et al. 2009). Later the same process ultimately leading to the development of cancer can take place.

The current understanding about the relationship between *RB1* mutations and ROS production is that loss-of-function mutations in *RB1* lead to dysregulation of *E2F2*, a component of the transcription factor gene *E2F* involved in cell-cycle regulation and DNA synthesis, which drives increased production of ROS (Bremner and Zacksenhaus 2010), and hence the rest of the same or similar process, possibly leading to cancer.

Based on the above discussion, one may speculate that the gradual accumulation of ROS over an extended period of time will ultimately lead to the constitutive activation of *NFκB* (Gloire et al. 2006), a master transcription regulator in response to ROS, which will ultimately lead to cancer as has been established (Karin et al. 2002). In addition, it has been well established that mitochondrial ROS triggers hypoxia-induced transcription (Chandel et al. 1998) and inflammation (Gupta et al. 2012). One can thus speculate that the same model discussed in Sect. 5.4 should essentially apply to the seven types of hereditary cancers, except that the initial trigger is increased ROS instead of persistent hypoxia. Potentially, this model may apply to most of the hereditary cancers for the same reason discussed here. A systematic study of this issue is clearly needed in order to work out the detailed mechanisms of how each of the seven mutations ultimately leads to cancer development and why the induced cancer tends to be organ-specific.

In this same vein, it is reasonable to speculate that aging-induced cancers may also follow this or a similar model as mitochondrial ROS accumulates and inflammatory cells increase, in addition to cellular senescence (Campisi et al. 2011) as one ages. At some point these cells may repress their mitochondrial activities with sufficiently high ROS levels accumulated (in steady state), leading to the reprogrammed energy metabolism and the associated phenomena discussed above. Figure 5.8 summarizes this driver model.

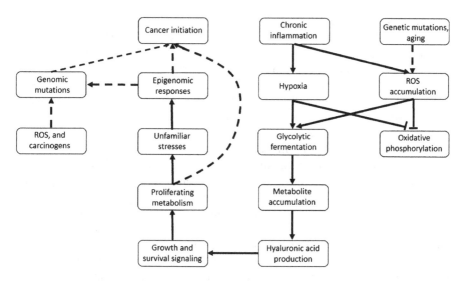

Fig. 5.8 A model for cancer initiation with two possible drivers: persistent hypoxia and ROS accumulation. Each *solid arrow* represents a strong causal relationship and a *dashed line* denotes a possible "lead to" relationship

5.5.2   Genetic Mutations in Sporadic Cancers

While genetic mutations are believed by many to be a primary reason for sporadic cancer development, recent studies are beginning to challenge this view as discussed in Chap. 4. A possible alternative is that genetic mutations may serve as facilitators rather than the primary drivers in sporadic cancers. More specifically, it has been suggested that loss-of-function or gain-of-function mutations in tumor suppressor genes or proto-oncogenes such as *P53* and *RAS*, respectively, are probably selected as "permanent" replacements for inhibitions or amplifications of the functions that are already being executed through functional regulations, post-translational modifications or other means. Such changes may be needed for sustained and efficient survival. While Chap. 9 provides an in-depth discussion on this issue, we use the following examples to illustrate the basic idea here.

The functional form of *PKM2* (pyruvate kinase isozyme M2) is a homo-tetramer that catalyzes the conversion from phosphoenolpyruvate to pyruvate and is also the rate-limiting step along the glycolytic pathway. It has been observed that the vast majority of advanced cancers have mutations in the *PKM2* gene (Mazurek et al. 2005), which inhibit the homo-tetramer formation. This finding indicates that there is strong evolutionary pressure for the affected cells to reduce their pyruvate production, possibly due to the glucose metabolite accumulation discussed earlier. It has been shown that oxidation of *PKM2* on specific residues by ROS can increase the possibility of the homo-tetramer's disassociation to dimers or monomers (Anastasiou et al. 2011), hence reducing its normal function. It is quite possible that disassociation of the *PKM2* tetramer occurs before the mutations in *PKM2*, as the first piece of data suggests that cells with fewer functional *PKM2* tetramers may have an advantage for survival. Also, the oxidation data suggest the possibility of other and less permanent means to accomplish the same loss of function.

Another example is the loss of contact inhibition of cells, a process that can terminate cell division when cells are in close proximity one to another (Sgambato et al. 2000). The increased hyaluronic acid synthesis and export, in response to glucose-metabolite accumulation (detailed in Chap. 6), will override contact inhibition, allowing for sustained proliferation by the cells. Mutations in genes responsible for activating the contact inhibition machinery, such as *ING4* (inhibitor of growth family, member 4) (Kim et al. 2004), are selected, thus ensuring the permanent loss of the inhibition capability, as observed in advanced cancer.

Overall, the selection of loss-of-function mutations of specific genes may represent a general mechanism in cancer-forming cells. That is, the surviving cells may require the repression or over-expression of specific genes to remain viable. Genetic mutations may prove to be the permanent replacement of the desired function, diminution or enhancement, initially accomplished through other means. We fully anticipate that a systematic analysis of all well documented cancer-related mutations, coupled with analyses of the matching transcriptomic and epigenomic data of cancer tissues at different developmental stages *versus* control tissues, could provide a detailed knowledge of which genetic mutations tend to serve as permanent replacements of on-going functions and which may be selected by cancer cells to start new functions. Such information should deepen, as well as broaden our current understanding of the process of tumorigenesis.

5.6 Exogenous Factors and Cancer

5.6.1 Microbial Infections

A number of cancers are known to be induced and closely associated with microbial infections. For example, human papilloma virus (HPV) is closely associated with cervical cancer (Walboomers et al. 1999; Crosbie et al. 2013); Hepatitis virus B and C (HBV and HCV) are with liver cancer (Perz et al. 2006); *H. pylori* is connected with gastric cancer; *Chlamydophila pneumonia* with lung cancer; and *Streptococcus bovis* is closely associated with colorectal cancer (Boleij et al. 2009). The current estimate is that 18 % of all the diagnosed cancer cases are related to infectious diseases, including viral, bacterial and parasitic infections (Anand et al. 2008). From the literature, the proposed mechanisms for different infection-related cancers vary considerably, but with one commonality: all these infections result in chronic inflammation (Shacter and Weitzman 2002) in the diseased lesions, suggesting a possibility that our model may apply to the tumorigenesis of the relevant cancers.

5.6.2 HBV, HCV and Hepatocellular Carcinoma

Hepatocellular carcinoma (HCC) is the third leading cancer worldwide, and 85 % of the liver cancers are associated with either HCV or HBV infection worldwide (Hiotis et al. 2012). Interestingly, the two types of viral infections have very little in common. HCV is a single-stranded RNA virus that does not have its genome integrated into the host genome, and HBV is a DNA virus whose genome does integrate into the host hepatocyte genome during early infection.

Previous studies have identified multiple integration sites of the HBV genes in the host hepatocyte genome, which can lead to the loss of heterozygosity for tumor suppressor genes such as *PRLTS* (Kahng et al. 2003). In addition, the HVB-x antigen is able to activate oncogenes, as well as interfere with the function of tumor suppressor genes such as *P53* (Feitelson and Duan 1997; Lian et al. 2003). In comparison, the current knowledge of the relationship between HCV and liver cancer is very limited. For example, it is not known whether the proliferating cells of liver cancer are HCV-infected cells or if they are responding to apoptosis induced by HCV in neighboring cells in order to maintain tissue homeostasis. A number of dysregulated cancer-associated genes, e.g., *RB* and *P53*, have been identified in such cancers (Lan et al. 2002; Munakata et al. 2007), but it is unclear whether these dysregulations are directly related to HCV infection or cancer development. One general observation has been that HCV-associated liver cancers tend to first develop cirrhosis that persists for an extended duration before the development of cancer.

Transcriptomic data analyses, coupled with statistical inference and guided by our model given in Sect. 5.5 and Chap. 6, should be able to help identify the key steps in initiation and development of HCV-induced cancer. Specifically, one can ask how different HCV genes, known to be relevant to liver cancer initiation and development, contribute to various significant events such as inflammation, hypoxia or ROS. Such studies have the potential of putting multiple and seemingly unrelated events into one coherent driver model, potentially leading to new and testable hypotheses about the initiation and development of the cancer. It is foreseeable that such studies can also reveal possible problems of our model that may require further refinement and expansion. The same strategy can be applied to studies of other infection-associated cancers as follows.

5.6.3 Human Papilloma Virus (HPV) and Cervical Cancer

Virtually all the reported cervical cancer cases are associated with the infection of human papilloma virus (Walboomers et al. 1999; Crosbie et al. 2013). Although HPV is essential to the transformation of cervical epithelial cells, studies have shown that it is not sufficient by viral infection itself for such a transformation. A number of cofactors are needed, including co-infection with cytomegalovirus or human herpes virus, while co-infection with adeno-associated virus can reduce the risk of cervical cancer. Interestingly, among 15 HPV strains known to be associated with cervical cancer, each acts independently when a person is infected with multiple strains of the virus, and infection with more HPV strains tend to increase the malignancy of the resulting cancer (Walboomers et al. 1999; Munoz et al. 2003). HPV16 and HPV18 have been identified as the most carcinogenic among all the strains (Ault 2006; Schiffman et al. 2009). Of note, long-term inflammation has been found in all cervical cancer cases. HPV infection leads to the up-regulation of 178 and the down-regulation of 150 genes in the host. The down-regulated genes are mainly those involved in the regulation of cell growth, and some are keratinocyte-specific genes and interferon-responsive genes (Chang and Laimins 2000). Unlike the situation in many other cancers, the *P53* gene in cervical cancer is typically not mutated. However, two proteins of HPV, *E6 and E7*, have a high binding affinity with both *P53* and *RB*, which disrupts the normal functions of the two proteins (Burd 2003; Oh et al. 2004), hence not requiring mutations of the genes.

5.6.4   *H. pylori and Gastric Cancer*

The relationship between *H. pylori* and gastric cancer is rather complex, as it is estimated that two thirds of the world's population is infected with *H. pylori* according to statistics from the US Center for Disease Control. Yet, most of the infected individuals do not develop gastric cancer. On the other hand, *H. pylori* is considered as a major risk factor of the disease, mainly because numerous studies have shown that the risk of developing gastric cancer is about six times higher for those infected with. *pylori* compared to those who are not (Helicobacter and Cancer Collaborative 2001). Recent studies have found that eradication of *H. pylori* does not substantially reduce the risk of gastric cancer development after chronic infection, suggesting important roles of *H. pylori* during the very early stage of the disease, possibly long before malignant transformation takes place. For example, it has been well established that *H. pylori* is a key contributor to the development of atrophic gastritis, an essential step leading to gastric cancer; consequently, atrophic gastritis induced by *H. pylori* may, quite early after infection, determine the disease development trajectories.

A number of *H. pylori* genes have been implicated to be relevant to the development of gastric cancer. For example, *vac-A* (vacuolatingcytotoxingene A) is a gene in some strains of *H. pylori* that can induce vacuolation in epithelial cells. It was observed that *cag-A*-containing *H. pylori* strains tend to be associated with patients having atrophic gastritis or gastric ulcer (Kuipers et al. 1995; Yamazaki et al. 2005). In addition, studies have shown that infection with *H. pylori* strains carrying the *cag-A* gene is associated with an increased risk of non-cardia gastric cancer (Nguyen et al. 2008). Here, *cag* refers to the pathogenicity island in an *H. pylori* genome, which consists of ~30 cytotoxin-associated genes, with other members of the island including *cag-C*, *cag-E*, *cag-L*, *cag-T*, *cag-V* and *cag-Gamma*. While these genes are believed to be directly relevant to the formation of atrophic gastritis, their roles in gastric carcinogenesis are yet to be understood.

While microbial infections have been found to be associated with the development of a variety of cancers, the detailed mechanisms of how these infections lead to cancer are unknown. It is, however, clear that all these infections lead to chronic inflammation, hence creating a hypoxic microenvironment of the infected tissue and possibly forming the initial pressure for the infected cells to evolve. Knowing that these microbial infections tend to be tissue specific, it is possible that specific biomolecules of the different microbes may target specific cell types and/or the specific microenvironment associated with these cell types. This would provide the needed vehicles for the affected cells to evolve to alleviate the pressures exerted on them by the microenvironment. Through a combination of high-throughput *omic* data analysis, computational prediction and statistical inference, it should be possible to make substantial progress in elucidating the detailed mechanisms of how a specific microbe may trigger the development of a particular cancer type.

5.6.5 *Radiation-Induced Cancer*

It is estimated that up to 10 % of the cancers is related to radiation exposure, including both ionizing radiation such as subatomic particles and non-ionizing radiation such as UV light (Anand et al. 2008). Ionizing radiation is known to induce leukemia among other cancer types. While the published studies have mostly focused on the damaging effects of ionizing radiation to DNA (Iliakis et al. 2003), the biochemical approaches have revealed that ionizing radiation can lead to the generation of ROS (Mikkelsen and Wardman 2003). For example, when water is exposed to ionizing radiation, a variety of ROS is generated, such as superoxide, hydroxyl radicals and hydrogen peroxide. If Warburg turns out to be correct in his hypothesis that the primary cause of cancer is a switch in energy metabolism, the ionizing radiation-induced ROS production may be a key reason for ionizing radiation-induced cancer development.

Multiple studies have recently shown that long-term exposure to non-ionizing radiation, including microwave radiation (Yakymenko et al. 2011) and UV light (Heck et al. 2003), leads to over-production of reactive oxygen and nitrogen species. Clearly, these results suggest that additional and more systematic analyses are needed to clarify the relevant mechanisms.

5.7 Concluding Remarks

Understanding the drivers of cancer initiation at the root level represents one of the most fundamental, most challenging and also most interesting problems in cancer research. Taking a very different approach to examine this problem, we focused on the intrinsic gap between energy demand and supply in human cells when oxidative phosphorylation is (partially) shut down, as mediated, for example, by hypoxia and/or accumulation of mitochondrial ROS. This energy gap, from our perspective, may represent an intrinsic "flaw" in our cellular system that allows cancer to develop since it can lead to the accumulation of glucose metabolites, forming the initial pressure for cells to evolve and rid themselves of the accumulation. In this context, genomic mutations probably serve as facilitators for the needed functional changes to take place, possibly in a sustained and efficient manner. The transcriptomic data analysis of energy metabolism in cancer *versus* control tissues has revealed that the current model is not complete since some cancers do not show increased uptake of glucose, at least based on the transcriptomic data. We speculate that the accumulation probably still takes place in those cancer-forming cells, possibly due to combined fluxes from glycolysis, amino acid and fatty acid metabolism, which overlaps with the glycolysis pathway. It is posited that glucose metabolite accumulation represents a most common driver for solid-tumor cancers and that the activation of hyaluronic acid synthesis pathway, as will be discussed in Chap. 6, represents the most essential step in malignant transformation of cells. Simply stated, persistent glucose metabolite accumulation necessitates cell proliferation since it provides a natural exit for the accumulated metabolites, and the natural link between glucose metabolite accumulation and hyaluronic acid synthesis serves as a facilitator for this action to take place.

Appendix

Table 5.1 A list of datasets used in analysis of Fig. 5.2

Tissue type	Data ID	Sample size	Platform
Pancreatic	GSE15471	78	GPL570
Kidney	GSE36895	76	GPL570
Colon	GSE21510	148	GPL570
Cervical	GSE6791	84	GPL570
Gastric	GSE13911	69	GPL570
Thyroid	GSE33630	105	GPL570
Brain	GSE50161	130	GPL570
Lung	GSE30219	307	GPL570
Breast	GSE42568	121	GPL570
Ovarian	GSE38666	45	GPL570
Metastatic prostate	GSE7553	43	GPL570
Metastatic melanoma	GSE7553	43	GPL570
Liver	GSE41804	40	GPL570
Primary melanoma	GSE7553	22	GPL570
Bladder	GSE31189	92	GPL570
Prostate primary	GSE3325	19	GPL570
Bcc	GSE7553	17	GPL570
Leukemia	GSE31048	221	GPL570

Table 5.2 A list of datasets used in analysis of Fig. 5.4

Tissue type	Data ID	Sample size	Platform
Inflammatory colon tissue *versus* normal colon	GSE11223	202	GPL1708
Inflammatory bowl disease/adenoma/ adenocarcinoma *versus* colon normal	GSE4183	53	GPL570
Colon adenoma *versus* colon normal	GSE8671	64	GPL570

Table 5.3 Gene expression data of different species under hypoxia and normoxia

GEO ID	Species	Number of samples	Description
GSE3537	Homo sapiens	69	Cell lines of human breast epithelial cell, renal proximal tubule epithelial cell, endothelial cell, and smooth muscle cell
GSE480	Mus musculus	20	Mouse brain, heart, lung and muscle cells
GSE3763	Blind mole rats	12	Muscle tissue
GSE1357	Rattus norvegicus	24	Hippocampus cell of hypoxia-sensitive and hypoxia-tolerant rat tissue
GSE30337	Naked mole rats	13	Transcriptome sequencing of naked mole rat tissue

Table 5.4 A list of ATP-consuming house-keeping genes in human used for the study of Sect. 5.4

Translation: GO_0006412
Sodium/potassium-exchanging ATPase activity: GO_0005391
Calcium-transporting ATPase activity: GO_0005388
Gluconeogenesis: GO_0006094
Urea cycle: *ASS1, ASL, NOS1, NOS2, NOS3, ARG1, ARG2, OTC*
Actin-dependent ATPase activity: GO_0030898

References

Anand P, Kunnumakkara AB, Sundaram C et al. (2008) Cancer is a preventable disease that requires major lifestyle changes. Pharmaceutical research 25: 2097-2116

Anastasiou D, Poulogiannis G, Asara JM et al. (2011) Inhibition of pyruvate kinase M2 by reactive oxygen species contributes to cellular antioxidant responses. Science 334: 1278-1283

Arai Y, Yoshiki T, Yoshida O (1997) c-erbB-2 oncoprotein: a potential biomarker of advanced prostate cancer. Prostate 30: 195-201

Assoian RK, Zhu X (1997) Cell anchorage and the cytoskeleton as partners in growth factor dependent cell cycle progression. Curr Opin Cell Biol 9: 93-98

Ault KA (2006) Epidemiology and natural history of human papilloma virus infections in the female genital tract. Infect Dis Obstet Gynecol 2006 Suppl: 40470

Bignell GR, Greenman CD, Davies H et al. (2010) Signatures of mutation and selection in the cancer genome. Nature 463: 893-898

Blasko I, Stampfer-Kountchev M, Robatscher P et al. (2004) How chronic inflammation can affect the brain and support the development of Alzheimer's disease in old age: the role of microglia and astrocytes. Aging Cell 3: 169-176

Boleij A, Schaeps RM, Tjalsma H (2009) Association between Streptococcus bovis and colon cancer. J Clin Microbiol 47: 516

Bremner R, Zacksenhaus E (2010) Cyclins, Cdks, E2f, Skp2, and more at the first international RB Tumor Suppressor Meeting. Cancer research 70: 6114-6118

Buchholz TA, Weil MM, Story MD et al. (1999) Tumor suppressor genes and breast cancer. Radiat Oncol Investig 7: 55-65

Budillon A (1995) Molecular genetics of cancer. Oncogenes and tumor suppressor genes. Cancer 76: 1869-1873

Burd EM (2003) Human papillomavirus and cervical cancer. Clin Microbiol Rev 16: 1-17

Buttgereit F, Brand MD (1995) A hierarchy of ATP-consuming processes in mammalian cells. The Biochemical journal 312 (Pt 1): 163-167

Campisi J, Andersen JK, Kapahi P et al. (2011) Cellular senescence: a link between cancer and age-related degenerative disease? Semin Cancer Biol 21: 354-359

Carracedo A, Cantley LC, Pandolfi PP (2013) Cancer metabolism: fatty acid oxidation in the limelight. Nature reviews Cancer 13: 227-232

Chandel N, Maltepe E, Goldwasser E et al. (1998) Mitochondrial reactive oxygen species trigger hypoxia-induced transcription. Proc Natl Acad Sci 95: 11715-11720

Chang YE, Laimins LA (2000) Microarray analysis identifies interferon-inducible genes and Stat-1 as major transcriptional targets of human papillomavirus type 31. J Virol 74: 4174-4182

Chen WY, Abatangelo G (1999) Functions of hyaluronan in wound repair. Wound repair and regeneration : official publication of the Wound Healing Society [and] the European Tissue Repair Society 7: 79-89

Chew V, Toh HC, Abastado JP (2012) Immune microenvironment in tumor progression: characteristics and challenges for therapy. J Oncol 2012: 608406

Coussens LM, Werb Z (2002) Inflammation and cancer. Nature 420: 860-867

Crosbie EJ, Einstein MH, Franceschi S et al. (2013) Human papillomavirus and cervical cancer. Lancet 382: 889-899

Donath MY, Shoelson SE (2011) Type 2 diabetes as an inflammatory disease. Nat Rev Immunol 11: 98-107

Edrey YH, Park TJ, Kang H et al. (2011) Endocrine function and neurobiology of the longest-living rodent, the naked mole-rat. Experimental gerontology 46: 116-123

Eltzschig HK, Carmeliet P (2011) Hypoxia and inflammation. N Engl J Med 364: 656-665

Estrella V, Chen T, Lloyd M et al. (2013) Acidity generated by the tumor microenvironment drives local invasion. Cancer research 73: 1524-1535

Evans VC, Barker G, Heesom KJ et al. (2012) De novo derivation of proteomes from transcriptomes for transcript and protein identification. Nature methods 9: 1207-1211

Fearon ER, Vogelstein B (1990) A genetic model for colorectal tumorigenesis. Cell 61: 759-767

Feitelson MA, Duan LX (1997) Hepatitis B virus X antigen in the pathogenesis of chronic infections and the development of hepatocellular carcinoma. Am J Pathol 150: 1141-1157

Firestein R, Bass AJ, Kim SY et al. (2008) CDK8 is a colorectal cancer oncogene that regulates beta-catenin activity. Nature 455: 547-551

Frey RS, Ushio-Fukai M, Malik AB (2009) NADPH oxidase-dependent signaling in endothelial cells: role in physiology and pathophysiology. Antioxidants & redox signaling 11: 791-810

Fujiwara S, Nakagawa K, Harada H et al. (2007) Silencing hypoxia-inducible factor-1alpha inhibits cell migration and invasion under hypoxic environment in malignant gliomas. Int J Oncol 30: 793-802

Gloire G, Legrand-Poels S, Piette J (2006) NF-kappaB activation by reactive oxygen species: fifteen years later. Biochemical pharmacology 72: 1493-1505

Greenman C, Stephens P, Smith R et al. (2007) Patterns of somatic mutation in human cancer genomes. Nature 446: 153-158

Grivennikov SI, Greten FR, Karin M (2010) Immunity, inflammation, and cancer. Cell 140: 883-899

Gupta SC, Hevia D, Patchva S et al. (2012) Upsides and downsides of reactive oxygen species for cancer: the roles of reactive oxygen species in tumorigenesis, prevention, and therapy. Antioxidants & redox signaling 16: 1295-1322

Hanahan D, Weinberg RA (2011) Hallmarks of cancer: the next generation. Cell 144: 646-674

He X, Ni Y, Wang Y et al. (2011) Naturally occurring germline and tumor-associated mutations within the ATP-binding motifs of PTEN lead to oxidative damage of DNA associated with decreased nuclear p53. Hum Mol Genet 20: 80-89

Heck DE, Vetrano AM, Mariano TM et al. (2003) UVB light stimulates production of reactive oxygen species: unexpected role for catalase. The Journal of biological chemistry 278: 22432-22436

Helicobacter, Cancer Collaborative G (2001) Gastric cancer and Helicobacter pylori: a combined analysis of 12 case control studies nested within prospective cohorts. Gut 49: 347-353

Hiotis SP, Rahbari NN, Villanueva GA et al. (2012) Hepatitis B vs. hepatitis C infection on viral hepatitis-associated hepatocellular carcinoma. BMC gastroenterology 12: 64

Hirayama A, Kami K, Sugimoto M et al. (2009) Quantitative metabolome profiling of colon and stomach cancer microenvironment by capillary electrophoresis time-of-flight mass spectrometry. Cancer research 69: 4918-4925

Hochachka PW, Buck LT, Doll CJ et al. (1996) Unifying theory of hypoxia tolerance: molecular/metabolic defense and rescue mechanisms for surviving oxygen lack. Proceedings of the National Academy of Sciences of the United States of America 93: 9493-9498

Huang LE, Bindra R, Glazer P et al. (2007) Hypoxia-induced genetic instability—a calculated mechanism underlying tumor progression. J Mol Med 85: 139-148

Hudson TJ, Anderson W, Artez A et al. (2010) International network of cancer genome projects. Nature 464: 993-998

Ibragimova I, Ibanez de Caceres I, Hoffman AM et al. (2010) Global reactivation of epigenetically silenced genes in prostate cancer. Cancer Prev Res (Phila) 3: 1084-1092

Iliakis G, Wang Y, Guan J et al. (2003) DNA damage checkpoint control in cells exposed to ionizing radiation. Oncogene 22: 5834-5847

Isaacs W, Kainu T (2001) Oncogenes and tumor suppressor genes in prostate cancer. Epidemiol Rev 23: 36-41

Itano N, Atsumi F, Sawai T et al. (2002) Abnormal accumulation of hyaluronan matrix diminishes contact inhibition of cell growth and promotes cell migration. Proceedings of the National Academy of Sciences of the United States of America 99: 3609-3614

Jiang D, Liang J, Noble PW (2007) Hyaluronan in tissue injury and repair. Annual review of cell and developmental biology 23: 435-461

Kahng YS, Lee YS, Kim BK et al. (2003) Loss of heterozygosity of chromosome 8p and 11p in the dysplastic nodule and hepatocellular carcinoma. J Gastroenterol Hepatol 18: 430-436

Kalo E, Kogan-Sakin I, Solomon H et al. (2012) Mutant p53R273H attenuates the expression of phase 2 detoxifying enzymes and promotes the survival of cells with high levels of reactive oxygen species. Journal of Cell Science 125: 5578-5586

Karin M, Cao Y, Greten FR et al. (2002) NF-kappaB in cancer: from innocent bystander to major culprit. Nature reviews Cancer 2: 301-310

Khansari N, Shakiba Y, Mahmoudi M (2009) Chronic inflammation and oxidative stress as a major cause of age-related diseases and cancer. Recent patents on inflammation & allergy drug discovery 3: 73-80

Kim S, Chin K, Gray JW et al. (2004) A screen for genes that suppress loss of contact inhibition: identification of ING4 as a candidate tumor suppressor gene in human cancer. Proceedings of the National Academy of Sciences of the United States of America 101: 16251-16256

Knudson AG, Jr. (1971) Mutation and cancer: statistical study of retinoblastoma. Proceedings of the National Academy of Sciences of the United States of America 68: 820-823

Kosaki R, Watanabe K, Yamaguchi Y (1999) Overproduction of hyaluronan by expression of the hyaluronan synthase Has2 enhances anchorage-independent growth and tumorigenicity. Cancer research 59: 1141-1145

Kubasiak LA, Hernandez OM, Bishopric NH et al. (2002) Hypoxia and acidosis activate cardiac myocyte death through the Bcl-2 family protein BNIP3. Proc Natl Acad Sci U S A 99: 12825-12830

Kuipers EJ, Perez-Perez GI, Meuwissen SG et al. (1995) Helicobacter pylori and atrophic gastritis: importance of the cagA status. J Natl Cancer Inst 87: 1777-1780

Lan KH, Sheu ML, Hwang SJ et al. (2002) HCV NS5A interacts with p53 and inhibits p53-mediated apoptosis. Oncogene 21: 4801-4811

Larson J, Peterson BL, Romano M et al. (2012) Buried Alive! Arrested Development and Hypoxia Tolerance in the Naked Mole-Rat. Frontiers in Behavioral Neuroscience:

Lian Z, Liu J, Li L et al. (2003) Upregulated expression of a unique gene by hepatitis B x antigen promotes hepatocellular growth and tumorigenesis. Neoplasia 5: 229-244

Liao D, Johnson RS (2007) Hypoxia: a key regulator of angiogenesis in cancer. Cancer Metastasis Rev 26: 281-290

Liaw D, Marsh DJ, Li J et al. (1997) Germline mutations of the PTEN gene in Cowden disease, an inherited breast and thyroid cancer syndrome. Nature genetics 16: 64-67

Lin EY, Jones JG, Li P et al. (2003) Progression to malignancy in the polyoma middle T oncoprotein mouse breast cancer model provides a reliable model for human diseases. Am J Pathol 163: 2113-2126

Liu Y, Zuckier LS, Ghesani NV (2010) Dominant uptake of fatty acid over glucose by prostate cells: a potential new diagnostic and therapeutic approach. Anticancer research 30: 369-374

Lonser RR, Glenn GM, Walther M et al. (2003) von Hippel-Lindau disease. Lancet 361: 2059-2067

Lorusso G, Ruegg C (2008) The tumor microenvironment and its contribution to tumor evolution toward metastasis. Histochem Cell Biol 130: 1091-1103

Lu T, Gabrilovich DI (2012) Molecular pathways: tumor-infiltrating myeloid cells and reactive oxygen species in regulation of tumor microenvironment. Clinical cancer research : an official journal of the American Association for Cancer Research 18: 4877-4882

Lunt SY, Vander Heiden MG (2011) Aerobic glycolysis: meeting the metabolic requirements of cell proliferation. Annual review of cell and developmental biology 27: 441-464

Manov I, Hirsh M, Iancu TC et al. (2013) Pronounced cancer resistance in a subterranean rodent, the blind mole-rat, Spalax: in vivo and in vitro evidence. BMC biology 11: 91

Martinez-Outschoorn UE, Balliet R, Lin Z et al. (2012) BRCA1 mutations drive oxidative stress and glycolysis in the tumor microenvironment: implications for breast cancer prevention with antioxidant therapies. Cell cycle 11: 4402-4413

Martinez MAR, Francisco G, Cabral LS et al. (2006) Genética molecular aplicada ao câncer cutâneo não melanoma. Anais Brasileiros de Dermatologia 81: 405-419

Mazurek S, Boschek C, Hugo F et al. (2005) Pyruvate kinase type M2 and its role in tumor growth and spreading. Semin Cancer Biol 15: 300-308

Mehta HH, Gao Q, Galet C et al. (2011) IGFBP-3 is a metastasis suppression gene in prostate cancer. Cancer research 71: 5154-5163

Mikkelsen RB, Wardman P (2003) Biological chemistry of reactive oxygen and nitrogen and radiation-induced signal transduction mechanisms. Oncogene 22: 5734-5754

Moeller BJ, Cao Y, Vujaskovic Z et al. (2004) The relationship between hypoxia and angiogenesis. Semin Radiat Oncol 14: 215-221

Morin PJ, Sparks AB, Korinek V et al. (1997) Activation of beta-catenin-Tcf signaling in colon cancer by mutations in beta-catenin or APC. Science 275: 1787-1790

Munakata T, Liang Y, Kim S et al. (2007) Hepatitis C virus induces E6AP-dependent degradation of the retinoblastoma protein. PLoS pathogens 3: 1335-1347

Munoz N, Bosch FX, de Sanjose S et al. (2003) Epidemiologic classification of human papillomavirus types associated with cervical cancer. N Engl J Med 348: 518-527

Murdoch CE, Zhang M, Cave AC et al. (2006) NADPH oxidase-dependent redox signalling in cardiac hypertrophy, remodelling and failure. Cardiovascular research 71: 208-215

Murphree AL, Benedict WF (1984) Retinoblastoma: clues to human oncogenesis. Science 223: 1028-1033

Mwenifumbo JC, Marra MA (2013) Cancer genome-sequencing study design. Nature Reviews Genetics 14: 321-332

Nathaniel TI, Otukonyong E, Abdellatif A et al. (2012) Effect of hypoxia on metabolic rate, core body temperature, and c-fos expression in the naked mole rat. International journal of developmental neuroscience : the official journal of the International Society for Developmental Neuroscience 30: 539-544

Nauseef WM (2008) Biological roles for the NOX family NADPH oxidases. The Journal of biological chemistry 283: 16961-16965

Nguyen LT, Uchida T, Murakami K et al. (2008) Helicobacter pylori virulence and the diversity of gastric cancer in Asia. J Med Microbiol 57: 1445-1453

Nishi H, Nakada T, Kyo S et al. (2004) Hypoxia-inducible factor 1 mediates upregulation of telomerase (hTERT). Molecular and cellular biology 24: 6076-6083

Nishisho I, Nakamura Y, Miyoshi Y et al. (1991) Mutations of chromosome 5q21 genes in FAP and colorectal cancer patients. Science 253: 665-669

Noble PW (2002) Hyaluronan and its catabolic products in tissue injury and repair. Matrix Biology 21: 25-29

Nowell P, Hungerford D (1960) A minute chromosome in human chronic granulocytic leukemia. Science 142:

Oh ST, Longworth MS, Laimins LA (2004) Roles of the E6 and E7 proteins in the life cycle of low-risk human papillomavirus type 11. J Virol 78: 2620-2626

Paterson S, Vogwill T, Buckling A et al. (2010) Antagonistic coevolution accelerates molecular evolution. Nature 464: 275-278

Pedersen PL, Greenawalt JW, Chan TL et al. (1970) A Comparison of Some Ultrastructural and Biochemical Properties of Mitochondria from Morris Hepatomas 9618A, 7800, and 3924A. Cancer research 30: 2620-2626

Perz JF, Armstrong GL, Farrington LA et al. (2006) The contributions of hepatitis B virus and hepatitis C virus infections to cirrhosis and primary liver cancer worldwide. J Hepatol 45: 529-538

Pfeiffer T, Schuster S, Bonhoeffer S (2001) Cooperation and competition in the evolution of ATP-producing pathways. Science 292: 504-507

Polyak K, Xia Y, Zweier JL et al. (1997) A model for p53-induced apoptosis. Nature 389: 300-305

Ramachandran K, Gopisetty G, Gordian E et al. (2009) Methylation-mediated repression of GADD45alpha in prostate cancer and its role as a potential therapeutic target. Cancer research 69: 1527-1535

Reitzer LJ, Wice BM, Kennell D (1979) Evidence that glutamine, not sugar, is the major energy source for cultured HeLa cells. The Journal of biological chemistry 254: 2669-2676

Roberts NJ, Vogelstein JT, Parmigiani G et al. (2012) The predictive capacity of personal genome sequencing. Sci Transl Med 4: 133ra158

Roche-Lestienne C, Soenen-Cornu V, Grardel-Duflos N et al. (2002) Several types of mutations of the Abl gene can be found in chronic myeloid leukemia patients resistant to STI571, and they can pre-exist to the onset of treatment. Blood 100: 1014-1018

Rolfe D, Brown G (1997) Cellular energy utilization and molecular origin of standard metabolic rate in mammals. PHYSIOLOGICAL REVIEWS 77: 731-758

Sawyers CL, Hochhaus A, Feldman E et al. (2002) Imatinib induces hematologic and cytogenetic responses in patients with chronic myelogenous leukemia in myeloid blast crisis: results of a phase II study: Presented in part at the 43rd Annual Meeting of The American Society of Hematology, Orlando, FL, December 11, 2001. Blood 99: 3530-3539

Schaffer J (2003) Lipotoxicity: when tissues overeat. Curr Opin Lipidol 14: 281-287

Schiffman M, Clifford G, Buonaguro FM (2009) Classification of weakly carcinogenic human papillomavirus types: addressing the limits of epidemiology at the borderline. Infect Agent Cancer 4: 8

Scholl S, Beuzeboc P, Pouillart P (2001) Targeting HER2 in other tumor types. Ann Oncol 12 Suppl 1: S81-87

Segal NH, Cohen RJ, Haffejee Z et al. (1994) BCL-2 proto-oncogene expression in prostate cancer and its relationship to the prostatic neuroendocrine cell. Arch Pathol Lab Med 118: 616-618

Sgambato A, Cittadini A, Faraglia B et al. (2000) Multiple functions of p27Kip1 and its alterations in tumor cells: a review. J Cell Physiol 183: 18-27

Shacter E, Weitzman SA (2002) Chronic inflammation and cancer. Oncology (Williston Park) 16: 217-226, 229; discussion 230-212

Sloma I, Mitjavila-Garcia MT, Feraud O et al. (2013) Whole Genome Sequencing Of Chronic Myeloid Leukemia (CML)-Derived Induced Pluripotent Stem Cells (iPSC) Reveals Faithful Genocopying Of Highly Mutated Primary Leukemic Cells. Blood 122: 514

Sounni NE, Noel A (2013) Targeting the tumor microenvironment for cancer therapy. Clin Chem 59: 85-93

Srivastava S, Zou Z, Pirollo K et al. (1990) Germ-line transmission of a mutated p53 gene in a cancer-prone family with Li–Fraumeni syndrome. Nature 348: 747-749

St-Pierre J, Brand MD, Boutilier RG (2000) The effect of metabolic depression on proton leak rate in mitochondria from hibernating frogs. The Journal of experimental biology 203: 1469-1476

Stehelin D, Varmus HE, Bishop JM et al. (1976) DNA related to the transforming gene (s) of avian sarcoma viruses is present in normal avian DNA.

Stern R, Asari AA, Sugahara KN (2006) Hyaluronan fragments: an information-rich system. European journal of cell biology 85: 699-715

Stratton MR, Campbell PJ, Futreal PA (2009) The cancer genome. Nature 458: 719-724

Sudarshan S, Sourbier C, Kong HS et al. (2009) Fumarate hydratase deficiency in renal cancer induces glycolytic addiction and hypoxia-inducible transcription factor 1alpha stabilization by glucose-dependent generation of reactive oxygen species. Molecular and cellular biology 29: 4080-4090

Sunaga N, Kohno T, Kolligs FT et al. (2001) Constitutive activation of the Wnt signaling pathway by CTNNB1 (beta-catenin) mutations in a subset of human lung adenocarcinoma. Genes Chromosomes Cancer 30: 316-321

Sundaresan M, Yu ZX, Ferrans VJ et al. (1996) Regulation of reactive-oxygen-species generation in fibroblasts by Rac1. The Biochemical journal 318 (Pt 2): 379-382

The-Cancer-Genome-Atlas-Research-Network (2008) Comprehensive genomic characterization defines human glioblastoma genes and core pathways. Nature 455: 1061-1068

Thery M, Bornens M (2006) Cell shape and cell division. Curr Opin Cell Biol 18: 648-657

Toole BP (2002) Hyaluronan promotes the malignant phenotype. Glycobiology 12: 37R-42R

Toro JR, Nickerson ML, Wei MH et al. (2003) Mutations in the fumarate hydratase gene cause hereditary leiomyomatosis and renal cell cancer in families in North America. Am J Hum Genet 73: 95-106

Valen LV (1973) A new evolutionary law. Evolutionary Theory 1: 1-30

Vigetti D, Ori M, Passi. A (2010) The Xenopus model for evaluating hyaluronan during development,.

Vogler M, Vogel S, Krull S et al. (2013) Hypoxia modulates fibroblastic architecture, adhesion and migration: a role for HIF-1alpha in cofilin regulation and cytoplasmic actin distribution. PloS one 8: e69128

Vojtek AB, Der CJ (1998) Increasing complexity of the Ras signaling pathway. Journal of Biological Chemistry 273: 19925-19928

Walboomers JM, Jacobs MV, Manos MM et al. (1999) Human papillomavirus is a necessary cause of invasive cervical cancer worldwide. J Pathol 189: 12-19

Warburg O (1956) On the origin of cancer cells. Science 123: 309-314

Warburg O (1967) The Prime Cause and Prevention of Cancer. Triltsch, Würzburg, Germany: 6-16

Warburg O, Posener K, Negelein E (1924) U¨ber den Stoffwechsel der Tumoren [On metabolism of tumors]. Biochem Z 152: 319-344.

Weinhouse S, Warburg O, Burk D et al. (1956) On Respiratory Impairment in Cancer Cells. Science 124: 267-272

Wilson WR, Hay MP (2011) Targeting hypoxia in cancer therapy. Nature reviews Cancer 11: 393-410

Wiseman H, Halliwell B (1996) Damage to DNA by reactive oxygen and nitrogen species: role in inflammatory disease and progression to cancer. The Biochemical journal 313 (Pt 1): 17-29

Witz IP, Levy-Nissenbaum O (2006) The tumor microenvironment in the post-PAGET era. Cancer Lett 242: 1-10

Yakymenko I, Sidorik E, Kyrylenko S et al. (2011) Long-term exposure to microwave radiation provokes cancer growth: evidences from radars and mobile communication systems. Experimental oncology 33: 62-70

Yamazaki S, Yamakawa A, Okuda T et al. (2005) Distinct diversity of vacA, cagA, and cagE genes of Helicobacter pylori associated with peptic ulcer in Japan. J Clin Microbiol 43: 3906-3916

Zhao M, Sun J, Zhao Z (2013) TSGene: a web resource for tumor suppressor genes. Nucleic Acids Res 41: D970-976

Chapter 6
Hyaluronic Acid: A Key Facilitator of Cancer Evolution

Otto Warburg made a seminal speculation in the 1960s that <u>the switch in cellular energy metabolism from aerobic respiration to glycolytic fermentation is the driving force for cancer development</u>. While increasingly more cancer researchers tend to agree with Warburg, it remains unknown, even five decades after his hypothesis, how this switch is linked to cell proliferation. We have discussed in Chap. 5 how chronic hypoxia and mitochondrial ROS accumulation will lead to continual accumulation of glucose metabolites, possibly resulting in cell death if not removed. This buildup imposes a persistent pressure on the host cells to evolve in order to survive, and cell proliferation represents a most feasible way for these cells to remove the accumulation of metabolites in a sustained manner. We will discuss in this chapter how this pressure can trigger a cellular program to facilitate cell proliferation, hence providing an exit for the accumulated glucose metabolites and a pathway to survival through proliferation.

The traditional view has been that genomic mutations, particularly mutations in proto-oncogenes or tumor suppressor genes, drive and facilitate cancer evolution. This may be true once the disease has reached a certain developmental stage, specifically after a few encoded constraints related to tissue development can be bypassed, as observed from the cancer genome data, but it is very unlikely for them to be the initial driver or even facilitator as discussed in Chap. 4. Here a model, based on the analysis results of large-scale transcriptomic data, is presented to offer a detailed explanation of how persistent hypoxia and ROS-induced stresses can lead to cell proliferation through activation and utilization of the tissue-repair system encoded in the human genome. Under normal conditions, signals for tissue repair come from the fragmented ECM of a damaged tissue; however, under stressful conditions as defined above, the cells can produce such fragments, specifically those of hyaluronic acid, from the accumulated glucose metabolites. This situation creates (or mimicks) all the signals needed for tissue repair, such as those for inflammation, anti-apoptosis, cell proliferation, cell survival and angiogenesis, leading to cell proliferation. Clearly, cell division provides relief for the stressed cells by consuming some of the accumulated metabolites for DNA and lipid synthesis. It is worth noting

© Springer Science+Business Media New York 2014
Y. Xu et al., *Cancer Bioinformatics*, DOI 10.1007/978-1-4939-1381-7_6

that such molecular accumulation will continue, even in new cells, due to the hypoxia and ROS condition in the microenvironment, which may become increasingly more stressful due to cell proliferation, possibly creating a vicious cycle for cell division.

6.1 Hyaluronic Acid and Its Physiological Functions

The glycosaminoglycan, *hyaluronic acid* (also known as hyaluronate or hyaluronan), is a key component of the extracellular matrix. It consists of a long chain of a repeating disaccharide, each consisting of one D-glucuronic acid (GlcUA) and one D-*N*-acetylglucosamine (GlcNAc). Each hyaluronic acid molecule may comprise up to 2×10^5 disaccharides with a total molecular weight approaching 10^7 Da. The negatively charged glycosaminoglycan binds various cations, e.g. Na^+, K^+ and Ca^{2+}, and forms an extended left-handed helix. Three enzymes, *HAS1-3* (hyaluronic acid synthases 1–3), are known to synthesize hyaluronic acid by lengthening the molecule through repeated addition of one glucuronic acid and one *N*-acetylglucosamine, derived from UDP-GlcUA and UDP-GlcNAc, respectively, to the nascent polydisaccharide as it is extruded via *ABC* transporters into the extracellular space. The exported hyaluronic acid, if not incorporated into extracellular matrices, will be degraded by at least six enzymes, hyaluronidases *HYAL1-6*, or by ROS into fragments of different sizes. Not until the late 1970s was the molecule found to play a significant role in cell migration during cardiac cushion development (Bernanke and Markwald 1979). Since then, a substantial amount of information has been learned about the amazingly wide range of functions of this molecule and its fragments.

6.1.1 Hyaluronic Acid Synthesis and Its Regulation

The synthesis of hyaluronic acid is tightly controlled at the transcriptional level of the three synthases, *HAS1-3*, because of its unique signaling roles in tissue development, remodeling and repair (Toole 2004). The current understanding is that *HAS1-2* tends to synthesize longer hyaluronic acid chains while *HAS3* produces shorter ones. *HAS2* is probably responsible for the production of most of the hyaluronic acid in human tissues while, interestingly, *HAS3* has been found to be up-regulated in cancer in general (Liu et al. 2001a, b; Tammi et al. 2011a; Teng et al. 2011). Figure 6.1 shows the synthetic pathway of this molecule from glucose 6-phosphate (G6P), the first glucose intermediate of the glycolytic pathway. The following table gives a list of the nine enzymes in the synthetic pathway of hyaluronic acid and their associated reactions as this information is important to our discussion in this chapter (Table 6.1).

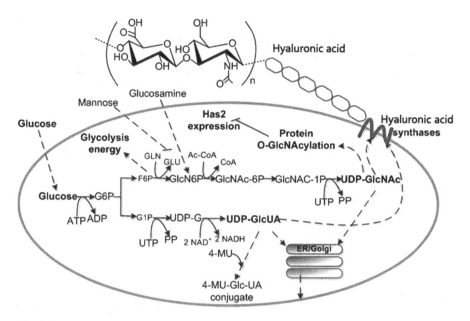

Fig. 6.1 The biosynthetic pathway of hyaluronic acid from UDP-GlcUA and UDP-GlcNAc, both derived from the common precursor glucose 6-phosphate (adapted from (Tammi et al. 2011a))

Table 6.1 Enzymes and reactions in the hyaluronic acid biosynthesis pathway

Enzyme name	Reaction catalyzed
Phosphoglucose isomerase	G6P → F6P
Glutamine-fructose-6-phosphate transaminase	F6P + GLN → GlcN6P + GLU
Glucosamine-phosphate *N*-acetyltransferase	GlcN6P + Ac-CoA → GlcNAc-6P + CoA
Phosphoacetyl glucosamine mutase	GlcNAc-6P → GlcNAc-1P
UDP-*N*-acetylglucosamine pyrophosphorylase	GlcNAc-1P + UTP → UDP-GlcNAc + PP
Phosphoglucomutase	G6P → G1P
UDP-glucose pyrophosphorylase	G1P + UTP → UDP-G + PP
UDP-glucose dehydrogenase	UDP-G + 2 NAD+→ UDP-GlcUA + 2 NADH
Hyaluronic acid synthase	UDP-GlcNAc + UDP-GlcUA → hyaluronic acid

Among the nine enzymes, (1) *GPI* (phosphoglucose isomerase) is part of the glycolytic pathway and is activated whenever glycolysis is activated; (2) the following three enzymes: *GNPNAT* (glucosamine phosphate *N*-acetyltransferase), *PGM3* (phosphoacetyl glucosamine mutase) and *UAP1* (UDP-N acetylglucosamine pyrophosphorylase), are shared with the hexosamine pathway (Fantus et al. 2006), which is positively regulated by the concentration of glucosamine (Patti et al. 1999) and by hypoxia (Guillaumond et al. 2013); (3) *PGM* (phosphoglucomutase) and *GFPT2* (glutamine-fructose-6-phosphate transaminase) are up-regulated by hypoxia

(Pelletier et al. 2012; Guillaumond et al. 2013) and by *PAK1(P21* protein-activated kinase) (Gururaj et al. 2004); (4) *UGP2* (UDP-glucose pyrophosphorylase) is up-regulated by hypoxia (Pescador et al. 2010); (5) *UGDH* (UDP-glucose dehydrogenase) is positively regulated by *TGFβ* but negatively regulated by hypoxia, which is different from all the other genes discussed above; and (6) the hyaluronic acid synthases can be regulated by various growth factors (Tammi et al. 2011a) such as *PDGF*, *KGF* (keratinocyte growth factor), *FGF2*, *EGF*, *TGFβ*, *IL1β* and *TNFα*. In addition, the level of UDP-GlcNAc has been found to control the expression of *HAS2*(Tammi et al. 2011a). The above information suggests that the high possibility of synthesis of hyaluronic acid when the cell contains ample G6P under hypoxic conditions in a *TGFβ*-containing microenvironment.

6.1.2   Functions of Hyaluronic Acid

The most relevant function of hyaluronic acid, in the current context, is its role in ECM, which also consists of collagen fibrils and numerous linker proteins. The other physiological functions of hyaluronic acid include: (1) its role as a major component of human skin, where it mainly serves as part of the tissue-repair machinery (Jiang et al. 2007); (2) the formation of a coat around each cell in articular cartilage (Holmes et al. 1988); and (3) a possible role in the development of brain as a previous study suggested (Margolis et al. 1975).

Studies in the past decade have discovered that a number of key interaction partners of hyaluronic acid have essential functional roles in tissue development and immunity. Among these partners, *CD44* and *RHAMM* (hyaluronic acid-mediated motility receptor) are the most significant ones since hyaluronic acid mediates many of its functions through interactions with these two cell-surface receptors. In addition, *CD44* has long been known to be closely associated with cancer development (Toole 2009). Other partners include *EMMPRIN* (extracellular matrix metalloproteinase inducer), *LYVE1* (lymphatic vessel endothelial hyaluronic acid receptor 1) and *HARE* (hyaluronic acid receptor for endocytosis), all being cell-surface receptors. Their detailed functions are as follows.

The wild-type *CD44* has three functional domains, an ectodomain, a transmembrane domain and a cytoplasmic domain, where the ectodomain binds hyaluronic acid and the cytoplasmic domain can bind with numerous regulatory molecules such as *NFκB* and *RAS* (Isacke 1994; Okamoto et al. 1999; Thorne et al. 2004; Misra et al. 2011). The majority of these regulatory molecules are involved in changing the key functional states of the host cell, such as proliferation, survival, differentiation, migration, production of cytokines and chemokines, and angiogenesis. *CD44*-hyaluronic acid interactions have been found to play critical roles in all these processes (Ahrens et al. 2001; Alaniz et al. 2002; Bourguignon et al. 2006; Bourguignon et al. 2009; Bourguignon et al. 2011; Park et al. 2012).

CD44 has 32 known functional splicing variants (Roca et al. 1998; Brown et al. 2011), suggesting the diversity of its function. The transcription of *CD44* is regulated in part by β-catenin and the *WNT* signaling pathway (Zeilstra et al. 2008; Ishimoto et al. 2013), and the functional state of the protein depends largely on its binding with hyaluronic acid (Toole 2009). Numerous post-translational modifications add another layer of the functional diversity to this protein, including its well-studied sialofucosylated glycoform, *HCELL* (Jacobs and Sackstein 2011), which serves as P-, L- and E-selectin ligands and fibrin receptors, where selectins are involved in chronic and acute inflammation, as well as in constitutive lymphocyte homing.

EMMPRIN (also known as *CD147*) is a cell-surface glycoprotein, and its primary function is an inducer of metalloproteinase (Guo et al. 2000; Attia et al. 2011). In addition, the protein can regulate, or at least mediate, a variety of cellular processes such as mono-carboxylate transporters and responsiveness of lymphocytes. It can bind with immunosuppressants such as cyclophilins A and B (*CYCA* and *CYCB*) and with integrins, which mediate the attachment of the host cell to the ECM. The protein interacts with hyaluronic acid indirectly through interactions with *CD44* or *LYVE1*. Interactions between *CD44* and *EMMPRIN* have been found to be a key player in the multi-drug resistance pathway (Toole and Slomiany 2008; Slomiany et al. 2009). Consequently, cancers harboring this molecular interaction tend to have poor clinical outcomes due to the cell survival capability induced by the interactions.

RHAMM (also known as *CD168*) functions both extra- and intra-cellularly. For its intracellular functions, *RHAMM* interacts with numerous signaling proteins such as tyrosine kinases (e.g., focal adhesion kinase or mitogen-induced protein kinase), *NFκB*, *RAS*, *ERK1* (extracellular regulated protein kinase 1) and actin cytoskeletal proteins. One key function relevant to the topic here is its involvement in the regulation of mitosis through interactions with *BRCA1* and *BARD1* (*BRCA1*-associated ring domain protein 1). For its extracellular functions, the protein binds with *CD44* on the cell surface. When the two proteins form a complex with hyaluronic acid, a variety of cellular processes will be activated, including the release of growth factors such as *PDGFBB*, *TGFβ2* and *FGF2* (Hamilton et al. 2007; Nikitovic et al. 2013), which (1) enhances the deposition of the hyaluronic acid chains to the ECM (Hall et al. 1994), (2) increases the locomotion in *RAS*-transformed cells (Hall et al. 1994; Hall and Turley 1995), (3) provides hyaluronic acid-mediated mobility (Hamilton et al. 2007; Nikitovic et al. 2013), and (4) sustains high basal motility when further bound with *ERK1-2* (mitogen-activated protein kinase) (Zhang et al. 1998; Lokeshwar and Rubinowicz 1999; Tolg et al. 2006).

The physiological function of *LYVE1* (also known as *XLKD1* (extracellular link domain containing 1)) is poorly understood, but its increased expression has been observed to be associated with lymph-node invasion. In addition, the protein exhibits expression patterns strongly correlated with those of *CD44* and *VEGFR3* (vascular endothelial growth factor receptor 3) during lymph-node invasion. While the detailed biochemical functions of *LYVE1* are not known, a large number of gene-expression datasets containing this protein are publicly available. With this information readily available, it is possible to conduct in-depth statistical analyses of its expression patterns and associations with other genes, potentially revealing causal relationships between this gene and other biological processes, and hence gaining an improved understanding of its functions.

Similar to *LYVE1*, very little is known about the function of *HARE*. A recent study found that *HARE* can activate *NFκB*-mediated gene expressions in response to low molecular-weight hyaluronic acid fragments (in the range of 40–400 kDa) (Pandey et al. 2013). A hypothesis was formulated based on this observation regarding its possible role in monitoring the homeostasis of ECM turnover. Again, this could be another case where computational data mining and statistical inference can lead to important new information about its functional roles.

6.1.3   Hyaluronic Acid Fragments as Signaling Molecules

The realization that hyaluronic acids have signaling roles in reporting tissue injury marks a major breakthrough in our understanding of the physiological functions of this glycosaminoglycan and the significant pathophysiological implications to cancer research. Initially this knowledge came from studies on tissue injury and repair (Noble 2002). When assaulted, an injured tissue releases ECM fragments, among which those derived from hyaluronic acid serve as signals for repairing the injured tissue. Most interestingly, hyaluronic acid fragments of different sizes have been found to serve as signals for different purposes, including the induction of inflammation, anti-apoptosis, cell survival, cell-cycle activation, cell proliferation, activation of angiogenesis and cell motility, all related to injury response, maintenance of tissue integrity, tissue repair and remodeling (Jiang et al. 2007).

The identified functions of these hyaluronic acid fragments include:(1) fragments consisting of four disaccharides typically serve as signals for suppression of apoptosis, up-regulation of *MMPs*, *HSF1* (heat shock factor-1) and *FASL* (a member of the *TNF* family); (2) fragments of four to six disaccharides signal for induction of cytokine synthesis; (3) six disaccharides function as signals for activation of *HAS2*, nitric oxide and *MMPs*; (4) 10-disaccharide fragments serve as signals for displacement of proteoglycans from cell surfaces; (5) fragments of 12 disaccharides function as signals for up-regulation of *PTEN* and endothelial cell differentiation; (6) 8–32 disaccharides serve as signals for angiogenesis stimulation; (7) 10–40 disaccharides, overlapping with the aforementioned, signal for induction of *CD44* cleavage and promotion of tumor migration; (8) ~1,000 disaccharides function as signals for production of inflammatory chemokines and stimulation of *PAI1* (plasminogen activator inhibitor) and *UPA* (urokinase); and (9) fragments of 1,000–5,000 disaccharides tend to signal for immune suppression and suppression of hyaluronic acid synthesis, i.e., providing a negative feedback for its synthesis (Stern et al. 2006; Duan and Kasper 2011).

One can see from this list that some hyaluronic acid fragments each may serve as signals for multiple biological functions, suggesting that they require additional partners to carry out their specific functions. Overall, these hyaluronic acid-derived signaling molecules promote cell survival, proliferation, angiogenesis and mobility, plus pro- or anti-immune responses depending upon their sizes. Hence they generate a microenvironment having the majority, if not all, of the essential ingredients needed for cancer (tissue) development, as discussed in the following section.

Therefore, these fragments become a high risk factor for cancer development if persistently present.

It is noteworthy that some of these signals interact with immune cells. For example, some hyaluronic acid fragments can induce the maturation of dendritic cells and the activation of allogenetic and antigen-specific T-cells (Jiang et al. 2007, 2011). Other fragments have been found to be capable of stimulating the development of the CD34+ progenitor cells into mature eosinophil cells (Hamann et al. 1995). There are also reports that these fragments can stimulate the production and release of cytokines such as *IL1β*, *TNFα* and *IL8* from the pericellular fibroblasts (Kobayashi and Terao 1997; Wilkinson et al. 2004), and some other fragments can induce proliferation of endothelial cells (West et al. 1985; Slevin et al. 1998). Many of these functional capabilities play key roles in cancer development, particularly in the early stages and during the metastatic transformation (see Chap. 10 for details).

Overall, evolution has selected hyaluronic acid fragments as signaling cues for maintaining tissue integrity and homeostasis, their release triggering the tissue repair system. Cancer or cancer-forming cells seem have learned to take full advantage of this capability, allowing them to survive the stress discussed in Chap. 5, through cell proliferation facilitated by hyaluronic acid synthesis, export and degradation.

6.2   Hyaluronic Acid: Its Links with Cancer Initiation and Development

Knowing the functional roles of hyaluronic acid and its fragments in response to tissue injury, specifically their roles in the induction of inflammation, anti-apoptosis, cell survival, proliferation, motility and angiogenesis, it is only natural to ask: *What roles do these functional capabilities have during cancer development, particularly in its early stage?*

Multiple roles by hyaluronic acid and fragments derived therefrom have been implicated in cancer development across different cancer types, particularly related to cancer metastasis (Hall and Turley 1995; Savani et al. 2001; Yoshihara et al. 2005; Bharadwaj et al. 2007, 2009; Ouhtit et al. 2007; Naor et al. 2008; Pandey et al. 2013). Our own study suggests that hyaluronic acid may have active roles throughout the entire process of cancer development and particularly in cancer initiation. The following summarizes the functions known to be exhibited by these molecules during tumorigenesis, which serves as a starting point for developing a hyaluronic acid-facilitated cancer initiation model in Sect. 6.3.

6.2.1   Inflammatory Signaling

Chronic inflammation is known to be closely related to cancer initiation and early development (Lu et al. 2006; Rakoff-Nahoum 2006; Colotta et al. 2009), as discussed in the earlier chapters. The current understanding is that cancer-inducing microenvironments are largely orchestrated by inflammatory cells, which are an integral part of a neoplastic process as they foster cell proliferation, survival and migration (Coussens and Werb 2002). Low molecular-weight hyaluronic acid fragments have been found to promote inflammation by up-regulating the expression of a number of pro-inflammatory genes such as *MIP1α* (macrophage inflammatory protein 1α), *MIP1β*, *KC* (keratinocyte chemo-attractant), *MCP1* (macrophage chemo-attractant protein1), *IFIT10* (interferon induced protein 10), *TNFα* and a few other cytokines, and by down-regulating the anti-inflammatory gene *A2AR* (adenosine A2a receptor) (Collins et al. 2011; Black et al. 2013).

6.2.2 Cell Survival Signaling

A key characteristic of cancer cells is that they can survive conditions that should lead to apoptosis of normal cells. Interactions between hyaluronic acid and *CD44* are known to have key roles in this distinct capability as they have been widely observed to be associated with the activation of survival pathways in both cancer and normal cells under stressful conditions. While detailed discussions on survival pathways are given in Chap. 7, a model of how hyaluronic acid and its interaction partners can activate survival pathways is introduced here.

The *PI3K/AKT* signaling pathway is at the core of a number of survival pathways. A recent study has found that constitutive synthesis and export of hyaluronic acid can activate a *PI3K/AKT*-mediated survival pathway (Ghatak et al. 2002). The mechanism for this activation involves the binding of hyaluronic acid to *CD44*, leading to the activation of *ERBB2* (also known as *HER2*), a receptor tyrosine kinase that can activate the *PI3K/AKT* signaling pathway. Specifically, an activated *PI3K* can phosphorylate *AKT,* which alters its structural conformation to enable the protein to be activated by the *PDK1/PRK* (phosphoinositide-dependent kinase 1 and phosphoribulokinase) complex (Datta et al. 1999). Then the active form of *PI3K/AKT* activates β-catenin, a cell–cell adhesion regulator that, in turn, up-regulates and activates *COX2* (cyclooxygenase-2), an enzyme that has been found to inhibit apoptosis in multiple cancers (Ding et al. 2000; Nzeako et al. 2002; Basu et al. 2004; Kern et al. 2006). *COX2* leads to the production of *PGE2* (prostaglandin E2), that activates the *RAS-MAPK-ERK* pathway via *EP4* (prostaglandin E receptor 4). This activated pathway will then up-regulate the expression of the anti-apoptotic protein *BCL2* via *CREB* (*CAMP*-response element-binding protein) (May 2009).

Hyaluronic acid can also activate cell-survival pathways in a *CD44*-independent manner. For example, the molecule can lead to the retention and concentration of *IL6* near its site of secretion (Vincent et al. 2001b), an event that promotes cell survival through the activation of *STAT3* (signal transducer and activator of transcription 3), this being an essential step for cell survival for a number of cancer types (Calame 2008; Diehl et al. 2008; Avery et al. 2010; Lin et al. 2011). A study has recently reported that an activated *STAT3* can enhance cell survival through up-regulation of *OX40* (also known as *CD134*), a member of the *TNFR* superfamily, and *BCL2*, as well as down-regulation of *FASL* and *BAD* (*BCL2*-associated agonist of cell death), a pro-apoptotic member of the *BCL2* family (Malemud 2013).

6.2.3   Mediating the Cell Cycle

Tight coupling between cell-cycle progression and cell polarity is crucial for cell division (Budirahardja and Gonczy 2009; Noatynska et al. 2013). The main regulators, the cyclins and *CDKs*, act on both cell-cycle progression and development of cell polarity (Drubin and Nelson 1996). While this relationship was established in the 1990s, only recently were the main targets of cyclins and *CDKs* identified, these being the *RHO GTPases* such as *CDC42* (Croft and Olson 2006; Yoshida and Pellman 2008). Interestingly, previous studies have shown that hyaluronic acid-CD44 interactions can activate various *GTPases* such as *RHOA* (*RAS* homolog gene A), *RAC1* and *CDC42* (also see earlier discussion on hyaluronic acid synthesis pathway), as well as cytoskeletal functions in cancer (Bourguignon et al. 2005; Bourguignon 2008), strongly suggesting the possibility that hyaluronic acid can modulate the cell cycle. A recent study has shown that the activation of hyaluronidase *HYAL1* increases the cell-doubling rate (Bharadwaj et al. 2009), hence providing indirect evidence to the above speculation.

6.2.4   Insensitivity to Anti-growth Signals

A hallmark of cancer is that the transformed cells become insensitive to anti-growth signals (Hanahan and Weinberg 2000, 2011), as introduced in Chap. 1. Cancer cells have evolved various mechanisms to achieve this, one being discussed here. The anti-growth factor *TGFβ* is known to be activated by different cellular stressors such as inflammation, destruction of the ECM, high ROS level, tissue injury and high intracellular acidity. The activated *TGFβ* inhibits a dividing cell by blocking its advancement through the G_1 phase of the cell cycle. However, this anti-growth role by *TGFβ* can be converted to a pro-growth role as has been widely observed in advanced cancers (Tang et al. 2003). It has been shown that the concentration of hyaluronic acid determines the anti- or pro-growth function of *TGFβ* (Porsch 2013). It is noteworthy that while the anti-growth role of *TGFβ1* is *CD44* independent, its switch to a pro-growth role is mediated by *CD44* in conjunction with an increased concentration of hyaluronic acid (Meran et al. 2011). We speculate here that the activation of *CD44* is also induced by the increased concentration of hyaluronic acid. Overall, upon binding *TGFβ*, the receptor *TGFR* up-regulates *EGF*, which binds to and activates its cognate receptor, *EGFR*, hence promoting cell growth. An environment rich in hyaluronic acid synthesized by *HAS2*, which tends to be longer than those synthesized by other synthases, has been found to promote binding between *CD44* and *EGFR*. The formation of this complex leads to the activation of the *MEK* (*MAPK/ERK1-2* kinases) and then the activation of *ERK1-2* (Meran et al. 2011), which are known to be cancer promoting (Hamilton et al. 2007).

6.2.5   Cell Proliferation and Anchorage-Independent Growth

It has been established that hyaluronic acid and a versican-rich ECM are required for cell proliferation, at least in smooth muscle cells (Evanko et al. 1999). Furthermore, it has been observed that an increased hyaluronic acid concentration can lead to cell proliferation through hyaluronic acid-*CD44* interactions (Hamann et al. 1995; Ghatak et al. 2002). While the detailed mechanism of how hyaluronic acid-*CD44* interactions promote cell proliferation are yet to be fully understood, a number of studies have revealed how such interactions may be associated with proliferation in specific cancers. For example, it was found that hyaluronic acid-*CD44* interactions with *PRKCE* (protein kinase C epsilon type) promote oncogenic signaling via *NANOG* (a transcription factor critical to self-renewal of stem cells) and the production of microRNA-21, leading to down-regulation of the tumor suppressor protein *PDCD4* in breast cancer (Bourguignon et al. 2009). In another study, increased hyaluronic acid production was found to promote coupling between *CD44* and *EGFR*, which will lead to *CD44*-dependent activation of *EGFR*-mediated growth signaling in head and neck cancers (Wang and Bourguignon 2011). Figure 6.2 shows a model of how hyaluronic acid may be linked to cell proliferation.

Another unique role of hyaluronic acid in cancer growth is that it facilitates anchorage-independent growth. In normal human tissues, cells require a surface on which to flatten and divide, that being the basement membrane. This process, termed the *anchorage dependence of growth*, is a mechanism used to prevent persistent and unregulated cell division. This is executed by a requirement of having companion ECM signals when signaled to grow. Studies have shown that over-production of hyaluronic acid, in conjunction with over-expression of *EMMPRIN*, allows cell growth on soft agar or even in suspension (Marieb et al. 2004), although the detailed mechanism remains to be fully understood.

A related capability gained by cancer cells is their loss of contact inhibition, another mechanism encoded in human cells for preventing over-growth during normal tissue development and remodeling. All cancer cells appear to have lost this preventive machinery or have gained a capability to override the processes leading to contact inhibition. The over-production of *HAS2*-synthesized hyaluronic acid has been found to allow cells to escape from the contact-inhibition constraint through formation of a large hyaluronic acid matrix regulated by *PI3K* (Itano et al. 2002).

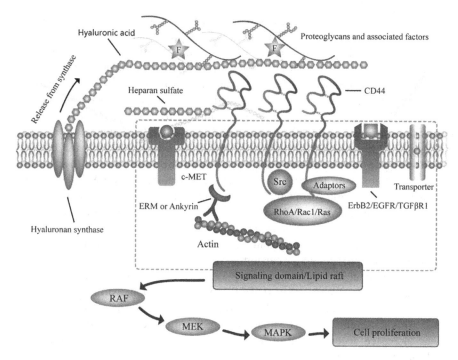

Fig. 6.2 A model for linkages between hyaluronic acid-*CD44* interactions and cell proliferation (adapted from (Toole 2009))

6.2.6   Tumor Angiogenesis

As discussed in Sect. 6.1, hyaluronic acids can provide signals for angiogenesis during tissue repair. The activation of the angiogenesis process requires a high concentration of *HIF1α*, in addition to hyaluronic acids (Pugh and Ratcliffe 2003; Stern et al. 2006; Toole 2009), both of which are available in a typical neoplastic environment. The actual formation of tumor blood vessels requires *MMPs* to partially degrade the basement membrane, thus making connections with existing blood vessels possible. The *MMPs* are also available in the environment under discussion (see below).

6.2.7   Invasiveness, EMT and Metastasis

Tissue invasion typically refers to cell growth across the basement membrane (a type of ECM), which underlies epithelial cells and surrounds blood vessels. A key step in tumor invasion, to be detailed in Chap. 10, is ECM proteolysis by *MMPs*.

It has been well established that low molecular-weight hyaluronic acid fragments can induce *MMP3* (Fieber et al. 2004). The current understanding is that ECM proteolysis is regulated by the balance between the concentrations of *MMPs* and their inhibitors *TIMPs* (tissue inhibitors of metalloproteinase) (Maeso et al. 2007), i.e., a higher *MMP* concentration generally indicates higher invasion rates. This relationship can explain the observation that interactions between *HAS2*-synthesized hyaluronic acid and *CD44* are important in determining the level of invasiveness of a cancer (Zoltan-Jones et al. 2003), since *HAS2* is known to be a suppressor of *TIMP* expression (Bernert et al. 2011), and hence its activation can shift the balance towards a higher *MMP* concentration.

An essential step in tumor metastasis is the activation of the EMT (epithelial-mesenchymal transition) pathway. Recent studies have found that hyaluronic acid has a central role in EMT activation. Specifically, the accumulation of extracellular low molecular-weight hyaluronic acids can create cyclic mechanical stretches when the innate immune adaptor protein *MYD88* (myeloid differentiation primary response gene) is present, which can induce the activation of EMT. Mechanistically, *WISP1* (*WNT*-inducible signaling protein 1) is significantly up-regulated in hyaluronic acid-stretched cells in a *MYD88*-dependent fashion, while inhibition of *WISP1* has been found to prevent the activation of EMT in these cells (Heise et al. 2011). This represents the best explanation to date that mechanical forces, created by the increased accumulation of hyaluronic acids, can lead to the activation of EMT via the innate immune system.

Hyaluronic acid has also been found to have other important roles during the metastatic processes such as intravasation, circulation in blood, extravasation of the blood vessels into new locations and reactivation from dormancy as the colonizing cells become established in the new location (s). Detailed discussions of these issues are given in Chaps. 10 and 11.

6.2.8 Evasion of Immune Detection

There is a close relationship between hyaluronic acids and immune signaling (Taylor et al. 2004; Scheibner et al. 2006; Jiang et al. 2007; Shirali and Goldstein 2008; Jiang et al. 2011; Erickson and Stern 2012), as discussed in Sect. 6.1. Specifically, hyaluronic acids can serve as endogenous activators of the innate immune system. For example, hyaluronic acid fragments can activate dendritic cells and prime T-cell alloimmunity via a *TLR4* (toll-like receptor 4)/*TIRAP*-dependent pathway (Shirali and Goldstein 2008). Furthermore, *CD44* is able to modulate these immune responses through augmenting the regulatory T-cell functions and enhancing the expression of the negative regulators of *TLR*-signaling (Shirali and Goldstein 2008). Knowing the intimate relationship between immune signaling and hyaluronic acid, it is reasonable to speculate that cancer cells may have evolved ways to alter the interactions between hyaluronic acid and its key receptors such as *CD44* to avoid detection by the immune system.

A recent study on the evolution of hyaluronic acid and associated genes suggests that the original function of the molecule was to provide a protective shield of cells for two situations where survival is vitally important: (1) in the cumulus mass that surrounds the ovum, and (2) in the stem cell niche (Salustri et al. 1999; Haylock and Nilsson 2006; Schraufstatter et al. 2010; Csoka and Stern 2013). For these situations, it is necessary that hyaluronic acid provide a shield for the cells to avoid detection by surveillance machineries. More generally, hyaluronic acids, particularly in their high molecular weight forms, have been shown to be intrinsically immuno-suppressive (McBride and Bard 1979; Delmage et al. 1986). For example, hyaluronic acid suppresses septic responses to lipopolysaccharides and acts to maintain immune tolerance. In addition, it can induce the production of immunosuppressive macrophages. Possibly through repeated trial-and-error for survival, cancer cells have evidently adapted to create favorable environmental conditions that promote the production of hyaluronic acids of the right sizes and take advantage of their immunosuppressive capability.

Knowing that hyaluronic acids of different sizes serve as signaling molecules for different aspects of tissue repair, one can hypothesize that upon tissue injury the hyaluronic acid fragments must have the "appropriate" size distribution to facilitate tissue repair being done in a **coordinated** manner, a response that has presumably been perfected through millions of years of evolution. In the following section, we present how cancer cells have adapted to take advantage of this powerful signaling capability by creating an environment that promotes the synthesis, export and degradation of hyaluronic acid to mimic the hyaluronic acid-based signals for their survival. It is worth noting that there is one fundamental difference between injury-induced *versus* neoplastic cell-induced hyaluronic acid-based signaling, as presented in Sect. 6.3. That is, the tissue-repair signaling induced by persistent hypoxia is done in an **uncoordinated** manner, which may be the key reason for needing genomic mutations to assist these cells' survival as discussed in Chap. 4. *[N.B. As a point of clarification, it should be stressed that the cells continue evolving with natural selection, favoring those that can best survive and proliferate. Many of course, probably the vast majority, fail to evolve appropriately and are destroyed by the normal defenses of the body.]*

6.3 A Model for Hyaluronic Acid-Facilitated Cancer Initiation and Development

A recent study reported that elevated cellular glucose concentrations can lead to the production of hyaluronic acid (Yevdokimova 2006). In addition, other investigations have shown that hyaluronic acid processing and interaction with the host cells can lead to proliferation (Kosaki et al. 1999; Vincent et al. 2001a). From these observations and discussions in the above two sections, one can see that hyaluronic acid can be synthesized from glucose metabolites, leading to the generation of signals needed for tissue repair. By integrating all this information, one can hypothesize that <u>the accumulation of glucose metabolites in inflammation-induced</u> <u>hypoxic cells will lead to the production of hyaluronic acid, which ultimately</u> <u>provides an exit for the accumulated glucose metabolites out of the cells through</u> <u>cell division</u>.

A model, based on extensive analyses of transcriptomic data of multiple cancer types, is now presented in support of this hypothesis. Through this model, the following will be demonstrated:(1) how the accumulation of glucose metabolites in inflammation-induced hypoxic cells can trigger the activation of the synthesis, export and degradation of hyaluronic acid; (2) how the fragments of this molecule lead to cell proliferation, thus providing an exit for the accumulated glucose metabolites and a (temporary) relief of the pressure forced on the relevant cells; and (3) how this process may continue as long as the hypoxic condition persists; some components of the model may ultimately be replaced by genetic mutations to accomplish the same functions but with better sustainability and possibly energy efficiency. While no statistical significance analysis is presented of the model against the available *omic* data, the model is nonetheless highly statistically significant, reaching a high level of consistence with the available transcriptomic data.

6.3.1 Activation of Synthesis, Export and Degradation of Hyaluronic Acid

One may recall from Chap. 5 that chronic hypoxia leads to a switch in cellular energy metabolism from aerobic respiration to (anaerobic) glycolysis, ultimately resulting in the accumulation of glucose metabolites. From the earlier discussion in this chapter, hypoxia, in conjunction with a high concentration of cellular G6P and *TGFβ* in the microenvironment, can lead to the activation of seven out of the nine enzymes in the hyaluronic acid synthetic pathway, namely *GPI, GFPT2, GNPNAT1, PGM3* and *UAP1* in the upper part of the pathway and *PGM2* and *UGP2* in the lower part of the pathway (Fig. 6.1). All the three triggering conditions should be satisfied under inflammation-induced hypoxic conditions, as discussed in Sect. 6.1. For the two other genes, both *UGDH* and *HAS* can be up-regulated by *TGFβ* if its active form is available, which should be the case in chronic inflammatory sites (Clarkin et al. 2011; Tammi et al. 2011b).

Gene expression data have been examined on the same set of colon precancerous adenoma and adenocarcinoma samples used for Fig. 5.4 of Chap. 5, with a focus on the nine enzyme-encoding genes involved in hyaluronic acid synthesis, along with a number of related genes, namely *UAP1L1* (a homolog of *UAP1*), three hyaluronidase genes *HYAL1-3*, the hyaluronic acid exporter gene *CFTR* (cystic fibrosis transmembrane conductance regulator) (Schulz et al. 2010), two anti-apoptosis genes (*BCL2A1* and *BCL2L1*), one heat shock gene (*HSF1*), *TGFβ* and *MYC* (v-myc avian myelocytomatosis viral oncogene homolog). Figure 6.3 shows the expression data of these genes except for *UGP2*, which is missing in the gene-expression datasets used here.

Fig. 6.3 Expression data of the hyaluronic acid synthetic pathway and related genes. The six columns represent the tissue types, going from *left to right*: (1) tissues of inflammatory bowel disease, (2) colon adenoma, and (3–6) colon adenocarcinoma stages 1–4. The 18 rows represent 18 relevant genes. The data shown here strongly suggest that hyaluronic acid is being produced and degraded into fragments

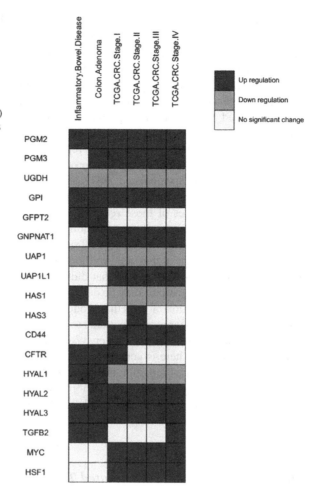

From the figure, one can see the following: all genes in the pathway of Fig. 6.1 are up-regulated in adenomas except for *UAP1* and *UGDH*, which are consistently down-regulated across all the samples. Interestingly for the down-regulated *UAP1* (EC2.7.7.23), its homolog *UAP1L1* is consistently up-regulated across all the adenocarcinoma samples, whose function is only partially determined to be in the EC2.7.7 enzyme class with the last digit of the EC (enzyme classification) class undetermined. Based on a KEGG pathway analysis, GlcNAc-1P, the substrate of *UAP1*, can only be metabolized to UDP-GlcNAc by an EC2.7.7.23 enzyme without other exits. Hence, it is reasonable to posit that *UAP1L1* is used to make this conversion. Regarding the other down-regulated enzyme, *UGDH,* the gene has 13 known splicing variants. Interestingly multiple splicing variants are up-regulated across the adenocarcinoma samples, raising the possibility that some of these splicing variants may have the same enzymatic function for the following reason: the rate-limiting gene (*GFPT2*) of hyaluronic acid synthesis pathway, the hyaluronic acid exporter gene *CFTR* and multiple hyaluronidase genes (*HAYL1-3*) are all up-regulated. Interestingly, *TGFβ* is up-regulated in the precancerous stage, and its expression then returns to the background level once the downstream genes of hyaluronic acid synthesis such as *HSF1* and *MYC* are up-regulated. This indicates that once the tissue becomes cancerous, cell proliferation may be driven by factors other than hypoxia-induced hyaluronic acid production and fragmentation, such as over-expression of certain proto-oncogenes. Clearly further analyses are needed to infer which oncogenes or other genes may have replaced the roles played by hyaluronic acid fragments to drive the tumorigenesis process.

The observation of up-regulated *HAYL1-3* strongly suggests that synthesized hyaluronic acid is exported and degraded into fragments, some of which, by chance, will be of the same sizes of those required to be tissue-repair related signals, as discussed in Sect. 6.2. Since all sizes of short hyaluronic acids serve as signals for inflammation induction, anti-apoptosis, cell survival, cell proliferation or angiogenesis, it is reasonable to posit that cell proliferation will be initiated. In addition, it is reasonable to further posit that cells that did not produce the necessary combinations of signals for tissue repair may be destroyed. That is, the natural selection has selected the sub-population of cells that produce the right combinations of signals.

Highly similar gene-expression patterns to those in Fig. 6.3 are observed in various other precancerous and early cancer types, strongly suggesting that the driver model being developed here is generally applicable to other cancers, at least solid cancers. One interesting observation is that while *ABCC5* has been found to encode the main exporter for hyaluronic acids in fibroblasts cells (Schulz et al. 2007), epithelial cells tend to use *CFTR* (Schulz et al. 2010), which is indeed up-regulated as shown in Fig. 6.3.

6.3.2   Hyaluronic Acid-Facilitated Tissue Development

The exported hyaluronic acid is degraded by hyaluronidases or through partial de-polymerization by ROS (Ågren et al. 1997). Multiple factors can activate different hyaluronidases. For example, inflammation and necrosis-related signals such as *TNFα* and *IL1β* can activate *HYAL2* (Monzón et al. 2008), which should be available under the conditions being considered here. *HYAL1* can be up-regulated by the binding of *EGR1* (early growth response protein 1) to the promoter region of the gene. The same study also found that while the binding of *NFκB* is not necessary to activate the gene, it enhances its expression.

A number of cell proliferation signaling pathways can be activated by hyaluronic acids, including genes in the *ERK1-2* pathway, namely *RAF1*, *MAP* (mitogen-activated protein), *ERK1* (Slevin et al. 1998) and heat-shock receptor binding protein *HSF1* (Xu et al. 2002). The link between hyaluronic acids and *HSF1* is particularly interesting since *HSF1* is known to orchestrate a large network of core cellular functions including proliferation, survival, protein synthesis and glucose

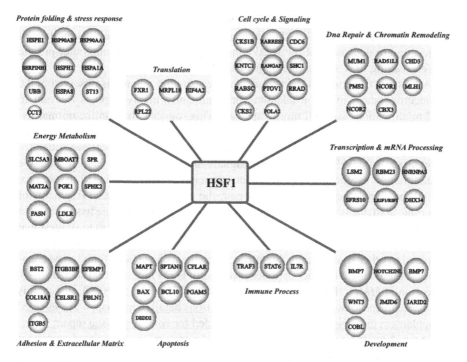

Fig. 6.4   *HSF1* and the gene network it regulates (adapted from (Mendillo et al. 2012))

metabolism (Dai et al. 2007) and is up-regulated in adenocarcinoma as shown in Fig. 6.3. Figure 6.4 shows the genes that are regulated by *HSF1*. Clearly multiple genes involved in cell proliferation such as cell cycle control, transcription and protein synthesis are up-regulated by this protein. From the above discussion, it is reasonable to postulate that growth signals and various regulatory signals are made available to the evolving cells under consideration. We will now examine how these signals are related to the development of cancer.

As discussed in Chap. 4, growth factors alone are not sufficient to initiate cell proliferation in a tissue environment as a number of conditions must be met. For a cell to divide or a tissue to grow via hyperplasia, at least three classes of extracellular signals are needed: (1) *mitogens* such as *PDGF* or *EGF* that stimulate cell division by activating intracellular growth signaling proteins such as *RAS* and the *MAPK* cascade and triggering cell-cycle progression; (2) *growth factors* such as *PI3K* and *PDGF* that stimulate cells to increase their mass by up-regulating genes involved in cell metabolism and macromolecular syntheses, typically through up-regulation of the *MYC* gene *[N.B. growth factors should not be confused with mitogens although some genes may serve in both roles]*; and (3) *survival factors* such as the anti-apoptotic members of the *BCL2* family, which inhibit apoptosis during tissue development (Alberts et al. 2002). Certain growth signals such as *FGF* and *PDGF* are communicated into the cells through ECM-cell adhesions (also called *focal adhesions*). Specifically, such signals alter the physical properties of an

ECM, which leads to changes in actin cytoskeletal structures through interactions between ECM proteins such as laminins and fibronectins and integrins on the cell surface, and to activation of the focal adhesion kinase (*FAK*) among other protein kinases, leading to signaling of cell growth.

Recall from Chap. 4 that the majority of the genomic mutations observed in the precancerous stage of colon adenoma predominantly involves genes relevant to ECM modifications and cell morphogenesis. This suggests that, unlike normal tissue development, the process of tumorigenesis, as triggered by the signals discussed above, may not have all of the needed signals or else lack sufficient quantities for cancer tissue to develop based on the signals alone. This is not hard to imagine since, although hyaluronic acid fragments, generated from a damaged tissue, provide all the signals needed for tissue repair, the randomly generated fragments of hyaluronic acid due to persistent hypoxia may not necessarily enable tissue repair in a coordinated manner. The aforementioned genomic mutations seem to suggest that signals related to changes in ECM and cell morphology may be relatively too weak compared to other signals, even though such fragments have been found to be capable of up-regulating a number of genes involved in ECM modification such as over-expression of *MMP*s (Fieber et al. 2004) and *UPA*s (Horton et al. 2000). Hence, it is reasonable to speculate that the observed mutations detailed in Chap. 4 are selected to supplement the missing or weak signals needed for continual tissue repair, which is triggered by the released hyaluronic acid fragments as a result of persistent hypoxia in an inflammatory site.

In addition, hyaluronic acid and fragments provide other signals needed for tissue repair. Specifically, a number of tumor angiogenesis-related genes are activated by hyaluronic acid (Slevin et al. 2002; Takahashi et al. 2005); similarly a number of survival genes such as *HSPA2* (heat shock 70 kDa protein 2) are over-expressed by hyaluronic acid (Xu et al. 2002). Furthermore, hyaluronic acid can also facilitate cells to overcome the constraints of anchorage-dependent growth and contact inhibition (Itano et al. 2002), as discussed in Sect. 6.2. Overall, it is the combination of hyaluronic acid, hyaluronic acid fragments and mutations in genes involved in changing the composition and hence the physical properties of the ECM, as well as in altering cell morphology, that facilitates cell proliferation in a tissue without actual signals for tissue repair.

Clearly, when cells divide, their accumulated glucose metabolites can be used as the building blocks for DNA and lipid synthesis, hence providing an exit for the accumulation and relieving the pressure as discussed in Chap. 5. This, however, is only temporary, as all the cells, including the newly synthesized ones, will become hypoxic again because of the local environment, thus again leading to the accumulation of glucose metabolites, and ultimately continuous cell proliferation. As this process continues and the biomass grows due to cell division, the microenvironment will become increasingly more hypoxic. One can expect that this process continues as long as the hypoxic condition persists, while the vast majority of the proliferated cells will die due to various reasons, including programmed cell death and cell-cell competition as discussed in Chap. 8, in the early phase of the tumorigenesis.

This process may go on for years without visible tumor growth, but fundamental changes are taking place inside these cells.

It is noteworthy that it was recently discovered that inhibition of tumor growth by the high molecular weight hyaluronic acid produced in naked mole rat tissues is reversed by treatment that removed the accumulated glycosaminoglycan (Tian et al. 2013). This observation provides strong supporting evidence for the key role played by hyaluronic acid synthesis in hypoxia-induced cell proliferation.

6.3.3   General Roles of Hyaluronic Acid During Tumorigenesis

It is known that extracellular hyaluronic acid can positively regulate lactate efflux (Slomiany et al. 2009), suggesting a possibility that its synthesis may serve as an overflow buffer for glycolytic metabolites and that the exported hyaluronic acid functions as a signal for increasing the exit flux of glycolysis by increasing the lactate efflux. Interestingly, lactate has also been found to serve as a stimulator for the increased production of hyaluronic acid in some cell types such as fibroblasts (Stern et al. 2002). These two pieces of data suggest a possible vicious cycle between hyaluronic acid production and lactate efflux, which may continuously generate cell proliferation signals. Clearly, this possibility requires experimental validation.

As this cell-division process continues, mutations in some genes, not limited to proto-oncogenes and tumor suppressor genes as discussed in Chap. 4, may be selected to allow constitutive activation or inhibition of functionalities to facilitate cell proliferation in a more sustained and efficient manner. It is foreseeable that, as a cancer evolves, the signaling roles of hyaluronic acids may be gradually replaced by the constitutive activation of the relevant processes of tissue repair, made possible via selection of certain genomic mutations such as oncogenes. When all the signaling roles of hyaluronic acid and its fragments are replaced by genomic mutations, some cancers may select, at some developmental stage, to cease the biosynthesis of hyaluronic acid, hence terminating their facilitator's role for cancer initiation. This hypothesis is clearly supported by the decreased expressions of hyaluronic acid synthase genes as the cancer advances (see Fig. 6.3).

We speculate that hyaluronic acid production is essential to cancer initiation not only because it provides signals for tissue growth, but also because it generates cell-survival signals. These may prove to be essential for the underlying cells to select mutations in genes with essential functions without being destroyed by apoptosis or tissue-level surveillance (see Chap. 8). The reason is that, by promoting anti-apoptotic activities, survival pathways may keep alive those cells harboring mutations in genes involved in essential cellular functions, which otherwise will be killed by apoptosis. A good example is that of the proto-oncogene *EGFR* since it can be constitutively activated due to mutations (Okabe et al. 2007) or oxidation of specific residues in the absence of its ligand *EGF* (discussed in Chap. 5).

During the entire course of neoplastic development, hyaluronic acid and their fragments are known to exhibit a variety of other regulatory actions, particularly at

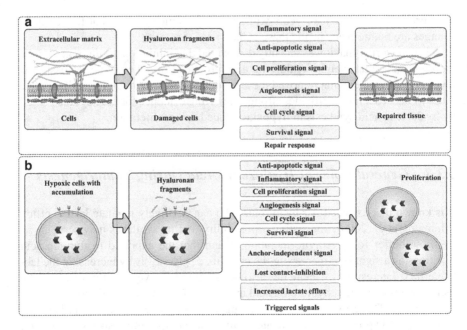

Fig. 6.5 A schematic illustration of roles played by hyaluronic acid (hyaluron) and its fragments in tissue injury and repair *versus* their roles in cell proliferation. (**a**) A description of the process of going from normal tissue to damaged tissue and the role(s) of hyaluronic acid and fragments in triggering tissue repair. (**b**) A description of the process of going from hypoxic cells with accumulated glucose metabolites to synthesis, export and fragmentation of hyaluronic acid in the extracellular space to triggering tissue repair and then proliferation

key transition points throughout cancer development, including cancer metastasis and exiting dormancy in the metastatic locations as discussed in Chap. 10. The up-regulated *TGFβ* gene expression at stage 4 cancers in Fig. 6.3 seems to suggest that the gene may have a key role in cancer metastasis as its activation will trigger the increased production of hyaluronic acid, which is essential for the initiation of metastasis (see Chap. 10). Figure 6.5 shows a model for the proposed hypoxia-driven, hyaluronic acid-facilitated cell proliferation in the early phase of cancer development.

6.4   Bioinformatics Opportunities and Challenges

The model given in Sect. 6.3 represents, to the best of our knowledge, the first model that links cancer initiation to inflammation-induced hypoxia and hyaluronic acid production at the molecular level. A number of opportunities present themselves in this model for further study of the important issue of cancer initiation through computational analyses of large-scale *omic* data of cancer and statistical inference in a systematic manner in terms of model validation, refinement and expansion. It is expected that such computational studies may lead to fundamental and novel insights of cancer initiation, particularly through this type of comparative analysis across multiple cancer types. From such data, one may be able to identify the most essential common characteristics in cancer initiation in general.

6.4.1 Completing the Details of the Model Based on Available Transcriptomic Data

The model provides a high-level conceptual framework of a mechanism by which cancer may be initiated, but numerous details are needed for specific cancer types. For example, from the transcriptomic data, strong correlations were observed between the increased glucose accumulation and the activation of hyaluronic acid synthesis, but the detailed signaling and regulatory processes may differ for different types of cancers. Another example relates to the sizes of hyaluronic acid fragments that are generated under specific cellular conditions. Intuitively, higher abundances of hyaluronidases are expected to produce shorter fragments due to the repeated degradations with higher frequencies. Some mathematical relationships between the expression levels of hyaluronidases and the size distributions of hyaluronic acids need to be established in order to address this issue. One approach involves the collection of data on the abundances and distributions of different hyaluronic acid fragments along with the corresponding gene-expression data under controlled conditions using cell lines. The results obtained could then be used to train a predictor for estimating the fragment size distributions based on the observed gene-expression data for specific cancer types. A similar approach could be taken to address issues such as identification of the: (1) specific signals that activate the genes responsible for hyaluronic acid synthesis, along with the required cellular environmental features such as the level of hypoxia and the availability of specific immune cells in the pericellular environment, and (2) detailed regulatory pathway between increased hyaluronic acid and lactate production and export.

6.4.2 Validation, Refinement and Expansion of the Model

The model can be checked against transcriptomic data of different cancer types for their validity. Specifically, one can check if each of the model-predicted associations holds against the available data for each cancer type. For example, if the model predicts transcription regulator X positively regulates the expression of gene Y, one can check if these two genes tend to show co-expression patterns under the relevant conditions, or determine if the up-regulation of gene Y tends to imply the up-regulation of gene X in the available data. With more advanced analyses, it may be possible to predict causal relations among genes found to be co-expressed.

Systematic analyses such as this on each model-predicted association can lead to identification of incorrect prediction or partially correct prediction, hence providing guidance for further refinement of the model.

6.4.3 Application of the Model

The above model allows one to study how hyaluronic acid may be functionally connected to other cancer-related activities during the early stage of tumorigenesis. For example, one can investigate if hyaluronic acid fragment patterns contribute to the clinical diversity of a specific cancer type across different patients using, say, cancer samples with different levels of differentiation (see Chap. 3), based on the knowledge that hyaluronic acid has key roles in cell differentiation of various cell types (Heldin 2003). Similarly, various mechanistic questions about hyaluronic acid and cancer can be asked and addressed (see below for examples).

– Information has been presented on the mechanism by which hyaluronic acid contributes to the initiation of cancer. In a similar vein, one can ask what roles hyaluronic acid has during the development of a cancer and have this question or related questions computationally studied in a fashion similar to the above. For example, one can determine how the expression patterns of hyaluronic acid synthesis, export and degradation genes change as a function of cancer progression, or how these patterns are statistically related to other environmental parameters such as the hypoxia level, ROS level, or to various cancer-related activities such as angiogenesis, activation of telomerase, signaling for angiogenesis and possibly other developments in cancer.
– Knowing the roles of hyaluronic acid in multiple aspects of cancer development, one can ask how the amount of hyaluronic acid generated and the resulting fragmentation patterns are linked to cancer mortality rates or the malignancy level. Such a study should be feasible using statistical correlation analyses between clinical parameters such as the average mortality rates and the average activation levels of hyaluronic acid-associated proteins. Similarly, studies can also be conducted between drug responses and hyaluronic acid-related proteins, knowing its role in the multi-drug resistance pathway (Misra et al. 2003).
– One can also investigate the possibility of using hyaluronic acid as a diagnostic or prognostic marker for different cancer types. For example, one can predict the size distribution of hyaluronic acid fragments based on the hyaluronic acid-related protein expression patterns and the expression levels of various cellular environmental parameters. Next, computational predictions can be undertaken to determine which of the abnormally expressed proteins is likely to be secreted (into the blood) and even excreted (into the urine) (see Chap. 12). Finally, such potential markers can be linked with different developmental stages of a cancer or with different cancers.

6.5   Concluding Remarks

Hyaluronic acid appears to represent a largely overlooked, albeit essential, element in cancer initiation because of its diverse roles as a facilitator in making multiple aspects of early cancer development possible. This includes: (1) serving as multiple types of signals by its fragmentation pattern, the products of which serve as signals for cell survival, cell-cycle control, cell proliferation, anti-apoptosis, angiogenesis, evasion of immune detection and induction of EMT; (2) facilitating anchorage-independent growth and overcoming contact inhibition through its interactions with the cell surfaces; (3) making intravasation and extravasation of blood vessels possible and (4) aiding migrating cells to become attached and reactivated from their dormancy states (see Chap. 10). Knowing the challenging nature of studying hyaluronic acid *in vivo* and the effectiveness in computational inference of new information as shown throughout this chapter, one can expect that computational techniques offer a unique and essential approach to better understand the functional roles of hyaluronic acid and its fragments, and particularly their interactions throughout cancer development. Various computational and statistical techniques are clearly needed to estimate the amounts, the fragment-size distributions and their dependence on various cellular environments.

References

Ågren UM, Tammi RH, Tammi MI (1997) Reactive oxygen species contribute to epidermal hyaluronan catabolism in human skin organ culture. Free Radical Biology and Medicine 23: 996–1001

Ahrens T, Assmann V, Fieber C et al. (2001) CD44 is the principal mediator of hyaluronic-acid-induced melanoma cell proliferation. J Invest Dermatol 116: 93–101

Alaniz L, Cabrera PV, Blanco G et al. (2002) Interaction of CD44 with Different Forms of Hyaluronic Acid. Its Role in Adhesion and Migration of Tumor Cells. Cell Communication and Adhesion 9: 117–130

Alberts B, Johnson A, Lewis J et al. (2002) Molecular Biology of the Cell, 4th edition. Available from: http://www.ncbi.nlm.nih.gov/books/NBK26877/, Extracellular Control of Cell Division, Cell Growth, and Apoptosis.

Attia M, Huet E, Delbe J et al. (2011) Extracellular matrix metalloproteinase inducer (EMMPRIN/CD147) as a novel regulator of myogenic cell differentiation. Journal of cellular physiology 226: 141–149

Avery DT, Deenick EK, Ma CS et al. (2010) B cell-intrinsic signaling through IL-21 receptor and STAT3 is required for establishing long-lived antibody responses in humans. The Journal of experimental medicine 207: 155–171

Basu GD, Pathangey LB, Tinder TL et al. (2004) Cyclooxygenase-2 inhibitor induces apoptosis in breast cancer cells in an in vivo model of spontaneous metastatic breast cancer. Molecular cancer research : MCR 2: 632–642

Bernanke DH, Markwald RR (1979) Effects of hyaluronic acid on cardiac cushion tissue cells in collagen matrix cultures. Texas reports on biology and medicine 39: 271–285

Bernert B, Porsch H, Heldin P (2011) Hyaluronan synthase 2 (HAS2) promotes breast cancer cell invasion by suppression of tissue metalloproteinase inhibitor 1 (TIMP-1). J Biol Chem 286: 42349–42359

Bharadwaj AG, Kovar JL, Loughman E et al. (2009) Spontaneous metastasis of prostate cancer is promoted by excess hyaluronan synthesis and processing. The American journal of pathology 174: 1027–1036

Bharadwaj AG, Rector K, Simpson MA (2007) Inducible hyaluronan production reveals differential effects on prostate tumor cell growth and tumor angiogenesis. J Biol Chem 282: 20561–20572

Black KE, Collins SL, Hagan RS et al. (2013) Hyaluronan fragments induce IFNbeta via a novel TLR4-TRIF-TBK1-IRF3-dependent pathway. Journal of inflammation 10: 23

Bourguignon LY (2008) Hyaluronan-mediated CD44 activation of RhoGTPase signaling and cytoskeleton function promotes tumor progression. Seminars in cancer biology 18: 251–259

Bourguignon LY, Gilad E, Rothman K et al. (2005) Hyaluronan-CD44 interaction with IQGAP1 promotes Cdc42 and ERK signaling, leading to actin binding, Elk-1/estrogen receptor transcriptional activation, and ovarian cancer progression. J Biol Chem 280: 11961–11972

Bourguignon LY, Ramez M, Gilad E et al. (2006) Hyaluronan-CD44 interaction stimulates keratinocyte differentiation, lamellar body formation/secretion, and permeability barrier homeostasis. J Invest Dermatol 126: 1356–1365

Bourguignon LY, Spevak CC, Wong G et al. (2009) Hyaluronan-CD44 interaction with protein kinase C(epsilon) promotes oncogenic signaling by the stem cell marker Nanog and the Production of microRNA-21, leading to down-regulation of the tumor suppressor protein PDCD4, anti-apoptosis, and chemotherapy resistance in breast tumor cells. J Biol Chem 284: 26533–26546

Bourguignon LY, Wong G, Earle CA et al. (2011) Interaction of low molecular weight hyaluronan with CD44 and toll-like receptors promotes the actin filament-associated protein 110-actin binding and MyD88-NFkappaB signaling leading to proinflammatory cytokine/chemokine production and breast tumor invasion. Cytoskeleton 68: 671–693

Brown RL, Reinke LM, Damerow MS et al. (2011) CD44 splice isoform switching in human and mouse epithelium is essential for epithelial-mesenchymal transition and breast cancer progression. The Journal of clinical investigation 121: 1064–1074

Budirahardja Y, Gonczy P (2009) Coupling the cell cycle to development. Development 136: 2861–2872

Calame K (2008) Activation-dependent induction of Blimp-1. Current opinion in immunology 20: 259–264

Clarkin CE, Allen S, Kuiper NJ et al. (2011) Regulation of UDP-glucose dehydrogenase is sufficient to modulate hyaluronan production and release, control sulfated GAG synthesis, and promote chondrogenesis. Journal of cellular physiology 226: 749–761

Collins SL, Black KE, Chan-Li Y et al. (2011) Hyaluronan fragments promote inflammation by down-regulating the anti-inflammatory A2a receptor. Am J Respir Cell Mol Biol 45: 675–683

Colotta F, Allavena P, Sica A et al. (2009) Cancer-related inflammation, the seventh hallmark of cancer: links to genetic instability. Carcinogenesis 30: 1073–1081

Coussens LM, Werb Z (2002) Inflammation and cancer. Nature 420: 860–867

Croft DR, Olson MF (2006) The Rho GTPase effector ROCK regulates cyclin A, cyclin D1, and p27Kip1 levels by distinct mechanisms. Molecular and cellular biology 26: 4612–4627

Csoka AB, Stern R (2013) Hypotheses on the evolution of hyaluronan: a highly ironic acid. Glycobiology 23: 398–411

Dai C, Whitesell L, Rogers AB et al. (2007) Heat shock factor 1 is a powerful multifaceted modifier of carcinogenesis. Cell 130: 1005–1018

Datta SR, Brunet A, Greenberg ME (1999) Cellular survival: a play in three Akts. Genes & development 13: 2905–2927

Delmage JM, Powars DR, Jaynes PK et al. (1986) The selective suppression of immunogenicity by hyaluronic acid. Ann Clin Lab Sci 16: 303–310

Diehl SA, Schmidlin H, Nagasawa M et al. (2008) STAT3-mediated up-regulation of BLIMP1 Is coordinated with BCL6 down-regulation to control human plasma cell differentiation. Journal of immunology 180: 4805–4815

Ding XZ, Tong WG, Adrian TE (2000) Blockade of cyclooxygenase-2 inhibits proliferation and induces apoptosis in human pancreatic cancer cells. Anticancer research 20: 2625–2631

Drubin DG, Nelson WJ (1996) Origins of cell polarity. Cell 84: 335–344

Duan J, Kasper DL (2011) Oxidative depolymerization of polysaccharides by reactive oxygen/nitrogen species. Glycobiology 21: 401–409

Erickson M, Stern R (2012) Chain gangs: new aspects of hyaluronan metabolism. Biochem Res Int 2012: 893947

Evanko SP, Angello JC, Wight TN (1999) Formation of hyaluronan- and versican-rich pericellular matrix is required for proliferation and migration of vascular smooth muscle cells. Arterioscler Thromb Vasc Biol 19: 1004–1013

Fantus IG, Goldberg H, Whiteside C et al. (2006) The Hexosamine Biosynthesis Pathway. In: Cortes P, Mogensen C (eds) The Diabetic Kidney. Contemporary Diabetes. Humana Press, pp 117–133

Fieber C, Baumann P, Vallon R et al. (2004) Hyaluronan-oligosaccharide-induced transcription of metalloproteases. Journal of cell science 117: 359–367

Ghatak S, Misra S, Toole BP (2002) Hyaluronan oligosaccharides inhibit anchorage-independent growth of tumor cells by suppressing the phosphoinositide 3-kinase/Akt cell survival pathway. J Biol Chem 277: 38013–38020

Guillaumond F, Leca J, Olivares O et al. (2013) Strengthened glycolysis under hypoxia supports tumor symbiosis and hexosamine biosynthesis in pancreatic adenocarcinoma. Proceedings of the National Academy of Sciences of the United States of America 110: 3919–3924

Guo H, Li R, Zucker S et al. (2000) EMMPRIN (CD147), an inducer of matrix metalloproteinase synthesis, also binds interstitial collagenase to the tumor cell surface. Cancer research 60: 888–891

Gururaj A, Barnes CJ, Vadlamudi RK et al. (2004) Regulation of phosphoglucomutase 1 phosphorylation and activity by a signaling kinase. Oncogene 23: 8118–8127

Hall CL, Turley EA (1995) Hyaluronan: RHAMM mediated cell locomotion and signaling in tumorigenesis. Journal of neuro-oncology 26: 221–229

Hall CL, Wang C, Lange LA et al. (1994) Hyaluronan and the hyaluronan receptor RHAMM promote focal adhesion turnover and transient tyrosine kinase activity. The Journal of cell biology 126: 575–588

Hamann KJ, Dowling TL, Neeley SP et al. (1995) Hyaluronic acid enhances cell proliferation during eosinopoiesis through the CD44 surface antigen. Journal of immunology 154: 4073–4080

Hamilton SR, Fard SF, Paiwand FF et al. (2007) The hyaluronan receptors CD44 and Rhamm (CD168) form complexes with ERK1,2 that sustain high basal motility in breast cancer cells. J Biol Chem 282: 16667–16680

Hanahan D, Weinberg RA (2000) The hallmarks of cancer. Cell 100: 57–70

Hanahan D, Weinberg RA (2011) Hallmarks of cancer: the next generation. Cell 144: 646–674

Haylock DN, Nilsson SK (2006) The role of hyaluronic acid in hemopoietic stem cell biology. Regen Med 1: 437–445

Heise RL, Stober V, Cheluvaraju C et al. (2011) Mechanical stretch induces epithelial-mesenchymal transition in alveolar epithelia via hyaluronan activation of innate immunity. J Biol Chem 286: 17435–17444

Heldin P (2003) Importance of hyaluronan biosynthesis and degradation in cell differentiation and tumor formation. Brazilian journal of medical and biological research = Revista brasileira de pesquisas medicas e biologicas / Sociedade Brasileira de Biofisica [et al] 36: 967–973

Holmes MW, Bayliss MT, Muir H (1988) Hyaluronic acid in human articular cartilage. Age-related changes in content and size. The Biochemical journal 250: 435–441

Horton MR, Olman MA, Bao C et al. (2000) Regulation of plasminogen activator inhibitor-1 and urokinase by hyaluronan fragments in mouse macrophages. American journal of physiology Lung cellular and molecular physiology 279: L707–715

Isacke CM (1994) The role of the cytoplasmic domain in regulating CD44 function. Journal of cell science 107 (Pt 9): 2353–2359

Ishimoto T, Sugihara H, Watanabe M et al. (2013) Macrophage-derived reactive oxygen species suppress miR-328 targeting CD44 in cancer cells and promote redox adaptation. Carcinogenesis:

Itano N, Atsumi F, Sawai T et al. (2002) Abnormal accumulation of hyaluronan matrix diminishes contact inhibition of cell growth and promotes cell migration. Proceedings of the National Academy of Sciences of the United States of America 99: 3609–3614

Jacobs PP, Sackstein R (2011) CD44 and HCELL: preventing hematogenous metastasis at step 1. FEBS letters 585: 3148–3158

Jiang D, Liang J, Noble PW (2007) Hyaluronan in tissue injury and repair. Annual review of cell and developmental biology 23: 435–461

Jiang D, Liang J, Noble PW (2011) Hyaluronan as an immune regulator in human diseases. Physiological reviews 91: 221–264

Kern MA, Haugg AM, Koch AF et al. (2006) Cyclooxygenase-2 inhibition induces apoptosis signaling via death receptors and mitochondria in hepatocellular carcinoma. Cancer research 66: 7059–7066

Kobayashi H, Terao T (1997) Hyaluronic acid-specific regulation of cytokines by human uterine fibroblasts. The American journal of physiology 273: C1151–1159

Kosaki R, Watanabe K, Yamaguchi Y (1999) Overproduction of hyaluronan by expression of the hyaluronan synthase Has2 enhances anchorage-independent growth and tumorigenicity. Cancer research 59: 1141–1145

Lin L, Liu A, Peng Z et al. (2011) STAT3 is necessary for proliferation and survival in colon cancer-initiating cells. Cancer research 71: 7226–7237

Liu N, Gao F, Han Z et al. (2001a) Hyaluronan synthase 3 overexpression promotes the growth of TSU prostate cancer cells. Cancer research 61: 5207–5214

Liu N, Gao F, Han Z et al. (2001b) Hyaluronan synthase 3 overexpression promotes the growth of TSU prostate cancer cells. Cancer research 61: 5207–5214

Lokeshwar V, Rubinowicz D (1999) Hyaluronic acid and hyaluronidase: molecular markers associated with prostate cancer biology and detection. Prostate cancer and prostatic diseases 2: S21

Lu H, Ouyang W, Huang C (2006) Inflammation, a key event in cancer development. Molecular cancer research : MCR 4: 221–233

Maeso G, Bravo M, Bascones A (2007) Levels of metalloproteinase-2 and -9 and tissue inhibitor of matrix metalloproteinase-1 in gingival crevicular fluid of patients with periodontitis, gingivitis, and healthy gingiva. Quintessence Int 38: 247–252

Malemud CJ (2013) Suppression of Pro-Inflammatory Cytokines via Targeting of STAT-Responsive Genes. Drug Discovery. InTech,

Margolis RU, Margolis RK, Chang LB et al. (1975) Glycosaminoglycans of brain during development. Biochemistry 14: 85–88

Marieb EA, Zoltan-Jones A, Li R et al. (2004) Emmprin promotes anchorage-independent growth in human mammary carcinoma cells by stimulating hyaluronan production. Cancer research 64: 1229–1232

May O (2009) COX-2/PGE2 Signaling: A Target for Colorectal Cancer Prevention.

McBride WH, Bard JB (1979) Hyaluronidase-sensitive halos around adherent cells. Their role in blocking lymphocyte-mediated cytolysis. The Journal of experimental medicine 149: 507–515

Mendillo ML, Santagata S, Koeva M et al. (2012) HSF1 drives a transcriptional program distinct from heat shock to support highly malignant human cancers. Cell 150: 549–562

Meran S, Luo DD, Simpson R et al. (2011) Hyaluronan facilitates transforming growth factor-beta1-dependent proliferation via CD44 and epidermal growth factor receptor interaction. J Biol Chem 286: 17618–17630

Misra S, Ghatak S, Zoltan-Jones A et al. (2003) Regulation of multidrug resistance in cancer cells by hyaluronan. J Biol Chem 278: 25285–25288

Misra S, Heldin P, Hascall VC et al. (2011) Hyaluronan-CD44 interactions as potential targets for cancer therapy. The FEBS journal 278: 1429–1443

Monzón ME, Manzanares D, Schmid N et al. (2008) Hyaluronidase Expression and Activity Is Regulated by Pro-Inflammatory Cytokines in Human Airway Epithelial Cells. Am J Respir Cell Mol Biol 39: 289–295

Naor D, Wallach-Dayan SB, Zahalka MA et al. (2008) Involvement of CD44, a molecule with a thousand faces, in cancer dissemination. Seminars in cancer biology 18: 260–267

Nikitovic D, Kouvidi K, Karamanos NK et al. (2013) The roles of hyaluronan/RHAMM/CD44 and their respective interactions along the insidious pathways of fibrosarcoma progression. BioMed research international 2013: 929531

Noatynska A, Tavernier N, Gotta M et al. (2013) Coordinating cell polarity and cell cycle progression: what can we learn from flies and worms? Open biology 3: 130083

Noble PW (2002) Hyaluronan and its catabolic products in tissue injury and repair. Matrix biology : journal of the International Society for Matrix Biology 21: 25–29

Nzeako UC, Guicciardi ME, Yoon JH et al. (2002) COX-2 inhibits Fas-mediated apoptosis in cholangiocarcinoma cells. Hepatology 35: 552–559

Okabe T, Okamoto I, Tamura K et al. (2007) Differential constitutive activation of the epidermal growth factor receptor in non-small cell lung cancer cells bearing EGFR gene mutation and amplification. Cancer research 67: 2046–2053

Okamoto I, Kawano Y, Tsuiki H et al. (1999) CD44 cleavage induced by a membrane-associated metalloprotease plays a critical role in tumor cell migration. Oncogene 18: 1435–1446

Ouhtit A, Abd Elmageed ZY, Abdraboh ME et al. (2007) In vivo evidence for the role of CD44s in promoting breast cancer metastasis to the liver. The American journal of pathology 171: 2033–2039

Pandey MS, Baggenstoss BA, Washburn J et al. (2013) The hyaluronan receptor for endocytosis (HARE) activates NF-kappaB-mediated gene expression in response to 40-400-kDa, but not smaller or larger, hyaluronans. J Biol Chem 288: 14068–14079

Park D, Kim Y, Kim H et al. (2012) Hyaluronic acid promotes angiogenesis by inducing RHAMM-TGF beta receptor interaction via CD44-PKC delta. Mol Cells 33: 563–574

Patti ME, Virkamaki A, Landaker EJ et al. (1999) Activation of the hexosamine pathway by glucosamine in vivo induces insulin resistance of early postreceptor insulin signaling events in skeletal muscle. Diabetes 48: 1562–1571

Pelletier J, Bellot G, Gounon P et al. (2012) Glycogen Synthesis is Induced in Hypoxia by the Hypoxia-Inducible Factor and Promotes Cancer Cell Survival. Frontiers in oncology 2: 18

Pescador N, Villar D, Cifuentes D et al. (2010) Hypoxia promotes glycogen accumulation through hypoxia inducible factor (HIF)-mediated induction of glycogen synthase 1. PLoS One 5: e9644

Porsch H (2013) Importance of Hyaluronan-CD44 Signaling in Tumor Progression : Crosstalk with TGFβ and PDGF-BB Signaling., Uppsala University., Digital Comprehensive Summaries of Uppsala Dissertations from the Faculty of Medicine.

Pugh CW, Ratcliffe PJ (2003) Regulation of angiogenesis by hypoxia: role of the HIF system. Nature medicine 9: 677–684

Rakoff-Nahoum S (2006) Why cancer and inflammation? The Yale journal of biology and medicine 79: 123–130

Roca X, Mate JL, Ariza A et al. (1998) CD44 isoform expression follows two alternative splicing pathways in breast tissue. The American journal of pathology 153: 183–190

Salustri A, Camaioni A, Di Giacomo M et al. (1999) Hyaluronan and proteoglycans in ovarian follicles. Human reproduction update 5: 293–301

Savani RC, Cao G, Pooler PM et al. (2001) Differential involvement of the hyaluronan (HA) receptors CD44 and receptor for HA-mediated motility in endothelial cell function and angiogenesis. J Biol Chem 276: 36770–36778

Scheibner KA, Lutz MA, Boodoo S et al. (2006) Hyaluronan fragments act as an endogenous danger signal by engaging TLR2. Journal of immunology 177: 1272–1281

Schraufstatter IU, Serobyan N, Loring J et al. (2010) Hyaluronan is required for generation of hematopoietic cells during differentiation of human embryonic stem cells. J Stem Cells 5: 9–21

Schulz T, Schumacher U, Prante C et al. (2010) Cystic Fibrosis Transmembrane Conductance Regulator Can Export Hyaluronan. Pathobiology 77: 200–209

Schulz T, Schumacher U, Prehm P (2007) Hyaluronan export by the ABC transporter MRP5 and its modulation by intracellular cGMP. J Biol Chem 282: 20999–21004

Shirali AC, Goldstein DR (2008) Activation of the innate immune system by the endogenous ligand hyaluronan. Curr Opin Organ Transplant 13: 20–25

Slevin M, Krupinski J, Kumar S et al. (1998) Angiogenic oligosaccharides of hyaluronan induce protein tyrosine kinase activity in endothelial cells and activate a cytoplasmic signal transduction pathway resulting in proliferation. Laboratory investigation; a journal of technical methods and pathology 78: 987–1003

Slevin M, Kumar S, Gaffney J (2002) Angiogenic oligosaccharides of hyaluronan induce multiple signaling pathways affecting vascular endothelial cell mitogenic and wound healing responses. J Biol Chem 277: 41046–41059

Slomiany MG, Grass GD, Robertson AD et al. (2009) Hyaluronan, CD44, and emmprin regulate lactate efflux and membrane localization of monocarboxylate transporters in human breast carcinoma cells. Cancer research 69: 1293–1301

Stern R, Asari AA, Sugahara KN (2006) Hyaluronan fragments: an information-rich system. European journal of cell biology 85: 699–715

Stern R, Shuster S, Neudecker BA et al. (2002) Lactate stimulates fibroblast expression of hyaluronan and CD44: the Warburg effect revisited. Experimental cell research 276: 24–31

Takahashi Y, Li L, Kamiryo M et al. (2005) Hyaluronan fragments induce endothelial cell differentiation in a CD44- and CXCL1/GRO1-dependent manner. J Biol Chem 280: 24195–24204

Tammi RH, Passi AG, Rilla K et al. (2011a) Transcriptional and post-translational regulation of hyaluronan synthesis. The FEBS journal 278: 1419–1428

Tammi RH, Passi AG, Rilla K et al. (2011b) Transcriptional and post-translational regulation of hyaluronan synthesis. FEBS J 278: 1419–1428

Tang B, Vu M, Booker T et al. (2003) TGF-beta switches from tumor suppressor to prometastatic factor in a model of breast cancer progression. The Journal of clinical investigation 112: 1116–1124

Taylor KR, Trowbridge JM, Rudisill JA et al. (2004) Hyaluronan fragments stimulate endothelial recognition of injury through TLR4. J Biol Chem 279: 17079–17084

Teng BP, Heffler MD, Lai EC et al. (2011) Inhibition of hyaluronan synthase-3 decreases subcutaneous colon cancer growth by increasing apoptosis. Anti-cancer agents in medicinal chemistry 11: 620–628

Thorne RF, Legg JW, Isacke CM (2004) The role of the CD44 transmembrane and cytoplasmic domains in co-ordinating adhesive and signalling events. Journal of cell science 117: 373–380

Tian X, Azpurua J, Hine C et al. (2013) High-molecular-mass hyaluronan mediates the cancer resistance of the naked mole rat. Nature:

Tolg C, Hamilton SR, Nakrieko KA et al. (2006) Rhamm-/- fibroblasts are defective in CD44-mediated ERK1,2 motogenic signaling, leading to defective skin wound repair. The Journal of cell biology 175: 1017–1028

Toole BP (2004) Hyaluronan: from extracellular glue to pericellular cue. Nature reviews Cancer 4: 528–539

Toole BP (2009) Hyaluronan-CD44 Interactions in Cancer: Paradoxes and Possibilities. Clinical cancer research : an official journal of the American Association for Cancer Research 15: 7462–7468

Toole BP, Slomiany MG (2008) Hyaluronan, CD44 and Emmprin: partners in cancer cell chemoresistance. Drug resistance updates : reviews and commentaries in antimicrobial and anticancer chemotherapy 11: 110–121

Vincent T, Jourdan M, Sy MS et al. (2001a) Hyaluronic acid induces survival and proliferation of human myeloma cells through an interleukin-6-mediated pathway involving the phosphorylation of retinoblastoma protein. The Journal of biological chemistry 276: 14728–14736

Vincent T, Jourdan M, Sy MS et al. (2001b) Hyaluronic acid induces survival and proliferation of human myeloma cells through an interleukin-6-mediated pathway involving the phosphorylation of retinoblastoma protein. Journal of Biological Chemistry 276: 14728–14736

Wang SJ, Bourguignon LY (2011) Role of hyaluronan-mediated CD44 signaling in head and neck squamous cell carcinoma progression and chemoresistance. The American journal of pathology 178: 956–963

West DC, Hampson IN, Arnold F et al. (1985) Angiogenesis induced by degradation products of hyaluronic acid. Science 228: 1324–1326

Wilkinson TS, Potter-Perigo S, Tsoi C et al. (2004) Pro- and anti-inflammatory factors cooperate to control hyaluronan synthesis in lung fibroblasts. Am J Respir Cell Mol Biol 31: 92–99

Xu H, Ito T, Tawada A et al. (2002) Effect of hyaluronan oligosaccharides on the expression of heat shock protein 72. J Biol Chem 277: 17308–17314

Yevdokimova NY (2006) Elevated level of ambient glucose stimulates the synthesis of high-molecular-weight hyaluronic acid by human mesangial cells. The involvement of transforming growth factor beta1 and its activation by thrombospondin-1. Acta biochimica Polonica 53: 383–393

Yoshida S, Pellman D (2008) Plugging the GAP between cell polarity and cell cycle. EMBO reports 9: 39–41

Yoshihara S, Kon A, Kudo D et al. (2005) A hyaluronan synthase suppressor, 4-methylumbelliferone, inhibits liver metastasis of melanoma cells. FEBS letters 579: 2722–2726

Zeilstra J, Joosten SP, Dokter M et al. (2008) Deletion of the WNT target and cancer stem cell marker CD44 in Apc(Min/+) mice attenuates intestinal tumorigenesis. Cancer research 68: 3655–3661

Zhang S, Chang MC, Zylka D et al. (1998) The hyaluronan receptor RHAMM regulates extracellular-regulated kinase. J Biol Chem 273: 11342–11348

Zoltan-Jones A, Huang L, Ghatak S et al. (2003) Elevated hyaluronan production induces mesenchymal and transformed properties in epithelial cells. The Journal of biological chemistry 278: 45801–45810

Chapter 7
Multiple Routes for Survival: Understanding How Cancer Evades Apoptosis

Apoptosis is a process of programmed cell-death encoded in all multicellular organisms. The system is designed to remove damaged, unhealthy or unneeded cells during development and under certain stresses. At the tissue level, it plays a key role in maintaining tissue homeostasis. For example, the typical human body produces approximately 50–70 billion new cells by mitosis each day (Karam 2009), and the same number of cells will be terminated by apoptosis to maintain total cell homeostasis, suggesting that there is a functional link between growth and cell death by apoptosis. Malfunctions of the apoptotic system, in either its regulators or effectors, have been linked to a variety of human diseases. Examples include: (1) human degenerative diseases, such as multiple sclerosis, which are known to be associated with abnormally high activities of apoptosis, and (2) cancer that is considered by some as a disease of abnormally low activities of apoptosis.

The effector component of the apoptosis system is relatively simple, consisting of a set of death substrates, whose release will kill the cell. In contrast, the regulatory component of apoptosis is rather extensive and complex. The activity level of the effector component is adjusted through changes in the concentration balance between pro- and anti-apoptotic proteins, as well as by enhancing or repressing the activities of specific proteins in response to external signals released to the extracellular space or to intracellular signals reflecting certain stresses. As discussed in the earlier chapters, cancer cells tend to accumulate a large number of genetic mutations, which should normally induce apoptosis and cell death. However, for reasons that are only partially understood, cancer cells have "learned" to be anti-apoptotic and remain viable through over-expression of their survival pathways, inhibiting the activities of their pro-apoptotic proteins or selecting genomic mutations that lose the connection to or the activity of the apoptotic effectors.

© Springer Science+Business Media New York 2014
Y. Xu et al., *Cancer Bioinformatics*, DOI 10.1007/978-1-4939-1381-7_7

The focus of this chapter is on that which can be learned through integrative analyses of *omic* data about how cancers may have become anti-apoptotic by creating a microenvironment through which they can advantageously use the encoded cellular mechanisms in normal cells for their survival.

7.1 The Basic Biology of Apoptosis

Apoptosis in Greek means "dropping off" of petals or leaves from plants. In cell biology, it refers to a programmed cell-death process, which involves a sequence of morphological changes such as cell shrinkage, membrane blebbing, chromatin condensation and DNA fragmentation, leading to cell death. This characteristic is considered to be the distinguishing property of apoptosis from that of the other programmed cell deaths such as necrosis, senescence, autophagy, paraptosis and mitotic catastrophe.

The discovery of the programmed natural cell-death process was first made by German scientist Carl Vogt in 1842, who introduced the designation *apoptosis* (Peter et al. 1997). Kerr, Wyllie and Currie raised the possibility in 1972 that apoptosis may participate in a major way during cancer development when they noted that the observed tumor-proliferation rate and the tumor size do not match, and hence inferred that more than 95 % of the tumor cells may have died due to apoptosis (Kerr et al. 1972). Later, Sydney Brenner, Robert Horvitz and John Sulston identified and characterized the genes that control the apoptotic process, for which they won the Nobel Prize in medicine in 2002. Over one hundred thousand scientific papers have been published on the topic of apoptosis since the early work by Kerr, Wyllie and Currie, indicating that apoptosis has been one of the most active research fields in modern biology.

7.1.1 The Apoptosis Execution System

The execution component of the apoptosis system consists of the following components: the activated *caspase-3* (a cysteine-aspartic protease) or *caspase-7* proteins will cleave the inhibitor of the *caspase*-activated deoxyribonuclease, which releases a number of death substrates such as LaminA, LaminB1, LaminB2, ICAD and D4-DGI, leading to the destruction of the cell through a sequence of well-defined steps: cell shrinkage, chromatin condensation, membrane blebbing, DNA fragmentation, nuclear collapse, apoptotic body formation and lysis of apoptotic bodies. Currently, a few hundreds of caspase substrates have been identified (Luthi and Martin 2007). Two well-studied signaling pathways, the intrinsic and extrinsic pathways, can activate *caspase-3* or *-7*, and hence activate the execution of apoptosis. The former activates *caspase-3* or *-7* through the release of cytochrome c molecules from mitochondria, which then activates *caspase-9*; and the latter activates it by activating *caspase-8* or *-10*, as shown in Fig. 7.1.

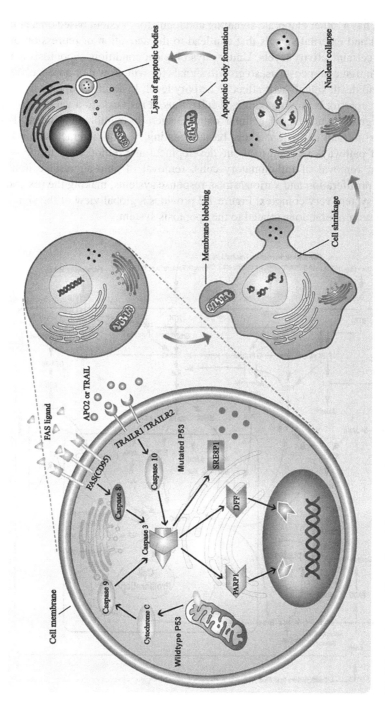

Fig. 7.1 A schematic illustration of the *caspase* signaling cascade and the execution of apoptosis

7.1.2   The Signaling and Regulatory System of Apoptosis

Apoptosis has a rather elaborate signaling and regulatory system based on a number of internal and external signals that can lead to the activation or repression of the system at certain activity levels. Under physiological conditions, apoptosis is used to maintain tissue homeostasis, so growth signals (or withdrawal or absence of such signals) and death signals will affect its activity level. In addition, apoptosis serves as a gatekeeper for removing damaged cells; thus, it responds to a variety of intracellular signals indicative of cellular damage such as membrane leakage, DNA damage or nutrient depletion. These basic signaling systems interact with a large number of pathways relevant to tissue development and remodeling, tissue injury and repair, removal of inflammatory cells, removal of auto-aggressive immune cells, cell proliferation and various stress-response systems, making the (extended) apoptosis system very complex. Figure 7.2 provides a global view of the signaling and regulatory interactions related to the apoptosis system.

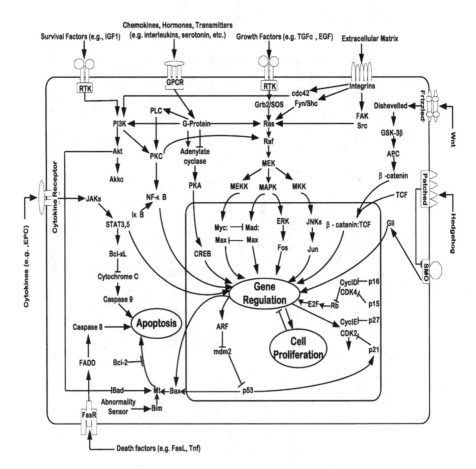

Fig. 7.2 A schematic of the signaling pathways of apoptosis, adapted from (Signal-transduction-pathways 2010)

The basic signaling pathways of apoptosis fall into three categories: (1) the *intrinsic signaling pathway* that is activated by intracellular stress signals followed by activation of the *caspase-3* or *-7* proteins, leading to the release of *cytochrome c* and *SMAC* (second mitochondrial activator of *caspases*), which bind with the *APAF1* (adaptor protein apoptotic protease-activating factor 1) protein, leading to the formation of apoptosomes and the activation of *caspase-9* proteins; (2) the *extrinsic signaling pathway* that is activated in response to external death signals such as *FAS*, and then activates *caspase-3* or *-7* through the activation of the *caspase-8* or *-10* proteins, which is induced through the formation of the death-induced signal complex (*DISC*); and (3) a number of non-canonical signaling pathways that do not fall into either of the first two groups and will be explained in detail later in this section. In addition, there is a family of apoptosis regulators, the *BCL2* (B-cell CLL/lymphoma 2) family, that can directly block or activate different proteins along the signaling pathways (see Fig. 7.2); these are often referred to as the main regulators of apoptosis.

Despite its name, *BCL2* genes are expressed in a variety of cell types, including epithelial cells that have been the focus of much of this book. As of now, 25 members of the *BCL2* family have been identified. Some of these members are pro-apoptotic, such as *BAX, BAD, BAK, BID, BIM, BOK, NOXA* and *PUMA*, while the other members are anti-apoptotic such as *BCL2, BCLB, BCLW, BCLXL* and *MCL1* (see Fig. 7.2 for the functional roles of these proteins). The overall apoptotic activity level is largely determined by the balance between the pro- and anti-apoptotic *BCL2* family members.

1. The intrinsic signaling pathway: The intrinsic pathway is activated when the cell is under severe intracellular stress, including extensive DNA damage, membrane damage, nutrient deprivation, hypoxia and viral infection, as well as withdrawal of growth factors, hormones and cytokines. These signals induce changes in the inner membrane of mitochondria, leading to an increase in mitochondrial permeability and the release of two groups of pro-apoptotic proteins from the intermembrane space of mitochondria to cytosol (Siskind 2005; Suen et al. 2008). One group of proteins includes *cytochrome c* and *SMAC*, which can lead to the formation of apoptosomes and activation of the caspase cascade, hence the activation of cell suicide. Since the initial signals to this pathway originate in mitochondria, the pathway is sometimes referred to as the *mitochondrial signaling pathway*. This may be one of the reasons why cancer cells tend to have dysfunctional mitochondria, i.e., so selected to prevent the production and release of the pro-apoptotic signals, as some have speculated (Gogvadze et al. 2008). Another group of released proteins includes *AIF* (apoptosis induced factor), endonuclease *G* and *CAD*, which are all translocated to the nucleus upon receiving signals for the activation of apoptosis to carry out the execution activities of cell death, including DNA fragmentation and chromatin condensation.

2. The extrinsic signaling pathway: The extrinsic signaling pathway is activated upon receiving death signals by their corresponding receptors on the cell surface, each of which has a cytoplasmic "death domain" that transmits the death signal to the intracellular signaling pathway. A number of such signals and the corresponding receptors have been identified, including *FAS* and *FASR, TNFα* and

TNFR1, APO3L and *DR3, APO2L* and *DR4,* and *APO2L* and *DR5,* all being in the superfamily of *TNFs,* also known as *cachectin,* and their corresponding receptors in the *TNFR* family. While many cell types can release death signals, macrophages represent the dominant cell type for the production and release of these signals, serving their supporting roles in determining the fate of their parenchymal cells based on their microenvironment. Upon binding a death signal such as *FAS,* the receptor protein *FASR* binds with the *TRADD* (tumor necrosis factor receptor type 1-associated death domain) protein, which in turn recruits the proteins, *FADD* (Fas-associated protein with death domain) and *RIP* (receptor-interacting serine/threonine protein). *FADD* then associates with *pro-caspase-8,* activating the caspase cascade and the execution of apoptosis. See Fig. 7.2 for the detailed relationships among these proteins, death signals and the execution of apoptosis.

3. The non-canonical signaling pathways: Studies in the past decade have found that the current classification between apoptosis and necrosis, both being programmed cell death, is probably an over-simplification. The two are probably part of a larger cell-death program, although each has its distinct morphological pattern during the respective cell-death process. Multiple investigations have reported different intermediates between the two, such as the apoptosis-like and necrosis-like death processes (Leist and Jaattela 2001; Jaattela 2004; Broker et al. 2005a; Qi and Liu 2006), which resemble only some aspects of the two canonical programs. A particularly interesting class of non-canonical apoptosis pathways is one referred to as *caspase-independent apoptosis* since it does not go through the canonical caspase cascade (Jaattela and Tschopp 2003), as its name suggested. The emergence of these newly identified signaling pathways has resulted in a number of studies with the aim to reclassify apoptosis, necrosis and the like. One study grouped all such pathways into four classes: (1) apoptosis, (2) apoptosis-like, (3) necrosis-like and (4) necrosis (Leist and Jaattela 2001), based on the cellular morphological differences during the death processes. Another study classified the signaling pathways based on their level of dependence on caspase proteins (Kolenko et al. 2000; Mathiasen and Jaattela 2002; Broker et al. 2005b).

A new model was recently proposed (see Fig. 7.3) suggesting that apoptosis is a part of a larger cell-death program that also covers necrosis and other newly discovered caspase-independent cell-death programs. This model expands the current apoptosis system that utilizes intracellular signals only from mitochondria to also include signals from lysosomes and ER. This recent model speculates that the different components of the cell death program may have distinct roles upon receiving stress-related signals from different organelles. The observed death phenotype is probably determined by the relative speeds in the execution of different death pathways under various conditions, and consequently only that by the fastest and most effective one is observed.

It should be possible to validate this model computationally by using the available *omic* data. For example, one can check for each observed activation based on gene-expression data whether the model-predicted associations indeed exhibit

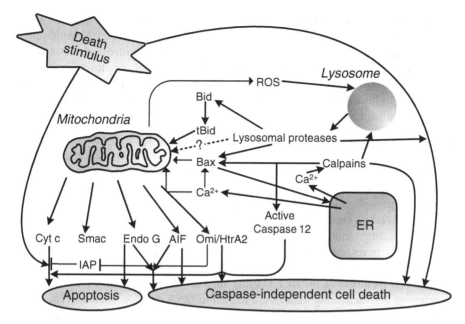

Fig. 7.3 A model for programmed cell death that integrates apoptosis, necrosis and newly discovered caspase-independent cell-death programs, adapted from (Desai 2013)

co-expression patterns for some cancer types (and at certain developmental stages). Such analyses can quickly identify incorrect associations predicted by the model or provide evidence for some model-predicted associations, hence validating or rejecting the model. For cases when time-course data are available, it should be possible to derive causal relationships among the identified associations.

7.1.3   The P53 Network

P53 is one of the best known tumor-suppressor genes. The fact that ~50 % of cancers have *P53* mutations across all cancer types reveals its dominating role in protecting cells from becoming cancerous. The current understanding is that *P53* is at the junction of a number of fundamental processes in cellular life, namely apoptosis, senescence, proliferation and immunity. According to (Levine and Oren 2009; Brady and Attardi 2010), the *P53* protein directly interacts with over one hundred different proteins under different cellular conditions, covering functions ranging from: *AKT/PKB* pathway regulator, adipogenesis, apoptosis, asymmetric cell division, cAMP pathway, cell cycle control, cell proliferation, chromatin proteins, chromosome condensation, development, differentiation, DNA damage and repair, DNA methylation, energy metabolism, extracellular matrix proteins, heat shock, hypoxia response, *MAPK*, *NFκB*, nucleocytoplasmic transport, nuclear receptors, proteasome degradation, protection against viral infection, ribosomal proteins to transforming activity and ubiquitination. From this list, one can imagine the very complex nature in the relationship between *P53* and apoptosis. The following summarizes a few major biological processes in which *P53* is involved, showing a framework for one to study both induction and inhibition of apoptosis via *P53*. It should be noted that this is clearly not the entire picture of the *P53* interaction network as the following four processes do not begin to cover every interaction partner of *P53* as outlined above.

1. Apoptosis: As an apoptosis regulator, *P53* responds to a large variety of cellular stresses as discussed earlier. Typically *P53* is inactive after being expressed because of binding to its negative regulator, *MDM2*, thus preventing it from functioning as a transcription factor. The activation of *P53* is accomplished through phosphorylation of its N-terminal domain by various protein kinases, which leads to a conformational change and the release of *MDM2*. One group of *P53*-activating kinases consists of members of the *MAPK* family such as *JNK1* and *ERK1* in response to cell-cycle abnormalities among other stresses. Under these conditions, the activated *P53* induces apoptosis by up-regulating pro-apoptotic members of the *BCL2* family such as *PUMA, BAX* and *BAK*. Another group of *P53*-activating kinases are related to DNA damage, such as *ATR, ATM* and *DNAPK*. The activated *P53* by these kinases will up-regulate a series of regulators relevant to *CDK* inhibition, growth arrest and the DNA damage-inducible gene *GADD45α* to repair the identified DNA damage. This process may lead to the execution of apoptosis if DNA repair fails.

2. Senescence: While an inactivated *P53* can increase the potential of cancer development, constitutive activation of *P53* will lead to tissue degeneration and premature aging as recently reported (Rodier et al. 2007; Feng et al. 2011). The speculation was that constitutive activation of *P53* can result in the annihilation of stem cells, which impairs the tissue-regeneration capability, in addition to apoptosis. Previous studies have shown that constitutive activation of *P53* can cause lifespan reduction (Dumble et al. 2007), providing supporting evidence to the above hypothesis. While research on the impact of the *P53* activity levels on tissue degeneration, as well as lifespan reduction, has emerged only in recent years and an understanding at the molecular level is still lacking, it is safe to state that the normal range of the *P53* activity level results from balancing tumor suppression and lifespan as determined by evolution.

3. Proliferation: *P53* can be activated by a number of oncogenes such as *E1A* (Debbas and White 1993; Lowe and Ruley 1993; Querido et al. 1997; Samuelson and Lowe 1997; Lowe 1999), *MYC* (Hermeking and Eick 1994) and *RAS* (Serrano et al. 1997), which requires the participation of the *P19* protein. This is probably part of a mechanism encoded in cells for prevention of over-growth. The key point to note here is that proliferation and death are tightly linked in the cellular mechanisms encoded in human cells.

4. Immunity: The connection between *P53* and the immune system is fairly extensive and is rooted at a fundamental level. It deserves mention that *NFκB* is a key regulator of both the innate and the adaptive immune systems. Published studies have established that *P53* and *NFκB* are opposing regulators in terms of apoptosis

versus survival, as the former promotes apoptosis while the latter enhances survival. Actually, the two proteins can directly inhibit each other (Schneider et al. 2010). As of now, ~30 immune-related genes have been found to be direct targets of *P53* regulation (Lowe et al. 2013).

The more detailed relationship between *P53* and the immune responses can be summarized as follows. It has been widely observed that *P53* is up-regulated in inflammation sites. While the detailed triggering mechanisms are not yet fully elucidated, the speculation has been that *P53* is induced by the increased ROS or reactive nitrogen species, both of which are produced during inflammation (Vousden and Prives 2009; Hafsi and Hainaut 2011). Conversely, *P53* has been found to inhibit autoimmune inflammation by suppressing the expression of inflammatory cytokine-encoding genes, possibly serving as a general negative regulator of inflammation (Santhanam et al. 1991; Pesch et al. 1996; Okuda et al. 2003; Takaoka et al. 2003; Zheng et al. 2005; Liu et al. 2009).

It has also been shown that an activated *P53* can serve as a co-stimulatory protein for the activation of T-cells (Gorgoulis et al. 2003; Lowe et al. 2013) and contribute to the initiation of adaptive immune responses. The link between *P53* and the innate immune system lies in the need for clearance of damaged or infected cells, which is typically done through cooperative actions between *P53* and the innate immune system (Martins et al. 2006; Ventura et al. 2007). There are clearly other links between *P53* and the innate immune system as it has been found that infections by several viruses, such as Epstein-Barr, adenovirus, influenza A and HIV-1, can activate *P53*. One potential activation mechanism is through *TLR* (toll-like receptor), which is a key regulator of the innate immune responses and provides a front-line protection against pathogens through recognition of their common features, referred to as the *pathogen-associated molecular patterns*. *TLR* can activate protein kinase *R*, which is capable of phosphorylating and activating *P53*. Interestingly, the activation of *P53* can lead to the expression of most *TLR* genes and the general response of the innate immune system (Menendez et al. 2010, 2011, 2013), suggesting a functional loop between *P53* and *TLR*.

Knowing the close and complex relationships between *P53* and the above pathways, it is logical to speculate that loss of *P53* function may overall be beneficial to cancer development in multiple ways as has been suggested for some time, in addition to just the loss of its inhibitory role of apoptosis. Specifically, cells without normal *P53* function will be more pro-inflammatory, more oxidative, less immune responsive, exhibit a lower level of counter-reaction to over-proliferation and be less prone to be eliminated due to fitness reasons (see Chap. 8), hence making the overall environment more cancer friendly. One may further speculate that it is these multiple beneficial factors to cancer development that have led to the high mutation rate in the *P53* gene in cancer tissues through natural selection. We believe that this hypothesis can be rigorously studied computationally by examining cancer tissue samples in terms of the activity levels of the relevant pathways *versus* the mutation rates of *P53*. To undertake such a project, one would need to statistically determine if there is a correlation between the *P53* mutation rates and the level of activity in each of the aforementioned processes such as inflammation response, production of antioxidant molecules or proliferation rate. With such data in hand, it may be possible to estimate the percentage of the *P53* mutations that are selected to directly benefit which particular processes.

7.2   Different Ways to Evade Apoptosis by Cancer

Since the seminal publication that links apoptosis to development and cancer by Kerr et al. in 1972 (Kerr et al. 1972), a large body of literature has been published on cancer apoptosis. Upon reviewing some of these publications, one quickly recognizes the considerable complexity of the topic. This involves close interactions and balances among multiple processes spanning a number of essential aspects of multicellular organisms as discussed in the previous section. In an effort to enable the reader to quickly grasp the essence of how different survival signaling pathways balance the apoptotic activities induced by cancer-associated environments and activities, the current knowledge is organized in the following fashion, focused on the fundamental (non-accidental) connections that link growth, apoptosis and survival, as well as genetic mutations that enhance such connections by making them constitutively active or repressed.

7.2.1   Growth and Apoptosis

There is a clear link between proliferation and apoptosis encoded in human genomes as briefly outlined in Sect. 7.1. Here the *MYC* gene is used as an example to explain the connection. *MYC* is an extensively studied oncogenic transcription regulator, whose overexpression can drive cellular proliferation in cancer tissue and has been widely observed in numerous cancer types. It has been well established that overexpression of *MYC* can directly induce apoptosis in the absence of survival factors in normal cells (Askew et al. 1991; Evan et al. 1992; Shi et al. 1992). Later studies have observed coordinated expression patterns between *MYC* and *BCL2* genes (Strasser et al. 1990; Bissonnette et al. 1992; Fanidi et al. 1992). Specifically the expression of *MYC* can inhibit the expression of *BCL2* (an anti-apoptotic member of the *BCL2* family) and induce the expression of *BAX* (a pro-apoptotic member of the *BCL2* family), hence establishing a functional link between growth and apoptosis. In addition, it has been shown that when *BCL2* is inhibited, the activation of *MYC* can induce cell death via apoptosis. The current understanding about these observations is that cellular systems have developed a mechanism to keep growth in check, i.e., when growth takes place, apoptosis is also activated at some level as a safety valve. It has been speculated that a large number of other oncogenes may have similar relationships with the apoptosis regulators as those between *MYC* and *BCL2*. We expect that carefully-designed data mining and statistical inference can lead to new discoveries of all the other oncogenes with the above properties, and even possibly provide information on the detailed mechanisms of how the oncogenes regulate *BCL2* and other anti-apoptotic genes.

7.2.2 Cell Cycle and Apoptosis

The cell cycle and apoptosis are, not surprisingly, tightly linked one with the other, mainly to ensure cellular integrity. For example, the cell cycle process consists of multiple checkpoints, such as those at the G_1/S phase transition, in the S phase and in the M phase. These checkpoints consist of a sequence of control steps that allow proliferation to proceed only when appropriate growth signals are present and when the DNA integrity is assured. If DNA damage is detected, then the necessary repair machinery will be activated. If the damage cannot be repaired, the cell will be eliminated through apoptosis. Interestingly, the cell cycle and apoptosis share a number of common genes, presumably to facilitate the close and efficient interactions between the two. For example, the cell cycle gene *cyclin D* binds with *CDK4* and *CDK6* to form a complex during the G_1 phase to facilitate their interaction and phosphorylation of the *RB* protein, a negative cell-cycle regulator. The hyperphosphorylated *RB* protein dissociates from the *E2F* protein, thus enabling it to function as a transcriptional activator for genes required in the S phase for DNA synthesis. It has been observed that the loss of *RB* function triggers the activation of the apoptosis pathway via *P53* (Morgenbesser et al. 1994; Macleod et al. 1996; Harbour and Dean 2000; Nevins 2001), indicating that cells are designed to be removed once their *RB* function is lost. Interestingly, in the majority of cancers, the *RB* gene is either repressed or mutated (Vandel et al. 2001; Du and Searle 2009; Engel et al. 2013), reflecting the strong anti-apoptotic role by *RB*.

The relationships between some cell-cycle genes and apoptosis seem to be condition-dependent. For example, some cyclin genes such as *cyclin G* can have either pro-apoptotic or anti-apoptotic roles dependent on the cellular conditions (Okamoto and Prives 1999; Russell et al. 2012). Some *CDK* genes seem to be required for the execution phase of apoptosis. For example, the complex of *cyclin A* and *CDK* is activated whenever the caspase cascade is activated (Levkau et al. 1998), but the detailed mechanism remains to be elucidated. This represents another fundamental biology problem to the study of which data mining and statistical inference could contribute.

7.2.3 Cancer-Associated Stresses and Apoptosis

As a cancer evolves, it will accumulate a variety of abnormalities such as increased hypoxia, ROS and (lactic) acidity in the microenvironment, as well as DNA damage and nutrient deprivation, which should trigger the activation of apoptosis under normal conditions. However, cancer cells have acquired capabilities to avoid such activation via using different mechanisms. The following summarizes the known alterations that cancer cells have adopted to avoid apoptosis.

1. Genetic mutations: as discussed in Chaps. 1 and 4, cancer genomes tend to accumulate a large number of mutations. While some of the mutations probably do not serve any purpose germane to cancer development, many single-point mutations

are selected to serve specific roles to the benefit of the cancer cells, such as those in tumor-suppressor genes. For example, close to 90 % of all colon cancers have mutations in their APC gene as discussed in Chap. 4; similarly, the majority of cancers have mutations in *P53* and/or *RB*. The advantage of such mutations in comparison with functional inhibition of the relevant genes is that the genetically modified functional state change is more sustainable and efficient for cancer development.

2. Epigenomic modifications: Genes can be silenced through epigenomic level changes such as DNA methylation or histone modification. A large number of tumor suppressor genes have been found to be highly methylated in cancer epigenomes. For example, the promoter region of the *GSTP1* (glutathione S-transferase P) gene is hyper-methylated in more than 90 % of prostate cancers (Cairns et al. 2001), and the *HPV16L1* (human papillomavirus 16 oncogene) gene is highly methylated in the majority of the cervical cancers (Clarke et al. 2012). Compared to the irreversible genomic mutations, epigenomic modifications are reversible, but clearly not as easy as reversion through transcriptional regulation.

3. Growth factors as survival factors: Previous studies have found that some growth factors can serve as survival factors and that their activation will lead to the inhibition of apoptosis under specific conditions. The best studied case is *IGF1* (insulin-like growth factor 1), the normal physiological function being that of a potent growth factor during early growth. The activation of this protein can inhibit apoptosis through the activation of the *PI3K/AKT* pathway (see the Sect. 7.2.4) (O'Connor 1998; Vincent and Feldman 2002; Kuemmerle 2003; Torres Aleman 2005). A literature search revealed that many known growth factors can serve as survival factors, including *EGF* (Rawson et al. 1991), *FGF* (Araki et al. 1990), *HDGF* (hepatoma-derived growth factor) (Tsang et al. 2008), *HGF* (hepatocyte growth factor) (Xiao et al. 2001), *IL3* and *IL4* (Collins et al. 1994), *NGF* (nerve growth factor) (Batistatou and Greene 1991), *PDGF1* (Barres et al. 1992) and *VEGF* (Harmey and Bouchier-Hayes 2002). Based on such information, it seems reasonable to speculate that all growth factors may serve as survival factors under certain conditions. This hypothesis could probably be validated computationally by mining cancer transcriptomic data through identifying coordinated relationships between the expression patterns of each growth factor and some survival pathway. If this turns out to be true, it may imply the existence of a common mechanism that links growth factor receptor activation and the induction of survival pathways. This would undoubtedly generate new information and knowledge about the canonical proliferation pathway(s) since, as discussed earlier, proliferation is typically accompanied with the activation of apoptosis at some level. With this new information, it may be possible to argue for the existence of an encoded mechanism to maintain the equilibrium between proliferation and apoptosis; specifically, overactive proliferation will trigger apoptosis while overactive apoptosis may enhance the effect of proliferation. It is likely that molecular level computational simulation among the relevant players may lead to interesting insights about how such a mutually inhibitory system works in detail.

4. <u>Functional-state changes at the protein level</u>: It has been observed that the activation of some oncogenes can inhibit apoptosis (see the Sect. 7.2.4); moreover, these oncogenes can be constitutively activated through genetic mutations. For example, it has been established that specific mutations in *EGFR* can activate the protein in the absence of ligand binding (Voldborg et al. 1997; Gazdar 2009). Specifically, these gain-of-function mutations can lead to conformational changes of the *EGFR* protein mimicking that induced by the natural ligand *EGF* (Dawson et al. 2005). More interestingly, oxidation of specific residues of *EGFR* (not the same residues with mutations) can accomplish exactly the same, i.e., having the protein constitutively activated, as we have recently discovered (Ji et al. 2014). This may be a general phenomenon, i.e., functional-state changes of key cancer-related genes may take place first through regulation or selection of accidental post-translational modification (e.g., oxidation) in response to the cellular environment. Such regulation-directed changes may then be gradually replaced by epigenomic or genomic level changes (see Chap. 9 for details), possibly to keep the evasion of apoptosis more sustained and more energetically efficient.

   It is worth mentioning that cancer cells have been found to release factors to either stabilize anti-apoptotic proteins or destabilize pro-apoptotic proteins, as another way to evade apoptosis. For example, cancer cells tend to activate post-translational modification factors to attenuate or abrogate the degradation of *MCL1* (Derouet et al. 2004; Zhong et al. 2005; Akgul 2009), an anti-apoptotic member of the *BCL2* family. Similar observations have been made that cancer cells tend to promote the degradation of pro-apoptotic members of the *BCL2* family, such as *BAX* and *BIM*, through the ubiquitin-proteasome pathway (Zhang et al. 2004; Meller et al. 2006; Brancolini 2008). Currently the detailed triggering mechanisms in releasing these factors are not understood, but it is likely that data mining and statistical inference can lead to new information and a better understanding of the possible triggering signals.

5. <u>Change through functional regulation</u>: In order to assess the impact of *P53* mutations, an examination of gene-expression data between cancer samples with *versus* those without such mutations has been conducted. Specifically, we searched for apoptosis-related genes that show consistent expression-level changes between the two sets of cancer samples. A wide range of changes were observed in the *BCL2* family members without any obvious consistent patterns among the tissues with *versus* those without *P53* mutations. The only apoptosis-related gene exhibiting consistent expression patterns was *MDM2*, the negative regulator of *P53*. Figure 7.4 shows consistent down-regulation of *MDM2* in samples without *P53* mutations *versus* those with *P53* mutations across nine cancer types. It is hypothesized that cancer may up-regulate the expression of *MDM2* to ensure that *P53* remains inactive before loss-of-function mutations appear in the gene. Once loss-of-function mutations occur in *P53*, there is no need to continue the up-regulation of *MDM2*, hence its expression level decreases. This observation is consistent with an earlier statement that mutations in tumor suppressor

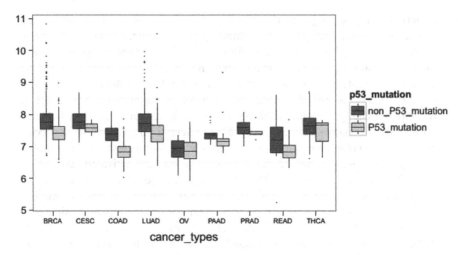

Fig. 7.4 A comparison of gene-expression levels of *MDM2* in samples without *P53* mutations *versus* those with *P53* mutations across nine cancer types: breast cancer (BRCA), cervical cancer (CESC), colon cancer (COAD), lung cancer (LUAD), ovarian cancer (OV), pancreatic cancer (PAAD), prostate cancer (PRAD), rectum cancer (READ) and thyroid cancer (THCA) (from *left* to *right*). The y-axis represents the gene-expression levels. The *light* and *dark grays* are for cancer samples with and without *P53* mutations, respectively

genes tend to occur in later stages during the development of cancer while the inhibition of its function may have taken place through other more reversible and less efficient means such as functional regulation in the early stage.

As indicated above, expression-level changes among the *BCL2* family members between the two groups of samples are clearly not as uniform as the *MDM2* gene (data not shown), suggesting that the impact of *P53* mutations is primarily on *MDM2* and less on the *BCL2* family members. One possible reason for this observation could be that the *BCL2* genes may need to be protected against mutations since they guard against the activation of apoptosis through other pathways as well.

IAP (inhibitor of apoptosis protein) is an interesting member of the apoptosis-inhibitor family as it can terminate apoptosis, even after the caspase genes are activated, by directly binding with *caspases-3, -7* and *-9* (Deveraux and Reed 1999; Shi 2004). Thus, the gene-expression pattern of the *IAP* gene was investigated and found to have highly elevated expression levels for a large fraction of the samples without elevated anti-apoptotic *BCL2* gene expression.

The above discussion depicts a plausible organization of the possible routes for apoptosis evasion. Generally, the apoptotic execution genes tend to remain at relatively low expression levels. *IAP* serves as the last safeguard against the execution of apoptosis, even after the activation of the *caspase* genes, through maintaining a high baseline expression-level of the protein. Together *P53* and *MDM2* serve as key regulators of the activity level of the apoptosis system, which responds to an intrinsic signaling pathway. The large set of *BCL2* genes and their splicing isoforms serve at the next layer of signaling and control. Recent studies have shown that not only do individual *BCL2* genes fall into two groups with opposing functions, i.e., pro- and anti-apoptosis, but also individual *BCL2* genes may have splicing isoforms with opposing functions. For example, *MCL1* has two known such isoforms (Boise et al. 1993; Craig 2002; Burlacu 2003; Youle and Strasser 2008), one being pro-apoptotic and the other anti-apoptotic. Numerous signaling pathways can then regulate these *BCL2* genes and their splicing isoforms in response to a large variety of survival signals (see the next section) or alterations at the protein, epigenomic or genomic levels. Clearly our understanding of the detailed mechanisms involved in apoptosis inhibition decreases as we move further away from the core apoptotic execution system.

We anticipate that carefully-designed data analyses and statistical inference will provide important new insights about the detailed mechanisms relevant to different survival pathways, as well as the mechanism(s) by which cancer cells have taken advantage of these encoded schemes to acquire new capabilities to avoid apoptosis. In time it may be possible to predict the evolutionary processes of cancer cells in selecting specific survival pathways for a fixed as well as a changing microenvironment.

To better understand the evolutionary trajectories of different cancers in terms of which survival pathways are selected and in what order, a detailed understanding of the survival mechanisms encoded in human cells is needed. These are of course used to stop cell death under severe conditions, which otherwise would lead to programmed cell death. Such survival pathways may have provided a basic framework through which cancer cells developed their initial anti-apoptotic capabilities. Later these capabilities may be further developed for more sustained and more efficient survival through the adoption of more permanent changes such as genomic or epigenomic alterations to replace the initial changes, either transcriptional or post-translational (see Chap. 9 for a more detailed discussion).

7.2.4 Survival Pathways and Apoptosis

To avoid destruction, survival pathways inject interference or termination signals to (damaged) cells that are destined or signaled to die. As of now, a number of survival pathways have been identified and found to be activated by different conditions. For example, the *KEAP1-NRF2-ARE* pathway is a survival pathway in response to severe oxidative stress that may cause injury to the cells (Kensler et al. 2007). The *NFκB*-dependent survival pathway can be activated by *TNFα* (Oeckinghaus et al. 2011). *MNK/EIF4E* is another survival pathway that can be induced by cytarabine, a chemotherapeutic that was discovered through analyses of cancers that had developed drug resistance (Altman et al. 2010). A number of other survival pathways have also been identified from studies of cancers that have developed drug resistance, including the well-studied multi-drug resistance pathway (Szakacs et al. 2006) in which hyaluronic acid plays a key regulatory role (Misra et al. 2003). At the core of this and other known survival pathways is that of the *PIK3/AKT* series of reactions (LoPiccolo et al. 2008).

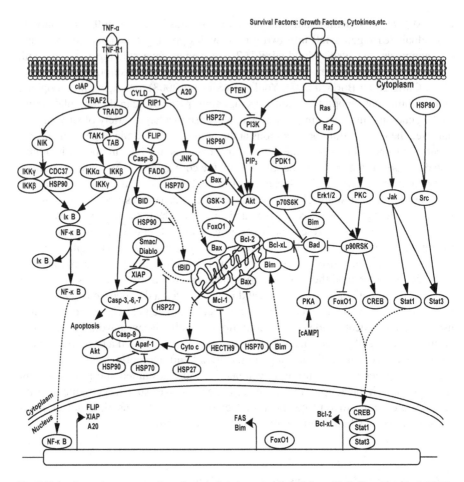

Fig. 7.5 A schematic representation of survival pathways, adapted from (Cell-Signaling-Tech 2011)

AKT is a centerpiece of multiple survival pathways. A number of proteins and signals can activate this protein kinase (see Fig. 7.5) through phosphorylation by the *PDK1/PRK* complex (see Chap. 6 for a more detailed introduction). It is speculated that the conformational change of the *AKT* protein, as induced by the growth factor-mediated *PI3K* activation, makes its phosphorylation possible. When activated, *AKT* can phosphorylate specific residues on a number of key proteins involved in signaling to apoptosis, thus inhibiting their function. Proteins so altered include the pro-apoptotic members *BAD, caspase-9*, transcription factors in the Forkhead family and the *NFκB* regulator *IKK* (Datta et al. 1999).

Computationally, one should be able to make informative inferences about the detailed relationships among the activation of survival pathways and the micro-environmental conditions through co-occurrence analyses of the activated survival pathways, the inactivated pro-apoptotic members of the *BCL2* family, and the activated *versus* inactivated apoptosis signaling pathways, in conjunction with the intracellular microenvironment such as oxidative state, hypoxic level and nutrient deprivation. The analyses should consider the following scenarios: (a) apoptosis signaling pathways that should have been activated, and (b) possible causes for "should be" but "not" activated signaling pathways. By making such analyses across multiple samples of the same cancer type at different developmental stages, one should be able to derive the occurrence order of the apoptosis-inhibition mechanisms adopted by cancer cells of the same cancer type. By conducting this analysis across multiple cancer types, it should be possible to derive different evolutionary trajectories by different cancer types in adopting different survival strategies. The same analytic approach can also be applied to examine possible relationships between the adopted apoptosis–inhibition schemes and clinical outcomes of different cancers, for example, by the cancer grade (see Chap. 3).

7.3   Cancer Characterization Through Resolution of How They Avoid Apoptosis

Our current understanding of the triggering mechanisms of apoptosis (see Fig. 7.2) and the known survival pathways (see Fig. 7.5) provides a framework to address several fundamental questions. For example, the possible relationships between the different routes for survival and the clinical outcome of specific cancer cases could potentially be derived through large-scale transcriptomic and other *omic* data for all cancer types in the public domain. Specifically one can ask and possibly address computationally the following questions regarding cancer samples:

- *Do different samples of the same cancer type tend to use consistent pathways to trigger their apoptosis system?*
- *Do different samples of the same cancer type tend to use consistent pathways to inhibit their apoptosis system?*
- *What are the major factors that may affect the selection of apoptosis activation and inhibition pathways?*
- *Is it possible to predict the activation and inhibition pathways used by a specific cancer based on its micro-environmental factors? If not, what other conditions need to be considered?*
- *Are there connections between the average survival rate of a cancer type and the activation and inhibition pathways of apoptosis?*

In addition to these general questions, one can also ask more detailed mechanistic questions about the activation and inhibition of apoptosis and possibly address them computationally. For example, it is plausible to address the following:

- *Why different cancer types have substantially different mutation rates in P53 or other cancer-related genes?*

– *Are functions of P53 or other cancer-related genes already inhibited or repressed in general before their loss-of-function mutations are selected in cancer-forming cells?*
– *If the answer to the above question is yes, then what are the triggering signals for functional inhibition or repression through regulation for a specific cancer type?*
– *Do epigenomic alterations such as DNA methylation tend to occur after functional inhibition through regulation and before genomic mutations are selected for the same gene as one would intuitively expect?*
– *Do cancers with P53 mutations tend to have higher benefits for sustained growth and survival than cancers without P53 mutations as measured using the terms discussed in the last paragraph of Section 7.1?*

Similarly, many other important questions can be asked regarding how different cancers evade apoptosis and then have them addressed by computationally mining the cancer *omic* data collected on tissue samples at different developmental stages (and possibly of different grades). These data are publicly available in databases such as the TCGA database (Collins and Barker 2007; Cancer-Genome-Atlas-Network 2012). Here we illustrate with a few examples how one can proceed towards solving some of the above questions by addressing simpler issues first and then integrating the solutions to provide answers to the questions posed.

What do cancers do when they are devoid of P53 mutations? Our current understanding is that loss of *P53* function is beneficial for cancer development. Intuitively one would thus expect that cancers without such mutations may need to functionally repress the activity of *P53*. To address this hypothesis, we have examined one set of *omic* data collected on 503 breast cancer samples in the TCGA database (Cancer-Genome-Atlas-Network 2012). 157 of these samples (31.2 %) have *P53* mutations according to the information provided on the sample set in TCGA. For the remaining samples, 149 have up-regulated expression levels of the *MDM2* gene, the negative regulator of *P53* that keeps *P53* in an inactive state; 145 have up-regulated *BCL2* expressions; 52 have up-regulated *BCL2L2*; and 77 have down-regulated *cytochrome c* (*CYC*) levels. Statistical analyses indicate that gene-expression level changes among *MDM2, BCL2* and *BCL2L2* are strongly correlated, suggesting that these genes are probably controlled through the same regulatory machinery. See Fig. 7.6 for the detailed information.

Analyses of other cancer types show similar trends in terms of up-regulated expression of anti-apoptotic genes and down-regulated expression of pro-apoptotic genes, but the actual genes with altered expression levels could be rather different. This leads to our next question.

What is the level of consistency in terms of up- and down-regulation of the apoptosis-associated genes across different cancer types? To answer this question, 15 cancer types were examined, namely bladder, brain, breast, colorectal, esophagus, gastric, head-neck, liver, lung, melanoma, ovary, pancreas, prostate, renal and thyroid cancers (see Appendix for the detailed names of the datasets used). The primary aim in analyzing the transcriptomic data of these cancer types was to identify the general triggers for apoptotic activation and apoptotic inhibition in different cancers. The

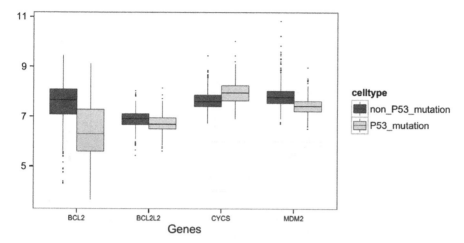

Fig. 7.6 Expression-level assessment of four genes in breast cancer samples (see designation along the x-axis) comparing those with *P53* mutations and those without. The y-axis represents the gene-expression levels. The *light* and *dark gray* are for cancer samples with and without *P53* mutations, respectively

rationale is based on the observation that the relative strengths of these two types of triggers determine the ultimate step of cell death or survival. Also, of course, one would like to know the evolutionary trajectories of these two competing processes, each having continuously increasing force and complexity.

From Fig. 7.7, it is first noted that the overall consistency level in any category of apoptotic genes is fairly low. This singular result suggests that different cancer types may have distinct driving forces for cell death and hence use different activation pathways of the apoptosis system, possibly due to the different microenvironments in different cancers. For example, it was noted that apoptosis is triggered by distinct stressors in different cancers, e.g., pancreatic cancer by oxidative stress, brain cancer via *P53* induced by DNA damage, gastric cancer often by natural killer cell-mediated cytotoxicity and thyroid cancer by ionic level changes. Interestingly, the toll-like receptor pathway was found to be consistently up-regulated across all the 15 cancer types, suggesting the essential role of this pathway in activating apoptosis.

The dominating survival signals in each of the 15 cancer types show a diverse range. For example, (1) bladder cancer tends to use the *ER* overload response to trigger the activation of *NFκB* and the associated survival pathway; (2) ovarian cancer triggers the replicative senescence system to curtail cancer rather than inducing cell death; (3) renal cancer tends to trigger the hyperosmotic response system for survival; and (4) head-neck cancer generally uses the "virus evasion of host immune system" mechanism for survival.

It is hypothesized that, as cancer continues to evolve, different types of cancers may utilize more than one survival pathway as their cellular environments diverge

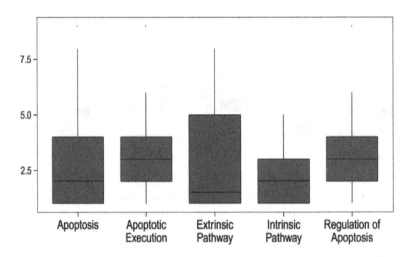

Fig. 7.7 Consistency levels in up- and down-regulated genes involved in: (1) the entire apoptosis system; (2) intrinsic signaling pathways; (3) extrinsic signaling pathways; (4) apoptotic execution; and (5) apoptosis regulation (from *left* to *right*), across 15 cancer types. For each differentially-expressed apoptotic gene, its consistency level is defined as the maximum number of cancer types that show consistent up- or down-regulation in this gene's expression (as defined in Fig. 7.5). The y-axis gives the distribution of the consistency levels of genes involved in one of the above categories

from normal. Such diversity will create more opportunities and pathways to ensure survival and overcome apoptosis. It is further hypothesized that mutations may be selected to make survival more sustainable and efficient as pressure from cell death mounts. Chapter 9 discusses these issues in a systematic manner.

7.4   Concluding Remarks

The current knowledge about induction and inhibition of apoptosis, as summarized in Figs. 7.2 and 7.5, provides a powerful framework to study inducers of apoptosis *versus* inducers of the counteraction, namely survival, across different cancer types. It is the availability of transcriptomic and genomic data of these cancer types that makes such delineation feasible. By carefully analyzing stress-related pathways that are consistently up-regulated and linked to apoptosis, it may be possible to identify the major triggers for apoptosis in the cancer type. A similar approach can provide information on the triggers for survival. Having such a capability, it is possible to study the evolutionary trajectories of the generation of signals for programmed cell death and for survival by analyzing *omic* data of cancers at different developmental stages. It is further anticipated that, when linking such analyses to cancers with high levels of malignancies, one could potentially derive useful insights about why certain cancer types result in greater mortalities than the others.

Appendix

Table 7.1 Transcriptomic datasets of 15 cancer types are collected from the GEO database and used in our data analysis

Bladder	GSE31676
Brain	GSE42906
Breast	GSE108106
Colorect	GSE209167
Esophagus	GSE20347
Gastric	GSE198264
Headneck	GSE98444
Liver	GSE143234
Lung	GSE198043
Melanoma	GSE31894
Ovary	GSE267128
Pancreas	GSE154717
Prostate	GSE69561
Renal	GSE156415
Thyroid	GSE36786

References

Akgul C (2009) Mcl-1 is a potential therapeutic target in multiple types of cancer. Cell Mol Life Sci 66: 1326-1336

Altman JK, Glaser H, Sassano A et al. (2010) Negative regulatory effects of Mnk kinases in the generation of chemotherapy-induced antileukemic responses. Mol Pharmacol 78: 778-784

Araki S, Shimada Y, Kaji K et al. (1990) Apoptosis of vascular endothelial cells by fibroblast growth factor deprivation. Biochem Biophys Res Commun 168: 1194-1200

Askew DS, Ashmun RA, Simmons BC et al. (1991) Constitutive c-myc expression in an IL-3-dependent myeloid cell line suppresses cell cycle arrest and accelerates apoptosis. Oncogene 6: 1915-1922

Barres BA, Hart IK, Coles HS et al. (1992) Cell death and control of cell survival in the oligodendrocyte lineage. Cell 70: 31-46

Batistatou A, Greene LA (1991) Aurintricarboxylic acid rescues PC12 cells and sympathetic neurons from cell death caused by nerve growth factor deprivation: correlation with suppression of endonuclease activity. J Cell Biol 115: 461-471

Bissonnette RP, Echeverri F, Mahboubi A et al. (1992) Apoptotic cell death induced by c-myc is inhibited by bcl-2. Nature 359: 552-554

Boise LH, Gonzalez-Garcia M, Postema CE et al. (1993) bcl-x, a bcl-2-related gene that functions as a dominant regulator of apoptotic cell death. Cell 74: 597-608

Brady CA, Attardi LD (2010) p53 at a glance. J Cell Sci 123: 2527-2532

Brancolini C (2008) Inhibitors of the Ubiquitin-Proteasome System and the cell death machinery: How many pathways are activated? Curr Mol Pharmacol 1: 24-37

Broker LE, Kruyt FA, Giaccone G (2005a) Cell death independent of caspases: a review. Clin Cancer Res 11: 3155-3162

Broker LE, Kruyt FAE, Giaccone G (2005b) Cell death independent of caspases: A review. Clinical Cancer Research 11: 3155-3162

Burlacu A (2003) Regulation of apoptosis by Bcl-2 family proteins. J Cell Mol Med 7: 249-257

Cairns P, Esteller M, Herman JG et al. (2001) Molecular detection of prostate cancer in urine by GSTP1 hypermethylation. Clin Cancer Res 7: 2727-2730

Cancer-Genome-Atlas-Network (2012) Comprehensive molecular portraits of human breast tumours. Nature 490: 61-70

Cell-Signaling-Tech (2011) Apoptosis Inhibition.

Clarke MA, Wentzensen N, Mirabello L et al. (2012) Human papillomavirus DNA methylation as a potential biomarker for cervical cancer. Cancer Epidemiol Biomarkers Prev 21: 2125-2137

Collins FS, Barker AD (2007) Mapping the cancer genome. Pinpointing the genes involved in cancer will help chart a new course across the complex landscape of human malignancies. Sci Am 296: 50-57

Collins MK, Perkins GR, Rodriguez-Tarduchy G et al. (1994) Growth factors as survival factors: regulation of apoptosis. Bioessays 16: 133-138

Craig RW (2002) MCL1 provides a window on the role of the BCL2 family in cell proliferation, differentiation and tumorigenesis. Leukemia 16: 444-454

Datta SR, Brunet A, Greenberg ME (1999) Cellular survival: a play in three Akts. Genes Dev 13: 2905-2927

Dawson JP, Berger MB, Lin CC et al. (2005) Epidermal growth factor receptor dimerization and activation require ligand-induced conformational changes in the dimer interface. Mol Cell Biol 25: 7734-7742

Debbas M, White E (1993) Wild-type p53 mediates apoptosis by E1A, which is inhibited by E1B. Genes Dev 7: 546-554

Derouet M, Thomas L, Cross A et al. (2004) Granulocyte macrophage colony-stimulating factor signaling and proteasome inhibition delay neutrophil apoptosis by increasing the stability of Mcl-1. Journal of Biological Chemistry 279: 26915-26921

Desai R (2013) Cell Death. Available at: http://drrajivdesaimd.com/,

Deveraux QL, Reed JC (1999) IAP family proteins–suppressors of apoptosis. Genes Dev 13: 239-252

Du W, Searle JS (2009) The rb pathway and cancer therapeutics. Curr Drug Targets 10: 581-589

Dumble M, Moore L, Chambers SM et al. (2007) The impact of altered p53 dosage on hematopoietic stem cell dynamics during aging. Blood 109: 1736-1742

Engel BE, Welsh E, Emmons MF et al. (2013) Expression of integrin alpha 10 is transcriptionally activated by pRb in mouse osteoblasts and is downregulated in multiple solid tumors. Cell Death Dis 4: e938

Evan GI, Wyllie AH, Gilbert CS et al. (1992) Induction of apoptosis in fibroblasts by c-myc protein. Cell 69: 119-128

Fanidi A, Harrington EA, Evan GI (1992) Cooperative interaction between c-myc and bcl-2 proto-oncogenes. Nature 359: 554-556

Feng Z, Lin M, Wu R (2011) The Regulation of Aging and Longevity: A New and Complex Role of p53. Genes Cancer 2: 443-452

Gazdar AF (2009) Activating and resistance mutations of EGFR in non-small-cell lung cancer: role in clinical response to EGFR tyrosine kinase inhibitors. Oncogene 28 Suppl 1: S24-31

Gogvadze V, Orrenius S, Zhivotovsky B (2008) Mitochondria in cancer cells: what is so special about them? Trends Cell Biol 18: 165-173

Gorgoulis VG, Zacharatos P, Kotsinas A et al. (2003) p53 activates ICAM-1 (CD54) expression in an NF-kappaB-independent manner. EMBO J 22: 1567-1578

Hafsi H, Hainaut P (2011) Redox control and interplay between p53 isoforms: roles in the regulation of basal p53 levels, cell fate, and senescence. Antioxid Redox Signal 15: 1655-1667

Harbour JW, Dean DC (2000) Rb function in cell-cycle regulation and apoptosis. Nat Cell Biol 2: E65-67

Harmey JH, Bouchier-Hayes D (2002) Vascular endothelial growth factor (VEGF), a survival factor for tumour cells: implications for anti-angiogenic therapy. Bioessays 24: 280-283

Hermeking H, Eick D (1994) Mediation of c-Myc-induced apoptosis by p53. Science 265: 2091-2093

Jaattela M (2004) Multiple cell death pathways as regulators of tumour initiation and progression. Oncogene 23: 2746-2756

Jaattela M, Tschopp J (2003) Caspase-independent cell death in T lymphocytes. Nature Immunology 4: 416-423

Ji F, Cao S, Kannan N et al. (2014) Oxidative stress counteracts intrinsic disorder in the EGFR kinase and promote receptor dimerization. In preparation.

Karam JA (2009) Apoptosis in Carcinogenesis and Chemotherapy. Netherlands: Springer:

Kensler TW, Wakabayashi N, Biswal S (2007) Cell survival responses to environmental stresses via the Keap1-Nrf2-ARE pathway. Annu Rev Pharmacol Toxicol 47: 89-116

Kerr JF, Wyllie AH, Currie AR (1972) Apoptosis: a basic biological phenomenon with wide-ranging implications in tissue kinetics. British journal of cancer 26: 239-257

Kolenko VM, Uzzo RG, Bukowski R et al. (2000) Caspase-dependent and -independent death pathways in cancer therapy. Apoptosis 5: 17-20

Kuemmerle JF (2003) IGF-I elicits growth of human intestinal smooth muscle cells by activation of PI3K, PDK-1, and p70S6 kinase. Am J Physiol Gastrointest Liver Physiol 284: G411-422

Leist M, Jaattela M (2001) Four deaths and a funeral: from caspases to alternative mechanisms. Nat Rev Mol Cell Biol 2: 589-598

Levine AJ, Oren M (2009) The first 30 years of p53: growing ever more complex. Nat Rev Cancer 9: 749-758

Levkau B, Koyama H, Raines EW et al. (1998) Cleavage of p21Cip1/Waf1 and p27Kip1 mediates apoptosis in endothelial cells through activation of Cdk2: role of a caspase cascade. Mol Cell 1: 553-563

Liu G, Park YJ, Tsuruta Y et al. (2009) p53 Attenuates lipopolysaccharide-induced NF-kappaB activation and acute lung injury. J Immunol 182: 5063-5071

LoPiccolo J, Blumenthal GM, Bernstein WB et al. (2008) Targeting the PI3K/Akt/mTOR pathway: effective combinations and clinical considerations. Drug Resist Updat 11: 32-50

Lowe J, Shatz M, Resnick M et al. (2013) Modulation of immune responses by the tumor suppressor p53. BioDiscovery 8:

Lowe SW (1999) Activation of p53 by oncogenes. Endocr Relat Cancer 6: 45-48

Lowe SW, Ruley HE (1993) Stabilization of the p53 tumor suppressor is induced by adenovirus 5 E1A and accompanies apoptosis. Genes Dev 7: 535-545

Luthi AU, Martin SJ (2007) The CASBAH: a searchable database of caspase substrates. Cell death and differentiation 14: 641-650

Macleod KF, Hu Y, Jacks T (1996) Loss of Rb activates both p53-dependent and independent cell death pathways in the developing mouse nervous system. EMBO J 15: 6178-6188

Martins CP, Brown-Swigart L, Evan GI (2006) Modeling the therapeutic efficacy of p53 restoration in tumors. Cell 127: 1323-1334

Mathiasen IS, Jaattela M (2002) Triggering caspase-independent cell death to combat cancer. Trends in Molecular Medicine 8: 212-220

Meller R, Cameron JA, Torrey DJ et al. (2006) Rapid degradation of Bim by the ubiquitin-proteasome pathway mediates short-term ischemic tolerance in cultured neurons. Journal of Biological Chemistry 281: 7429-7436

Menendez D, Inga A, Resnick MA (2010) Potentiating the p53 network. Discov Med 10: 94-100

Menendez D, Shatz M, Azzam K et al. (2011) The Toll-Like Receptor Gene Family Is Integrated into Human DNA Damage and p53 Networks. Plos Genetics 7:

Menendez D, Shatz M, Resnick MA (2013) Interactions between the tumor suppressor p53 and immune responses. Current Opinion in Oncology 25: 85-92

Misra S, Ghatak S, Zoltan-Jones A et al. (2003) Regulation of multidrug resistance in cancer cells by hyaluronan. Journal of Biological Chemistry 278: 25285-25288

Morgenbesser SD, Williams BO, Jacks T et al. (1994) p53-dependent apoptosis produced by Rb-deficiency in the developing mouse lens. Nature 371: 72-74

Nevins JR (2001) The Rb/E2F pathway and cancer. Hum Mol Genet 10: 699-703

O'Connor R (1998) Survival factors and apoptosis. Adv Biochem Eng Biotechnol 62: 137-166

Oeckinghaus A, Hayden MS, Ghosh S (2011) Crosstalk in NF-kappaB signaling pathways. Nature Immunology 12: 695-708

Okamoto K, Prives C (1999) A role of cyclin G in the process of apoptosis. Oncogene 18: 4606-4615

Okuda Y, Okuda M, Bernard CC (2003) Regulatory role of p53 in experimental autoimmune encephalomyelitis. J Neuroimmunol 135: 29-37

Pesch J, Brehm U, Staib C et al. (1996) Repression of interleukin-2 and interleukin-4 promoters by tumor suppressor protein p53. J Interferon Cytokine Res 16: 595-600

Peter ME, Heufelder AE, Hengartner MO (1997) Advances in apoptosis research. Proc Natl Acad Sci U S A 94: 12736-12737

Qi R, Liu XY (2006) New advance in caspase-independent programmed cell death and its potential in cancer therapy. Int J Biomed Sci 2: 211-216

Querido E, Teodoro JG, Branton PE (1997) Accumulation of p53 induced by the adenovirus E1A protein requires regions involved in the stimulation of DNA synthesis. Journal of Virology 71: 3526-3533

Rawson CL, Loo DT, Duimstra JR et al. (1991) Death of serum-free mouse embryo cells caused by epidermal growth factor deprivation. J Cell Biol 113: 671-680

Rodier F, Campisi J, Bhaumik D (2007) Two faces of p53: aging and tumor suppression. Nucleic Acids Res 35: 7475-7484

Russell P, Hennessy BT, Li J et al. (2012) Cyclin G1 regulates the outcome of taxane-induced mitotic checkpoint arrest. Oncogene 31: 2450-2460

Samuelson AV, Lowe SW (1997) Selective induction of p53 and chemosensitivity in RB-deficient cells by E1A mutants unable to bind the RB-related proteins. Proc Natl Acad Sci U S A 94: 12094-12099

Santhanam U, Ray A, Sehgal PB (1991) Repression of the interleukin 6 gene promoter by p53 and the retinoblastoma susceptibility gene product. Proc Natl Acad Sci U S A 88: 7605-7609

Schneider G, Henrich A, Greiner G et al. (2010) Cross talk between stimulated NF-kappaB and the tumor suppressor p53. Oncogene 29: 2795-2806

Serrano M, Lin AW, McCurrach ME et al. (1997) Oncogenic ras provokes premature cell senescence associated with accumulation of p53 and p16INK4a. Cell 88: 593-602

Shi Y (2004) Caspase activation, inhibition, and reactivation: a mechanistic view. Protein Sci 13: 1979-1987

Shi Y, Glynn JM, Guilbert LJ et al. (1992) Role for c-myc in activation-induced apoptotic cell death in T cell hybridomas. Science 257: 212-214

Siskind LJ (2005) Mitochondrial ceramide and the induction of apoptosis. J Bioenerg Biomembr 37: 143-153

Strasser A, Harris AW, Bath ML et al. (1990) Novel primitive lymphoid tumours induced in transgenic mice by cooperation between myc and bcl-2. Nature 348: 331-333

Suen DF, Norris KL, Youle RJ (2008) Mitochondrial dynamics and apoptosis. Genes Dev 22: 1577-1590

Szakacs G, Paterson JK, Ludwig JA et al. (2006) Targeting multidrug resistance in cancer. Nat Rev Drug Discov 5: 219-234

Takaoka A, Hayakawa S, Yanai H et al. (2003) Integration of interferon-alpha/beta signalling to p53 responses in tumour suppression and antiviral defence. Nature 424: 516-523

Torres Aleman I (2005) Role of insulin-like growth factors in neuronal plasticity and neuroprotection. Adv Exp Med Biol 567: 243-258

Tsang TY, Tang WY, Tsang WP et al. (2008) Downregulation of hepatoma-derived growth factor activates the Bad-mediated apoptotic pathway in human cancer cells. Apoptosis 13: 1135-1147

Vandel L, Nicolas E, Vaute O et al. (2001) Transcriptional repression by the retinoblastoma protein through the recruitment of a histone methyltransferase. Mol Cell Biol 21: 6484-6494

Ventura A, Kirsch DG, McLaughlin ME et al. (2007) Restoration of p53 function leads to tumour regression in vivo. Nature 445: 661-665

Vincent AM, Feldman EL (2002) Control of cell survival by IGF signaling pathways. Growth Horm IGF Res 12: 193-197

Voldborg BR, Damstrup L, Spang-Thomsen M et al. (1997) Epidermal growth factor receptor (EGFR) and EGFR mutations, function and possible role in clinical trials. Ann Oncol 8: 1197-1206

Vousden KH, Prives C (2009) Blinded by the Light: The Growing Complexity of p53. Cell 137: 413-431

Xiao GH, Jeffers M, Bellacosa A et al. (2001) Anti-apoptotic signaling by hepatocyte growth factor/Met via the phosphatidylinositol 3-kinase/Akt and mitogen-activated protein kinase pathways. Proc Natl Acad Sci U S A 98: 247-252

Youle RJ, Strasser A (2008) The BCL-2 protein family: opposing activities that mediate cell death. Nat Rev Mol Cell Biol 9: 47-59

Zhang HG, Wang J, Yang X et al. (2004) Regulation of apoptosis proteins in cancer cells by ubiquitin. Oncogene 23: 2009-2015

Zheng SJ, Lamhamedi-Cherradi SE, Wang P et al. (2005) Tumor suppressor p53 inhibits autoimmune inflammation and macrophage function. Diabetes 54: 1423-1428

Zhong Q, Gao W, Du F et al. (2005) Mule/ARF-BP1, a BH3-only E3 ubiquitin ligase, catalyzes the polyubiquitination of Mcl-1 and regulates apoptosis. Cell 121: 1085-1095

Chapter 8
Cancer Development in Competitive and Hostile Environments

From the perspective of cancer cells, they reside and must survive in a highly stressful and unfriendly environment. From the beginning, the cells have been under tremendous pressure to evolve in order to rid themselves of accumulated glucose metabolites and hence survival via cell division, as discussed in Chap. 5. In order to sustain their proliferation for long-term survival, they must continuously adjust their metabolism to adapt to the micro-environmental stresses and concurrently avoid the detection and destruction by myriad defense mechanisms that are present in humans to neutralize and destroy infectious organisms and aberrant cells. The adaptations of these cells to the challenging environment continue to drive their metabolism to become increasingly more irregular. As they continue to accumulate abnormalities, these cells must also continue to evolve to gain additional protection, thus ensuring their survival in the increasingly hostile environment. The ongoing processes of evolution and adaptation may drive the cellular metabolism to become even more anomalous, thus forming a vicious cycle and serving as a new driver for survival via proliferation. The defense systems that these evolving cells must overcome include: (1) the cellular surveillance and protection systems such as apoptosis and limited growth potential; (2) competition from the neighboring cells, a tissue-level mechanism for eliminating less fit cells and (3) the immune system. The main question addressed in this chapter is: *How have the cancer cells evolved to survive these obstacles and become increasingly more malignant?* The question of how the cancer cells respond to and overcome other microenvironment-induced stresses will be deferred to Chap. 9.

The first set of obstacles that these cells must overcome is intracellular, namely apoptosis, anti-growth signals and limited growth potential. As discussed in Chap. 6, hyaluronic acid fragments generated by the neoplastic cells can provide them the ability to avoid the activation of apoptosis and anti-growth mechanisms. In addition, the hypoxic environment can facilitate a gain in the capability for unlimited growth

© Springer Science+Business Media New York 2014
Y. Xu et al., *Cancer Bioinformatics*, DOI 10.1007/978-1-4939-1381-7_8

potential by activating telomerase to repair the shortened telomeres during cell division (Nishi et al. 2004). The focus of this chapter is on how such cells overcome challenges from their competing neighbors, the acidic environment and the immune system.

8.1   Cell-Cell Competition in Growing Tissues

Studies on *Drosophila* in the past decade, particularly in the past few years, have revealed that viable cells can die solely due to competition with their neighboring cells in growing tissues. Specifically during tissue growth, cell-cell communication plays a critical role in determining whether cells remain viable in the tissue or die. This is accomplished through cells sensing the fitness level of the neighboring cells and eliminating those determined to be less fit by signaling their apoptosis and then engulfing the debris (Vivarelli et al. 2012). This mechanism allows a tissue to keep the fittest cells during development, repair and remodeling when growth is involved. The available data suggest that competitive interactions among cells are regulated at the tissue level, i.e., competitions stay within boundaries of developmental compartments of organs (Johnston 2009). More specifically, super-competitors can eliminate less fit neighbors and replace them by their own off-spring, so the total tissue size is not affected by such competition. Interestingly, a counterbalance mechanism has also been discovered, namely the resistance of a competition-based elimination accomplished by adjusting the elimination threshold (Portela et al. 2010). Some details are summarized as follows.

Cell fitness level: Cell-cell competition in growing tissues was discovered when it was found that *Drosophila* cells with a specific mutation, *Minute*, are viable on their own, but are actively eliminated when growing in the same environment but containing wild-type cells (Simpson 1979; Simpson and Morata 1981). Further studies revealed that the expression level of *dmyc*, the *Drosophila* homolog of the human *MYC* gene, determines the winners and losers in such a competition (de la Cova et al. 2004; Moreno and Basler 2004). A later study suggested that a similar competition mechanism may exist in mice with *P53* serving as a sensor and reporter of cellular fitness, where an up-regulated *P53* puts the host cell in a growth disadvantage (Bondar and Medzhitov 2010a). Data from this study revealed that cells harboring loss-of-function *P53* mutants are unable to report their own fitness levels even when they are damaged, and hence can elude the machinery for eliminating them, which, otherwise, should have been competitive losers (Bondar and Medzhitov 2010b). This finding indicates that loss of *P53* function gives a survival advantage to the host cell in addition to the other growth advantages, as discussed in the earlier chapters. More recent studies further suggest that a number of other cancer-related genes such as *RAS* and *SRC* are also involved in the complex network in signaling and regulating cellular competition in growing tissues (Prober and Edgar 2000; Levayer and Moreno 2013).

The Flower code: A basic question in studying cell-cell competition is: *How do cells recognize each other's fitness level?* Again turning to *Drosophila,* recent studies have produced some exciting data about this issue. Specifically, the losing cells in a competition all express specific isoforms of the Flower (*fwe*) protein, one of its physiological functions being that of a Ca^{+2} channel (Yao et al. 2009). The protein is known to have three splicing isoforms: the predominant *fwe(ubi)* isoform and two *fwe(loss)* isoforms. The study found that, when putting wild-type wing-disc cells of *Drosophila* in the milieu containing cells of the same type but harboring overexpressed *dmyc* genes, two less frequently used isoforms, *fwe(loss)*, will be expressed in the disadvantaged cells. This happens only during competition, i.e., during growth with multiple sub-populations in the same environment (Rhiner et al. 2010). Similar observations have also been made about other competitive situations among *Drosophila* cells, suggesting that labeling less fit cells with *fwe(loss)* expression is a general mechanism for putting such cells in a queue to be eliminated.

Loser elimination mechanism: The expression of the *fwe(loss)* isoforms will ultimately lead to activation of the *caspase* cascade in the host cells (Baker 2011), and also activate an engulfment process in the neighboring cells expressing the winner *fwe(ubi)* isoform, ultimately leading to the engulfment of the loser cells by the winners (Johnston 2009).

Relevant signaling pathways: Two signaling pathways have been found to be involved in signal transduction among neighboring cells to inform their relative fitness: the *WNT* pathway and the *Hippo-Salvador-Warts* (*HSW*) pathway (Tamori and Deng 2011). Specifically, the *HSW* pathway, known for controlling an organ's size through its regulation of cell proliferation and apoptosis, controls the expression of *dmyc* via the *yki* transcription factor (Neto-Silva et al. 2010; Ziosi et al. 2010). The *WNT* pathway, known as the key regulator of cell survival, cell fate decision and tissue growth, serves as a signaling pathway for cell competition independent of *dmyc*.

A balancing force: A separate study has recently discovered that there is a counteracting mechanism in *Drosophila* that protects cells from being killed after they are labeled with *fwe(loss)* (Portela et al. 2010). It was found that *dsparc*, the *Drosophila* homolog of the human *SPARC* gene, serves as a regulator for protecting the losing cells by increasing the activation threshold of the *caspase* cascade. Interestingly, the human *SPARC* gene, long implicated in cancer development, has been found to serve multiple roles such as promotion of cancer development and metastasis (Puolakkainen et al. 2004; Sansom et al. 2007),as well as tumor suppression (Mok et al. 1996). These data suggest the possibility that a similar *SPARC-centric* mechanism may exist in human cells because of the multiple connections of the over-expressed *SPARC* gene with human cancers.

Extending the knowledge to humans: A substantial amount of information has been accumulated about the different roles played by *SPARC* in human cancer, suggesting the possibility that the gene may serve as some type of master regulator. It has, however, been challenging to piece together all this information into a cohesive *SPARC*-centric model from the published data since these reports are often conflicting (Arnold and Brekken 2009). If, however, all the data are reliable, this suggests a high complexity of the *SPARC*-associated mechanism. The discovery of the *dsparc* gene serving as a regulator for cell survival during cell-cell competition may provide a unifying scheme for human cells, which is consistent with all the published human *SPARC* data.

One can imagine the challenging nature of developing a logical framework of cancer-related *SPARC* data based on the above discussion, particularly when the relevant context information about *MYC* and the human homolog of *fwe* is not considered. The challenge arises from our knowledge that the effects of *SPARC* could be highly context-dependent, specifically depending on the relative fitness levels of the neighboring cells. This is based on the assumption that a similar mechanism indeed exists in human, which involves cell elimination through fitness competition and protection of such cells from elimination under some to-be-identified conditions. A recent study has found that deficiency in the mouse homolog of the *fwe* gene can reduce the susceptibility to skin papilloma formation in mice (Petrova et al. 2012), suggesting the possibility that when no cells are labeled as unfit by expressing *fwe(loss)*, there will be no trigger to activate *SPARC* for protecting such cells from being eliminated. This further suggests that the rescue provided by *SPARC* may represent a novel survival pathway that cancer cells may have possibly learned to use. Of course, to derive a *SPARC*-centric model in mouse or human similar to that in *Drosophila* may require integrated investigation of *omic* data generation, analysis and computational inference.

Despite reservations expressed about cell culture-based cancer studies in various chapters, this system has the advantage of enabling the experimentalist to study multiple sub-populations of cancer cells expressing different levels of *MYC* in order to elucidate the functional roles of *SPARC* in cancer. This will require the consideration of *MYC* and the human homolog of the *fwe* gene, along with genes identified to be relevant to the functionalities of *SPARC* as reported in the literature. Clearly, both RNA and DNA sequencing data will be needed to facilitate the inference of splicing isoforms of the key relevant genes, coupled with the available expression data on these genes in the public domain. It is worth emphasizing that while the mouse homolog of *fwe* has been identified (Petrova et al. 2012),the human homolog has not yet. Hence, the elucidation of the *SPARC*-regulated survival mechanism may represent a major undertaking in both bioinformatics analysis and *omic* data generation.

8.2   Cancer *Versus* Normal Cells in a Lactic Acidic Environment

As discussed in Chaps. 1 and 5, the reprogramming of energy metabolism is a hallmark of cancer. That is, glycolytic fermentation replaces aerobic respiration as the main form of glucose metabolism (at least in the early stage). A direct result of this change is that substantially more lactate, as the terminal receiver of electrons in glucose metabolism, is produced and transported extracellularly. To maintain the intracellular electro-neutrality when releasing lactate, the cells release one proton for each released lactate, the anionic form of lactic acid. This leads to increased acidity in the extracellular environment of the neoplastic cells, which will also affect the neighboring normal cells. The question addressed here is: *How do the neoplastic cells and the normal cells fare in the lactic acidosis environment?*

Normal cells: The typical extracellular pH for normal cells is 7.3–7.4 and the intracellular pH ranges between 6.99 and 7.20. In contrast, the extracellular pH of cancer cells tends to range between 6.2 and 6.9. Such an acidic extracellular pH, particularly when coupled with hypoxia, can lead to cell death via the activation of apoptosis, specifically through a direct activation of the *caspase* genes without going through the upstream regulators of apoptosis such as *P53* (Xu et al. 2013). Consequently, the acidic extracellular pH created by cancer cells provides these cells with an opportunity to encroach into the areas occupied by normal cells.

Neoplastic cells: While the acidic microenvironment will ultimately lead to the death of normal cells, this does not seem to be the case with cancer cells, at least not at the same level. The literature has suggested that such an acidic environment may be more favorable for cancer cells to thrive, as will be discussed later in this section.

8.2.1   Maintaining a Neutral or Slightly Alkaline Intracellular pH in Cancer Cells

Compared to normal cells, cancer cells tend to have higher intracellular pH levels, typically ranging between 7.12 and 7.56 (Calorini et al. 2012). This is consistent with the general understanding that a slightly alkaline environment is more ideal for cell proliferation. But this also creates a pressure on the neoplastic cells because of the pH difference, hence a proton gradient between the two sides of their cell membrane. Cancer cells seem to have found some effective responses to accommodate the proton gradient while maintaining the pH difference, which seems not to be available to normal cells. A comparative analysis of genome-scale transcriptomic data on six types of solid cancers has recently been carried out, namely breast, colon, liver, two lung (adenocarcinoma, squamous cell carcinoma) and prostate cancer, to understand how cancer cells, but not normal cells, have accomplished this (Xu et al. 2013). The following observations were made.

Cellular responses to increased acidity: In glycolysis, the degradation of each mole of glucose generates 2 lactates, 2 protons and 2 ATPs, detailed as follows:

$$glucose + 2NAD^+ + 2ADP + 2Pi \rightarrow 2\ ATP + 2NADH + 2H_2O + 2\ lactate + 2\ H^+$$

while in contrast the complete degradation of glucose through oxidative phosphorylation is pH neutral. The two extra protons generated by glycolytic fermentation must be removed or neutralized to avoid acidosis. Four monocarboxylate transporters, all in the *SLC16A* family, have been found to serve as exporters of the excess protons (Halestrap and Price 1999; Halestrap 2012), thus maintaining the intracellular pH neutrality (Casey et al. 2010). Among the four transporter genes, two are up-regulated in five out of the six cancers, suggesting that these transporters have been activated to export protons extracellularly. Specifically, *SLC16A1* is up-regulated in breast, colon, liver cancer and lung adenocarcinoma, and *SLC16A3* is up-regulated in colon and squamous cell lung cancer, as shown in the top part of Fig. 8.1. These data are consistent with previous reports that these transporter genes are

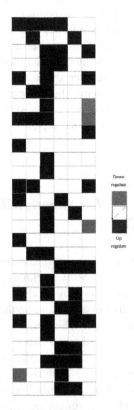

Fig. 8.1 A heat-map of expression-level changes in transporter genes and associated genes in six cancer types: breast, colon, liver, lung adenocarcinoma, squamous cell lung cancer, and prostate cancer, each represented by one column from *left* to *right*, in comparison with their matching control tissues. A total of 28 genes are included here, with each row representing one gene from top down: two genes for proton exporters: *SLC16A3, SLC16A1*; seven genes encoding the V_0 domain of *V-ATPase:ATP6V0E2, ATP6V0E1, ATP6V0B, ATP6V0A2, ATP6V0A1, ATP6AP2, ATP6AP1*;12 genes encoding the V_1 domain of *V-ATPase: ATP6V1H, ATP6V1G3, ATP6V1G2, ATP6V1G1, ATP6V1F, ATP6V1E2, ATP6V1E1, ATP6V1D, ATP6V1C2, ATP6V1C1, ATP6V1B1, ATP6V1A*;two genes for *mTORC1: GBL* and *FRAP1*; and five genes for *NHE* anti-porters: *SLC9A8, SLC9A7, SLC9A3R1, SLC9A3, SLC9A2*. Each entry is the log-ratio between a gene's expression levels in cancer and the matching control, averaged across all the samples for each cancer type, with *gray* representing down-regulation, white for no change and black for up-regulation as defined in the side bar on the right side of the figure. The detailed information about the datasets used here is given in the Appendix. Adapted from (Xu et al. 2013)

up-regulated in breast, colon, lung and ovarian cancer (Ganapathy et al. 2009; Pinheiro et al. 2010). Prostate cancer is the only exception here, as none of the four transporter genes were up-regulated, suggesting that the intracellular pH in prostate cancer may not be particularly low and hence the relevant genes are not needed. This is consistent with the fact that prostate cancer tends to use lipids as the major nutrient and thus may not generate as many protons as the other five cancer types do, reducing the need for the removal of excess protons (Liu et al. 2008).

SLC16A1 and *SLC16A3* are known to be partially regulated by intracellular hypoxia (Xu et al. 2013). In addition, hyaluronic acid has been found capable of regulating these genes (Slomiany et al. 2009). While these two conditions tend to be satisfied in neoplastic cells, neither of them will be met in normal cells in general, providing one possible explanation of why these genes are up-regulated only in cancer cells. In addition to the above proton exporters, other genes have also been examined that may be involved in the removal or neutralization of protons in cancer cells across multiple cancer types. Details follow.

Activation of V-ATPase: ATPases, transmembrane bidirectional transporters, can import many of the metabolites necessary for cellular metabolism and export toxins and wastes that can be detrimental to the health of the host cells (Perez-Sayans et al. 2009). *V-ATPase* is an ATPase that can transport solutes extracellularly and is fueled by ATP hydrolysis. For each proton it pumps out, it brings in one cation such as Na^+ or K^+ to maintain the electro-neutrality, and is found to be up-regulated in numerous cancer types (see Fig. 8.1). From Fig. 8.1, one can see that multiple *V-ATPase* genes are up-regulated in five cancer types except for prostate cancer, which is consistent with the earlier discussion regarding the *SLC16* genes in prostate cancers, suggesting that *V-ATPase* is being used to remove the excess protons in the five cancer types.

Now the question is: *What regulates V-ATPase?* A literature search revealed that *V-ATPase* can be regulated by *mTORC*, a key cell-growth regulator (Pena-Llopis et al. 2011). By examining the two genes encoding *mTORC1*, namely *GBL* and *FRAP1*, one can see that both genes are up-regulated in five out of six cancer types except for colon cancer (see Figure 8.1). Hence it is reasonable to speculate that it is the combined effect of decreased pH and up-regulation of *mTORC1* that activates *V-ATPase* to pump out the excess protons while normal cells, which are not proliferating and hence have no up-regulated *mTORC*-encoding genes, will not survive the increased acidity level.

Na⁺-H⁺ exchanger (NHE): *NHE* anti-porters represent another class of proteins that can transport protons out and exchange each for a cation to maintain intracellular electro-neutrality. They have an important role in the regulation of intracellular pH (Mahnensmith and Aronson 1985). An examination of the five genes encoding this group of transporters (see Fig. 8.1) found that these genes are highly up-regulated in the two lung cancer types among the six under consideration. Knowing that its main function is related to sodium homeostasis, one may speculate that the *NHE* system may be used here as a backup for removing protons, possibly under emergency conditions. It is interesting that the expression patterns are highly complementary between *NHE* and *V-ATPase* in five out of the six cancer types as shown in Fig. 8.1, suggesting that the *NHE* anti-porters may play a complementary role to that of the *V-ATPases* through coordinated regulation by an unknown mechanism.

The *NHE* genes have been found to be regulated by both growth factors and pH, among a few other factors (Donowitz and Li 2007). Again, the triggering condition of the system partially explains why it is more active in cancer than in normal cells. Based on the above results, one can expect that carefully-designed computational analyses of transcriptomic data of multiple cancer types may reveal detailed regulatory mechanisms regarding why some cancers use the *NHE* system while others use the *V-ATPase* system to export protons.

Carbonic anhydrases are important in pH neutralization in cancer cells: It has been previously suggested that carbonic anhydrases (*CAs*) have a role in neutralizing protons in cancer cells. For example, a model for proton export via the membrane-associated *CAs* has been reported (Swietach et al. 2007). Basically, the membrane-bound *CAs* reversibly catalyze the otherwise slow reaction from $CO_2 + H_2O$ to H_2CO_3 (carbonic acid), which dissociates into HCO_3^- (bicarbonate) and H^+ in an acidic extracellular environment, as detailed by the following.

$$HCO_3^- + H^+ \rightleftharpoons H_2CO_3 \rightleftharpoons CO_2 + H_2O$$

HCO_3^- is then imported across the membrane via an *NBC* transporter (Johnson and Casey 2009), where it reacts with a H^+ to form CO_2 and H_2O. Note that the membrane-permeability of CO_2 is regulated by the ratio between the content of membrane cholesterol and that of phospholipid (Itel et al. 2012), specifically the higher the ratio, the lower the permeability. As discussed in Chap. 11, cells in hypoxic conditions tend to have a lower membrane cholesterol content, hence making the membrane more CO_2 permeable (see Chap. 11). Thus, under hypoxic conditions, the high CO_2 efflux facilitates a cycle for removing excess H^+.

This model has been checked against gene-expression data of the six cancer types under consideration. Figure 8.2 shows expression changes of the relevant genes between cancer and the matching normal tissue. From the figure, one can see that: (1) at least one membrane-associated *CA* is up-regulated in five out of six cancer types (except for prostate cancer); and (2) at least one *NBC* gene is up-regulated in five cancer types. Interestingly the cytosolic *CAs* are generally down-regulated in all cancer tissues but prostate cancer. This observation indicates that (1) oxidative phosphorylation is not being used as actively as in five cancer types *versus* their matching normal tissues, hence producing less CO_2 in cytoplasmic matrices of cancer cells *versus* normal cells; and (2) prostate cancer relies on oxidative phosphorylation in its glucose metabolism.

Knowing that *CA9* and *CA12* are hypoxia-inducible in brain cancer (Proescholdt et al. 2005) and that the *NBC* genes are pH inducible (Chiche et al. 2010),one is tempted to predict that under hypoxic conditions with low pH, the membrane-associated *CA* genes and the *NBC* genes are activated to accomplish the functions of the above model.

Neutralization of acidity through decarboxylation reactions: A novel mechanism? Our recent study has led to the proposal of a novel de-acidification mechanism

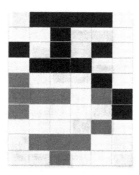

Fig. 8.2 Expression-levels of genes involved in carbonic anhydrases (*CAs*) pH regulation in six cancer types, each represented by one column: breast, colon, liver, lung adenocarcinoma, squamous cell lung cancer, and prostate cancer (from *left* to *right*). 10 genes are included here, with each row representing one gene from top down: three membrane-associated *CAs*: *CA9, CA14, CA12*;three *NBC*-transporter genes: *SLC4A7, SLC4A5, SLC4A4*; and four cytosolic *CAs*: *CA7, CA3, CA2, CA13*. Each entry is defined similarly to that in Fig. 8.1. Adapted from (Xu et al. 2013)

in cancer cells (Xu et al. 2013), based on a similar mechanism used in *Lactococcus lactis* for releasing lactic acid that the bacteria produce to the extracellular space. It has been reported that bacteria use the glutamate decarboxylases (*GAD*) to consume one (dissociable) H^+ during the decarboxylation reaction that *GAD* catalyzes (Cotter and Hill 2003) as presented, which converts glutamate to γ-aminobutyrate (GABA) plus carbon dioxide.

$$^-OOC–CH_2\_CH_2\_CH(NH_3^+)–COO^- + H^+ \rightarrow CO_2 + {}^-OOC–CH_2–CH_2–CH_2–NH^{3+}$$

Two human homologues of *GAD*(glutamate decarboxylase), *GAD1* and *GAD2*, have been found. A previous study has demonstrated that the activation of *GAD* leads to the synthesis of GABA in human brain (Hyde et al. 2011), suggesting that the human *GAD* genes have (or cover) the same function as the *GAD* gene in *Lactococcus lactis*, i.e., catalyzing the GABA-synthesis reaction.

Data analyses on the six cancer types have further revealed that *GAD1* is upregulated in three of the six cancer types, namely colon, liver and lung adenocarcinoma, and *GAD2* is up-regulated in prostate cancer. In addition, the concentration of glutamate, the main substrate of the above reaction catalyzed by *GAD*, is generally elevated in cancer (DeBerardinis et al. 2008). Hence, one can posit that the above reaction takes place in human cancer with the following supporting evidence, as depicted in Fig. 8.3: (a) at least one importer of glutamate is up-regulated in all six cancer types; and (b) at least one gene encoding the GABA exporter is up-regulated in four out of the six cancer types, indicating that the GABA is not used by cancer cells in general, but instead serves only as a vehicle for H^+ removal.

To elucidate what may have triggered the activation of the *GAD* genes, one can search the database for transcription regulations, Cscan (Zambelli et al. 2012), to identify ones for *GAD*. The search suggests that *GAD* is regulated by *FOS* (FBJ murine osteosarcoma viral oncogene homolog), a known oncogene (Wang et al. 2003). Analyses of transcriptomic data in the ENCODE database (Rosenbloom

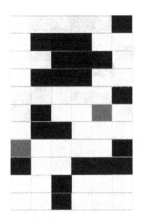

Fig. 8.3 Expression-levels of genes involved in the conversion of glutamate to GABA and CO_2, along with the genes encoding the GABA transporters in six cancer types, each represented as one column from *left* to *right*: breast, colon, liver, lung adenocarcinoma, squamous cell lung cancer, and prostate cancer. 11 genes are included in this figure, with each row representing one gene from top down: two *GAD* genes: *GAD2, GAD1*; four GABA transporter genes: *SLC6A8, SLC6A6, SLC6A11, SLC6A1*; and five glutamate transporter genes: *SLC1A7, SLC1A6, SLC1A4, SLC1A3, SLC1A2*. Each entry is defined similarly to that in Fig. 8.1. Adapted from (Xu et al. 2013)

et al. 2012) indicate that the expression of *GAD1* is positively correlated with that of *FOS* in the HUVEC cell-line. Based on this information, one can predict that *FOS*, in conjunction with some pH-associated regulator, controls *GAD*. This then would lead to the activation of GABA synthesis and remove one H⁺ for each GABA molecule synthesized, with the unneeded GABAs ultimately exported by the up-regulated GABA exporters from the cells. A more carefully designed analysis will likely provide more convincing data to support this novel possibility of a de-acidification mechanism used by cancers; further studies may lead to the identification of the pH-related regulatory factors responsible for *GAD* up-regulation in cancer cells.

A model for cancer to maintain intracellular pH in a normal range: A model for de-acidification is depicted in Fig. 8.4 based on the above discussion, which consists of six possible mechanisms for exporting or neutralizing excess protons. It is worth emphasizing that hypoxia and growth factors, in conjunction with the pH level, may serve as the main regulators of the de-acidification processes, thus making them available to cancer cells but not to normal cells in general. It is also notable that when dealing with acidity, cancer cells do not have to acquire new capabilities as all those genes and systems discussed above are encoded in normal cells. These capabilities are utilized by cancer cells, but not normal cells, simply because cancer cells have created the *appropriate* conditions that can trigger them, probably selected to give cancer cells a survival advantage.

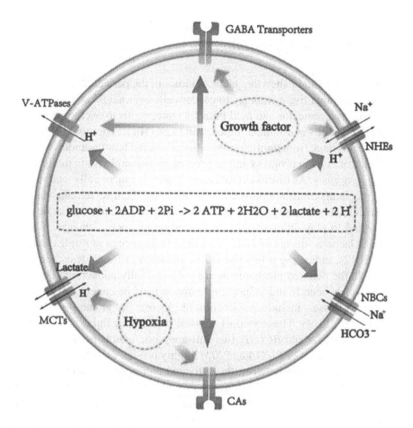

Fig. 8.4 A model for de-acidification in cancer cells using six sets of transporters, each represented as one component along the cell membrane (the *outer circle*), along with the triggering conditions for their activation, depicted using *arrows*. The intermediate, carbonic acid, in the reversible enzymatic reaction is not shown. Also, NAD$^+$ and NADH are omitted in the reaction shown for glycolysis

This model can be expanded to include a number of transcriptional regulators, tying the effector transporters and enzymes to a number of cancer-related genes. For example, a detailed examination of the above analysis shows that 44 genes are involved in the de-acidification process. Searches of these genes against the Cscan database indicate that 28 out of the 44 genes are predicted to be regulated by nine proto-oncogenes, namely *BCL3* (B-cell lymphoma 3-encoded protein), *ETS1* (protein C-ets-1), *FOS* (FBJ murine osteosarcoma viral oncogene homolog), *JUN*, *MXI1* (*MAX*-interacting protein 1), *MYC*, *PAX5* (paired box protein), *SPI1* (transcription factor PU.1) and *TAL1* (T-cell acute lymphocytic leukemia protein 1), and 17 genes are regulated by two tumor-suppressor genes, *IRF1* and *BRCA1*, suggesting that there is a strong connection between de-acidification and cancer growth.

8.2.2   Lactic Acidosis Facilitates Cancer Cells to Grow and Become More Malignant

Recent studies have established that lactic acidosis in the pericellular space does not only provide a competitive advantage to cancer cells over neighboring normal cells, but also serves as a protector and facilitator for cancer cells to overcome some challenges that the hostile environment imposes on them (Hirschhaeuser et al. 2011). We now examine this issue from multiple angles to understand how an acidic environment facilitates cancer cells to progress and to become increasingly more malignant.

Evading apoptosis: As discussed in Chaps. 1 and 7, cancer cells tend to lose the triggering mechanisms to apoptosis, hence allowing them to remain viable even when these cells have accumulated large numbers of abnormalities. Lactic acidosis has been found to play an essential role in facilitating such a change. It has been observed that lactic acidosis correlates with drug resistance in multiple cancer types (Wu et al. 2012), indicating a functional link between lactic acidosis and cell survival although the detailed mechanisms are yet to be fully understood. In addition, lactic acidosis has been found to have protective effects on cancer cells under multiple cytotoxic stresses, including starvation of glucose and glutamine (Ryder et al. 2012a; Wu et al. 2012). These studies showed that extracellular acidosis induces up-regulation of *BCL2* and *BCLXL*, two anti-apoptotic members of the *BCL2* family, and down-regulation of *PUMA* and *BIM*, highly pro-apoptotic relatives of *BCL2*. Hence, the increased ratio between the activated anti-apoptotic and pro-apoptotic members (or relatives) of the *BCL2* family provides a protection against apoptosis activation (Gross et al. 1999), where the acid-sensing enzyme *GPR65* (G-protein coupled receptor 65) serves as the mediator of the observed up-regulation of *BCL2* and *BCLXL* (Ryder et al. 2012b).

Becoming increasingly more malignant: As has been widely observed, extracellular lactic acidosis can enhance tumorigenesis. For example, lactic acidosis was recently found to directly reduce necrosis-induced cell death and moderately increase ROS production (Riemann et al. 2011), with the latter tending to be associated with increased survival (Trachootham et al. 2009). In addition, lactic acidosis can also trigger the activation of *ERK1-2* and *P38 MAP* kinases. Together they have essential roles in the survival of cancer cells. While the latter can phosphorylate *BCL2*, thus protecting the cells against apoptosis (Ruvolo et al. 2001), the former can increase the expressions of *BCL2*, *BCLXL* and *MCL1*, as well as stabilize the *MCL1* protein (Balmanno and Cook 2009). Furthermore, acidosis has been found to increase phosphorylation of transcription factor *CREB* via *P38*, suggesting its possible role in promoting cell proliferation, since inhibition of *P38/CREB* phosphorylation has been shown to have anti-proliferative effects. Overall, this study has demonstrated that acidosis leads to enhanced cell survival and that the *P38/CREB-mediated* transcriptional program can have lasting effects, which remain even after the cells leave the tumor environment (Riemann et al. 2011).

Facilitating invasiveness and metastasis: Degradation of the ECM represents the first step in tumor cell invasion and metastasis (see Chap. 10 for details). An acidic environment will promote this process in multiple ways, including the activation of

*MMP*s, the enzymes for matrix degradation. Specifically, it has been reported that low pH facilitates the redistribution of the activated *CTSB* (cathepsin B), a lysosomal aspartic proteinase, to the surfaces of malignant cells (Rozhin et al. 1994); and acid-activated Cathepsin L can promote amplification of the proteinase cascade through the activation of *UPA* (Goretzki et al. 1992), which is known to promote the conversion of *MMP*s to their active forms. It is noteworthy that another function of activated *MMP*s is to release various growth factors from the fragmented ECMs (Nagase et al. 2006).

Reducing the effectiveness of immune cells: Multiple mechanisms have been identified regarding how lactic acidity weakens immune responses to cancer development. The best understood mechanism is that of lactic acid directly interrupting the normal functions of T-cells (Fischer et al. 2007). Specifically, activated T-cells use glycolytic fermentation instead of oxygen respiration for ATP production, and normal functions of T-cells depend on the efficient removal of their glycolytic products, particularly lactic acid, into the extracellular space. Consequently a high concentration of lactic acid in the pericellular space will form a (partial) blockade of the lactate transporter *MCT1*, hence resulting in impaired T-cell function.

Another affected mechanism is related to the reduced responses by the natural killer (NK) cells of the innate immune system. Lactic acid can inhibit NK cell function via direct inhibition of their cytolytic function and by indirectly increasing the number of myeloid-derived suppressor cells that together inhibit NK cytotoxicity to cancer cells (Husain et al. 2013).

Overall, through their rapid evolution, cancer cells have learned to create an acidic microenvironment to facilitate them to thrive and expand, and to promote cell death in their non-cancerous neighboring cells. This environment also provides other factors for them to develop angiogenesis and the mobility to migrate (Hirschhaeuser et al. 2011). For example, a study has reported that membrane vesicles shed by cancer cells into the extracellular space contain *VEGF* and two different types of *MMP*s, and that the acidic environmental pH can lead to vesicle rupture, hence releasing these pro-angiogenesis factors that prompt angiogenesis (Taraboletti et al. 2006).

8.3 Cancer Development Under Immune Surveillance

The relationship between the immune system and cancer development is a very complex one, quite different from what one may intuitively speculate: <u>a healthy immune system will prevent cancer from happening</u>. While the current understanding about immunity and cancer is far from being complete, it is generally understood that different components of the immune system play different roles throughout cancer development, including both anti- and pro-cancer activities. One aspect is probably true: <u>without immune responses such as inflammation, most cancers will not develop</u>, knowing the deep connections between the immune system and the ECM (de Visser et al. 2006; Sorokin 2010; Lu et al. 2012). In addition, immune responses may have different roles at different stages during cancer development, including inflammatory responses in the beginning of tumorigenesis and providing growth factors by tumor-associated macrophages (*TAM*s) throughout cancer development.

A brief introduction to the human immune system: The human immune system consists of two layers, with some overlap, of defense mechanisms: the *innate immune* system and the *adaptive immune* system. The former is a non-specific and fast reacting system that can detect and destroy non-self cells, e.g., bacteria and viruses, and the latter can adapt its responses during the invasion of non-self pathogens with a more specific detection and elimination mechanism of the invading organism. The innate response is triggered when pathogens or altered self, sometimes referred to as *self pathogen* in the literature, are detected. Both systems can be triggered by inflammatory signals, the first response to foreign invasions or to disruption of tissue homeostasis (see Chap. 6), and by each other (Alberts et al. 2002).

The innate immune system consists of: (a) the inflammatory signals, (b) the complement pathway that attacks the surfaces of non-self pathogens and (c) multiple types of immune cells such as macrophages, neutrophils, dendritic cells, mast cells, eosinophils, basophils and natural killer cells. Each of these cell types serves distinct purposes and is often located in different sites, which together identify and eliminate non-self pathogens. The complement pathway serves as the command center of the innate immune system, with its key functions being to: (1) signal for inflammatory cells; (2) label pathogens for destruction by other immune cells; and (3) directly attack the cell membrane of pathogens. Another component of the innate system acts by engulfing and destroying invading organisms.

The adaptive immune system is comprised of lymphocytes, the majority being B-cells and T-cells, both of which carry receptor molecules that recognize specific targets. T-cells recognize a non-self-pathogen only after its cell-surface antigens (molecular fragments of the pathogen) have been processed and presented. Using membrane-associated immunoglobulins, B-cells bind intact antigens (e.g., from pathogens), then internalize and process (degrade) them before presenting the derived peptides via MHC (major histocompatibility complex) II to helper T cells. If the antigen is deemed foreign, the T cells secrete cytokines leading to a clonal expansion of the B cells producing the appropriate antibody. Gene switching then leads to secreted antibodies capable of binding intact pathogens and orchestrating their destruction. The main difference here is that the adaptive immune system retains an immunological memory through its B cells after its initial response to each pathogen type, thus allowing the system to have tailored a more effective response to individual pathogens.

Cancer development is essentially rooted in the overlap between tissue maintenance and immune responses for the following reasons: (1) immune systems respond to two types of signals, invasion of foreign pathogens and disruption of tissue homeostasis by recruitment of immune cells to the relevant sites; (2) signals reporting tissue damage are designed to come from damaged ECMs, but can also be generated directly from cells under persistent hypoxic conditions; (3) persistent low-grade immune responses, namely inflammation, will lead to persistent hypoxia; and (4) continuous interactions between innate and adaptive immune cells will lead to loss of tissue architecture (de Visser et al. 2006). By integrating this information with the information offered in Chaps. 5 and 6, one can see that: (1) persistent inflammation contributes to the initiation of a cancer; (2) the loss of tissue architecture induced by persistent interactions between innate and adaptive immune cells will lead to wound over-healing (de Visser et al. 2006), a characteristic of neo-plasms; and (3) the co-evolutionary relationships between the neoplastic cells and their pericellular innate and adaptive immune cells start from the very beginning, which must have contributed to the formation of the future conjugative relations between cancer cells and some immune cells such as TAMs (see below).

Knowing all this, one has to ask about the validity of cancer studies carried out on immune-deficient mouse models, particularly for fundamental mechanism stud-ies of cancer or drug treatments of cancer. This has been a common practice mainly to overcome the issue that active immune systems of mice will kill implanted for-eign cells but at the expense of losing a key, actually indispensable, ingredient in cancer development.

Furthermore, it is worth noting that while activated immune cells attack "patho-gens" based on their recognition of the non-self surface signals, cancer cells are self-cells, only altered. Consequently, it is more challenging for the (fresh) immune cells to recognize and attack these self-pathogens than foreign invaders such as bacteria or viruses. Even if the self-pathogens have distinct molecules on their sur-faces, they tend to be only weakly immunogenic, which may be the result of selec-tion by the immune system. Various cancer immunity studies using mouse models have found that T-cells tend to either ignore such weak signals or not attack them aggressively (Houghton and Guevara-Patino 2004). These studies have raised a number of questions about cancer immunity such as:(a) *Which types of the cancer-related genetic alterations may have their products recognized by a competent immune system?* And (b) *Why are some alterations more detectable by the immune system than others?* Answers to these questions may shed new light on our under-standing of cancer immunity.

Cancer immunity: A yet to be understood system: The hypothesis of cancer immuno-surveillance was initially proposed in 1982 by Lewis Thomas, which states that a human immune system has the capability to recognize and destroy nascent transformed cells (Thomas 1982). Since then the hypothesis has evolved to one much richer, now referred to as the *cancer immunoediting* system, which consists of three phases: *elimination, equilibrium* and *escape* (Dunn et al. 2002; Kim et al. 2007). Specifically, the elimination phase suggests that a competent immune system can recognize antigens associated with neoplastic cells, and then attack and elimi-nate them. The following molecules are believed to serve as "antigen" signals for neoplastic cells: uric acid, pro-inflammatory signals such as *TLR* (toll-like receptor) ligands, heat shock proteins and hyaluronic acid chains. γδ T-cells, αβ T-cells and killer T-cells form the basic components of the antitumor defense mechanism that can detect cancer-specific antigens. For example, CD4$^+$ and CD8$^+$ αβ T-cells have been found to serve such roles in mouse (Yusuf et al. 2008). The type II interferon, *IFN*γ, represents a key arsenal for attacking and killing the neoplastic cells by these T-cells (Dighe et al. 1994).

During the elimination phase, not all neoplastic cells are necessarily detected and eliminated. Then, interactions between cancer cells and immunity move to the equi-librium and escape phases. There are documented studies showing the possible existence of the hypothesized equilibrium phase when the surviving neoplastic cells remain inactive for an extended period of time. For example, two kidney-transplant recipients from the same donor both developed melanoma within 2 years of receiving the kidney, neither of whom had a history of cancer. Further investigation revealed that the donor was treated for primary melanoma 16 years before, and, at the time of donation, the donor was believed to be cancer free (MacKie et al. 2003).

During the equilibrium phase, new neoplastic cell populations may be generated if the initial elimination phase did not destroy all the neoplastic cells. The new cell populations are selected by the genetic disposition of the patient, particularly their immune systems, and hence may have the makeup for escaping immune detection. One can further carry out hypothesis-driven analyses of the available *omic* data in the TCGA database (The Cancer Genome Atlas Data Portal, 2010) to assess the plausibility and generality of the existence of such an equilibrium phase during cancer development across different types of cancers.

The pro-cancerous roles played by the immune system: It has been established that increased infiltrating lymphocytes, including both T-cells and B-cells, in cancer tissues generally correlate with favorable prognosis while, in contrast, increased infiltrating innate immune cells, such as macrophages and mast cells, correlate with increased tumor angiogenesis and poor prognosis (de Visser et al. 2006; Dirkx et al. 2006). Currently the detailed mechanisms are not well understood. Again, data mining and statistical inference could have important ramifications in identifying pathways and gene groups with distinct expression patterns between these two classes of cancer samples, and hence potentially provide a deeper understanding and new insights about the complex relationships between cancer and their infiltrating immune cells.

It has been well established that, when tissue homeostasis is disrupted, macrophages and mast cells are first recruited, and they, in turn, release multiple types of signals to recruit innate immune cells for tissue repair. The activation of the innate immune system will then lead to the activation of the adaptive immune system, which generally requires the creation of a pro-inflammatory environment (Charles 2001), indicating interactions between the two immune systems. As discussed earlier, this, if continues persistently, will lead to cycles of excessive tissue remodeling and repair, as well as loss of tissue architecture (de Visser et al. 2006). This, of course, increases the opportunity for cancer development.

In addition to the pro-cancer roles played by the innate immune system, published studies have also found that B-cells are essential for maintaining chronic inflammation (Hamel et al. 2008),known to be associated with pre-malignant progression. Therefore, potentially B-cells may also have indirect pro-cancer roles during cancer development.

Overall, while the detailed mechanisms and impact of adaptive immunity in cancer development are rather complex and not fully understood, chronic activation of the innate immune cells contributes to cancer development (de Visser et al. 2006). Multiple types of chronically activated immune cells therefore exert cancer-promoting effects directly by influencing proliferation and survival of neoplastic cells, as well as by indirectly modulating neoplastic microenvironments to favor cancer progression.

8.4   Elucidating Detailed Relationships Between Immunity and Cancer Evolution

Immunity seems to be most essential in cancer development based on the information provided above. Some important interactive relationships between immune responses and cancer development have been established. These include the conjugative relationship between cancer cells and TAMs (see Fig. 8.5) and immune cells serving a selection mechanism for more robust cancer cells. However, many detailed relationships between the two are yet to be elucidated. For example, a few questions about their relationships are listed below, which could benefit from well-designed computational data mining and associated statistical inference:

– It is known that multiple types of immune cells can infiltrate cancer tissues. One question is: *Can one possibly estimate the composition of the infiltrating immune cells, in terms of their types and population sizes in a tissue sample, based on the available transcriptomic data?* Along this line of questioning, one can further ask: *How do the sub-populations of different infiltrating immune cells change during the course of cancer evolution? Does the composition of such a cell popu-*

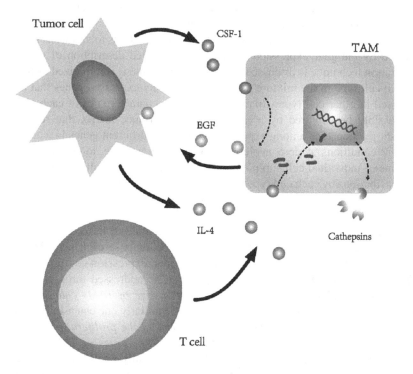

Fig. 8.5 A schematic illustration of the relationships between cancer cells and TAMs, where *arrows* showing the feeding relationships among cancer cells, TAMs and T-cells. The figure is adapted from (Wang and Joyce 2010)

lation at the early stage of a cancer have any predictive power for the cancer's evolutionary trajectory or 5-year survival rate? Or more specifically, *are there relationships between the composition of the infiltrating cell population and the clinical phenotype of a cancer?*

– Macrophages probably represent the most abundant immune-cell infiltrates into cancer tissues (Bingle et al. 2002). It is natural to ask: *How do the TAMs co-evolve with the cancer cells?* More specifically, *what are the common differences between TAMs across different cancer types and normal macrophages?* In the conjugative relationship between TAMs and cancer cells, *what other signals/ nutrients do these two types of cells exchange with each other in addition to the ones depicted in Fig. 8.5?*

– The same types of questions can also be asked about other types of infiltrating immune cells, including both the innate and the adaptive immune cells that have been detected in a cancer microenvironment.

Clearly it is very challenging to address these questions experimentally, not to mention the amount of time and expense needed. In contrast, computational techniques provide a unique and powerful approach in generating highly useful information and addressing some of these questions when the relevant transcriptomic data are available. A key tool needed is a computational capability for de-convoluting the transcriptomic data collected on a cancer tissue sample to contributions from individual cell types present in the cancer tissue, which will allow one to see the detailed sub-populations of different cell types and their contributions to various activities as reflected by the gene-expression data. While a detailed description of such a de-convolution technique is discussed in Chap. 2, an outline is summarized as follows.

Each cell type in the mixture has its own uniquely expressed genes that are not expressed in other cell types. In addition, the key characteristics of each cell type can possibly be captured by a (generalized) covariance matrix that reflects the invariant relationships among expression patterns of different genes in a cell under different conditions. A de-convolution process is, in essence, to determine (1) the fractions of individual cell sub-populations in a mixture of multiple cell types and (2) contributions by each cell type to the gene-expression data collected on the mixture, which are most consistent with the covariance matrix for each cell type derived based on clean cell line-data collected under multiple conditions, as well as with the estimated fractions of individual cell types in the collected tissue-cell population. Using a de-convolution tool like this, one can decompose each gene-expression dataset collected on cancer tissue samples into gene-expression contributions from different cell types. Example questions that can be addressed through such data analyses may include: *How does the population size of TAMs change over the course of development of a specific cancer type?* Similarly one can ask: *Is there any correlation between the immune cell population and the proliferation rate of the associated cancer?* Or *how do the activated functionalities of TAMs or T-cells change during the course of cancer evolution?* We fully expect that such studies could lead to fundamentally novel information about cancer initiation, development and metastasis.

8.5 Concluding Remarks

In order to understand the clinical phenotype and the evolutionary trajectory of a cancer, it is essential to study the interplay between cancer cells and other cell types in their pericellular environment such as immune cells *in vivo*, particularly along the developmental axis. Such investigations will allow one to examine the problem of cancer development in a more realistic setting since continuous and complex interactions between the neoplastic cells and their neighboring cells, including immune cells, probably define the key characteristics of a cancer. While traditional cancer studies in a controlled and isolated setting have clearly generated a great amount of information about the detailed molecular and cellular level mechanisms, they tend to suffer from being overly simplistic and from missing key information associated with essential interactions between cancer cells and their co-evolving microenvironments and the interaction results. Computational data analyses of cancer tissues allow one to integrate the rich molecular level, some cellular level and tissue level information with the enormously large pool of cancer *omic* data. Such an approach has the potential to identify key and subtle relationships among different players and provide highly informative models to guide experimental design for further studies in a highly rationalized manner. It is anticipated that tightly integrated studies between computational and experimental approaches will prove to be most effective in tackling complex cancer systems biology problems, hence possibly revealing the complex roles played by the two classes of immune cells in cancer development, as well as defining the key characteristics of individual cancers. An improved understanding about the relationships between cancer and immune responses has considerable potential in improving capabilities for cancer treatment.

Appendix

The gene-expression data for the six cancer types, (breast, colon, liver, lung adenocarcinoma, squamous cell lung, prostate), were downloaded from the GEO database (Edgar et al. 2002) of the NCBI. For each cancer type, the following criteria were applied in selecting the dataset used for this study: (1) all the data in each dataset were generated using the same platform by the same research group; (2) each dataset consists of only paired samples, i.e., cancer tissue sample and the matching adjacent noncancerous tissue sample; and (3) each dataset has at least ten pairs of samples. In the GEO database, only six cancer types have datasets satisfying these criteria. A summary of the 12 datasets, 2 sets for each cancer, is listed in the following table.

Table 8.1 Gene expression data for six cancer types used in transcriptomic data analysis in Sect. 8.2

Cancer type	Dataset 1	Dataset 2	Pairs
Breast cancer	GSE14999 (Uva et al. 2009)	GSE15852 (Pau Ni et al. 2010)	61/43
Colon cancer	GSE18105 (Matsuyama et al. 2010)	GSE25070 (Hinoue et al. 2012)	17/26
Liver cancer	GSE22058 (Burchard et al. 2010)	GSE25097 (Tung et al. 2011)	97/238
Lung adenocarcinoma	GSE31552 (Tan et al. 2012)	GSE7670 (Su et al. 2007)	31/26
Lung squamous cell carcinoma	GSE31446 (Hudson et al. 2010)	GSE31552 (Tan et al. 2012)	13/17
Prostate cancer	GSE21034 (Taylor et al. 2010)	GSE6608 (Chandran et al. 2007)	29/58

References

Alberts B, Johnson A, Lewis J et al. (2002) Molecular Biology of the Cell. 4th edn. New York and London: Garland Science,

Arnold SA, Brekken RA (2009) SPARC: a matricellular regulator of tumorigenesis. Journal of cell communication and signaling 3: 255-273

Baker NE (2011) Cell competition. Current Biology 21: R11-R15

Balmanno K, Cook SJ (2009) Tumour cell survival signalling by the ERK1/2 pathway. Cell death and differentiation 16: 368-377

Bingle L, Brown NJ, Lewis CE (2002) The role of tumour-associated macrophages in tumour progression: implications for new anticancer therapies. J Pathol 196: 254-265

Bondar T, Medzhitov R (2010a) p53-mediated hematopoietic stem and progenitor cell competition. Cell stem cell 6: 309-322

Bondar T, Medzhitov R (2010b) p53-mediated hematopoietic stem and progenitor cell competition. Cell Stem Cell 6: 309-322

Burchard J, Zhang C, Liu AM et al. (2010) microRNA-122 as a regulator of mitochondrial metabolic gene network in hepatocellular carcinoma. Mol Syst Biol 6: 402

Calorini L, Peppicelli S, Bianchini F (2012) Extracellular acidity as favouring factor of tumor progression and metastatic dissemination. Experimental oncology 34: 79-84

Casey JR, Grinstein S, Orlowski J (2010) Sensors and regulators of intracellular pH. Nat Rev Mol Cell Biol 11: 50-61

Chandran UR, Ma C, Dhir R et al. (2007) Gene expression profiles of prostate cancer reveal involvement of multiple molecular pathways in the metastatic process. BMC Cancer 7: 64

Charles AJP, T.; Mark, W.; Mark, J.S. (2001) Immunobiology: The Immune System in Health and Disease. 5th edn. New York: Garland Science,

Chiche J, Brahimi-Horn MC, Pouyssegur J (2010) Tumour hypoxia induces a metabolic shift causing acidosis: a common feature in cancer. J Cell Mol Med 14: 771-794

Cotter PD, Hill C (2003) Surviving the acid test: responses of gram-positive bacteria to low pH. Microbiol Mol Biol Rev 67: 429-453, table of contents

de la Cova C, Abril M, Bellosta P et al. (2004) Drosophila Myc Regulates Organ Size by Inducing Cell Competition. Cell 117: 107-116

de Visser KE, Eichten A, Coussens LM (2006) Paradoxical roles of the immune system during cancer development. Nature reviews Cancer 6: 24-37

DeBerardinis RJ, Lum JJ, Hatzivassiliou G et al. (2008) The biology of cancer: metabolic reprogramming fuels cell growth and proliferation. Cell Metab 7: 11-20

Dighe AS, Richards E, Old LJ et al. (1994) Enhanced in vivo growth and resistance to rejection of tumor cells expressing dominant negative IFN gamma receptors. Immunity 1: 447-456

Dirkx AE, Oude Egbrink MG, Wagstaff J et al. (2006) Monocyte/macrophage infiltration in tumors: modulators of angiogenesis. J Leukoc Biol 80: 1183-1196

Donowitz M, Li X (2007) Regulatory binding partners and complexes of NHE3. Physiol Rev 87: 825-872

Dunn GP, Bruce AT, Ikeda H et al. (2002) Cancer immunoediting: from immunosurveillance to tumor escape. Nat Immunol 3: 991-998

Edgar R, Domrachev M, Lash AE (2002) Gene Expression Omnibus: NCBI gene expression and hybridization array data repository. Nucleic Acids Res 30: 207-210

Fischer K, Hoffmann P, Voelkl S et al. (2007) Inhibitory effect of tumor cell-derived lactic acid on human T cells. Blood 109: 3812-3819

Ganapathy V, Thangaraju M, Prasad PD (2009) Nutrient transporters in cancer: relevance to Warburg hypothesis and beyond. Pharmacol Ther 121: 29-40

Goretzki L, Schmitt M, Mann K et al. (1992) Effective activation of the proenzyme form of the urokinase-type plasminogen activator (pro-uPA) by the cysteine protease cathepsin L. FEBS Lett 297: 112-118

Gross A, McDonnell JM, Korsmeyer SJ (1999) BCL-2 family members and the mitochondria in apoptosis. Genes & development 13: 1899-1911

Halestrap AP (2012) The monocarboxylate transporter family–Structure and functional characterization. IUBMB Life 64: 1-9

Halestrap AP, Price NT (1999) The proton-linked monocarboxylate transporter (MCT) family: structure, function and regulation. Biochem J 343 Pt 2: 281-299

Hamel K, Doodes P, Cao Y et al. (2008) Suppression of Proteoglycan-Induced Arthritis by Anti-CD20 B Cell Depletion Therapy Is Mediated by Reduction in Autoantibodies and CD4+ T Cell Reactivity. The Journal of Immunology 180: 4994-5003

Hinoue T, Weisenberger DJ, Lange CP et al. (2012) Genome-scale analysis of aberrant DNA methylation in colorectal cancer. Genome Res 22: 271-282

Hirschhaeuser F, Sattler UG, Mueller-Klieser W (2011) Lactate: a metabolic key player in cancer. Cancer research 71: 6921-6925

Houghton AN, Guevara-Patino JA (2004) Immune recognition of self in immunity against cancer. The Journal of clinical investigation 114: 468-471

Hudson TJ, Anderson W, Artez A et al. (2010) International network of cancer genome projects. Nature 464: 993-998

Husain Z, Huang Y, Seth P et al. (2013) Tumor-derived lactate modifies antitumor immune response: effect on myeloid-derived suppressor cells and NK cells. Journal of immunology 191: 1486-1495

Hyde TM, Lipska BK, Ali T et al. (2011) Expression of GABA signaling molecules KCC2, NKCC1, and GAD1 in cortical development and schizophrenia. J Neurosci 31: 11088-11095

Itel F, Al-Samir S, Oberg F et al. (2012) CO2 permeability of cell membranes is regulated by membrane cholesterol and protein gas channels. FASEB J 26: 5182-5191

Johnson DE, Casey JR (2009) Bicarbonate Transport Metabolons. In: Supuran CT, Winum JY (eds) Drug Design of Zinc-Enzyme Inhibitors: Functional, Structural, and Disease Applications. Wiley, Hoboken, New Jersey, pp 415–437

Johnston LA (2009) Competitive interactions between cells: death, growth, and geography. Science 324: 1679-1682

Kim R, Emi M, Tanabe K (2007) Cancer immunoediting from immune surveillance to immune escape. Immunology 121: 1-14

Levayer R, Moreno E (2013) Mechanisms of cell competition: themes and variations. The Journal of cell biology 200: 689-698

Liu Y, Zuckier L, Ghesani N (2008) Fatty acid rather than glucose metabolism is the dominant bioenergetic pathway in prostate cancer. J NUCL MED MEETING ABSTRACTS 49: 104P-a-

Lu P, Weaver VM, Werb Z (2012) The extracellular matrix: a dynamic niche in cancer progression. The Journal of cell biology 196: 395-406

MacKie RM, Reid R, Junor B (2003) Fatal melanoma transferred in a donated kidney 16 years after melanoma surgery. N Engl J Med 348: 567-568

Mahnensmith RL, Aronson PS (1985) The plasma membrane sodium-hydrogen exchanger and its role in physiological and pathophysiological processes. Circulation research 56: 773-788

Matsuyama T, Ishikawa T, Mogushi K et al. (2010) MUC12 mRNA expression is an independent marker of prognosis in stage II and stage III colorectal cancer. Int J Cancer 127: 2292-2299

Mok SC, Chan WY, Wong KK et al. (1996) SPARC, an extracellular matrix protein with tumor-suppressing activity in human ovarian epithelial cells. Oncogene 12: 1895-1901

Moreno E, Basler K (2004) dMyc Transforms Cells into Super-Competitors. Cell 117: 117-129

Nagase H, Visse R, Murphy G (2006) Structure and function of matrix metalloproteinases and TIMPs. Cardiovascular Research 69: 562-573

The Cancer Genome Atlas Data Portal (2010).

Neto-Silva RM, de Beco S, Johnston LA (2010) Evidence for a Growth-Stabilizing Regulatory Feedback Mechanism between Myc and Yorkie, the Drosophila Homolog of Yap. Developmental cell 19: 507-520

Nishi H, Nakada T, Kyo S et al. (2004) Hypoxia-inducible factor 1 mediates upregulation of telomerase (hTERT). Molecular and cellular biology 24: 6076-6083

Pau Ni IB, Zakaria Z, Muhammad R et al. (2010) Gene expression patterns distinguish breast carcinomas from normal breast tissues: the Malaysian context. Pathol Res Pract 206: 223-228

Pena-Llopis S, Vega-Rubin-de-Celis S, Schwartz JC et al. (2011) Regulation of TFEB and V-ATPases by mTORC1. The EMBO journal 30: 3242-3258

Perez-Sayans M, Somoza-Martin JM, Barros-Angueira F et al. (2009) V-ATPase inhibitors and implication in cancer treatment. Cancer Treat Rev 35: 707-713

Petrova E, Lopez-Gay JM, Rhiner C et al. (2012) Flower-deficient mice have reduced susceptibility to skin papilloma formation. Disease models & mechanisms 5: 553-561

Pinheiro C, Reis RM, Ricardo S et al. (2010) Expression of monocarboxylate transporters 1, 2, and 4 in human tumours and their association with CD147 and CD44. J Biomed Biotechnol 2010: 427694

Portela M, Casas-Tinto S, Rhiner C et al. (2010) Drosophila SPARC is a self-protective signal expressed by loser cells during cell competition. Developmental cell 19: 562-573

Prober DA, Edgar BA (2000) Ras1 Promotes Cellular Growth in the<i>Drosophila</i>Wing. Cell 100: 435-446

Proescholdt MA, Mayer C, Kubitza M et al. (2005) Expression of hypoxia-inducible carbonic anhydrases in brain tumors. Neuro Oncol 7: 465-475

Puolakkainen PA, Brekken RA, Muneer S et al. (2004) Enhanced growth of pancreatic tumors in SPARC-null mice is associated with decreased deposition of extracellular matrix and reduced tumor cell apoptosis. Molecular cancer research : MCR 2: 215-224

Rhiner C, Lopez-Gay JM, Soldini D et al. (2010) Flower forms an extracellular code that reveals the fitness of a cell to its neighbors in Drosophila. Developmental cell 18: 985-998

Riemann A, Schneider B, Ihling A et al. (2011) Acidic environment leads to ROS-induced MAPK signaling in cancer cells. PloS one 6: e22445

Rosenbloom KR, Dreszer TR, Long JC et al. (2012) ENCODE whole-genome data in the UCSC Genome Browser: update 2012. Nucleic Acids Res 40: D912-917

Rozhin J, Sameni M, Ziegler G et al. (1994) Pericellular pH affects distribution and secretion of cathepsin B in malignant cells. Cancer research 54: 6517-6525

Ruvolo P, Deng X, May W (2001) Phosphorylation of Bcl2 and regulation of apoptosis. Leukemia (08876924) 15:

Ryder C, McColl K, Zhong F et al. (2012a) Acidosis promotes Bcl-2 family-mediated evasion of apoptosis: involvement of acid-sensing G protein-coupled receptor Gpr65 signaling to Mek/Erk. The Journal of biological chemistry 287: 27863-27875

Ryder C, McColl K, Zhong F et al. (2012b) Acidosis Promotes Bcl-2 Family-mediated Evasion of Apoptosis: INVOLVEMENT OF ACID-SENSING G PROTEIN-COUPLED RECEPTOR GPR65 SIGNALING TO MEK/ERK. Journal of Biological Chemistry 287: 27863-27875

Sansom OJ, Mansergh FC, Evans MJ et al. (2007) Deficiency of SPARC suppresses intestinal tumorigenesis in APCMin/+mice. Gut 56: 1410-1414

Simpson P (1979) Parameters of cell competition in the compartments of the wing disc of Drosophila. Developmental biology 69: 182-193

Simpson P, Morata G (1981) Differential mitotic rates and patterns of growth in compartments in the Drosophila wing. Developmental biology 85: 299-308

Slomiany MG, Grass GD, Robertson AD et al. (2009) Hyaluronan, CD44, and emmprin regulate lactate efflux and membrane localization of monocarboxylate transporters in human breast carcinoma cells. Cancer research 69: 1293-1301

Sorokin L (2010) The impact of the extracellular matrix on inflammation. Nature reviews Immunology 10: 712-723

Su LJ, Chang CW, Wu YC et al. (2007) Selection of DDX5 as a novel internal control for Q-RT-PCR from microarray data using a block bootstrap re-sampling scheme. BMC Genomics 8: 140

Swietach P, Vaughan-Jones RD, Harris AL (2007) Regulation of tumor pH and the role of carbonic anhydrase 9. Cancer Metastasis Rev 26: 299-310

Tamori Y, Deng WM (2011) Cell competition and its implications for development and cancer. Journal of genetics and genomics=Yi chuan xue bao 38: 483-495

Tan XL, Marquardt G, Massimi AB et al. (2012) High-throughput library screening identifies two novel NQO1 inducers in human lung cells. Am J Respir Cell Mol Biol 46: 365-371

Taraboletti G, D'Ascenzo S, Giusti I et al. (2006) Bioavailability of VEGF in tumor-shed vesicles depends on vesicle burst induced by acidic pH. Neoplasia 8: 96-103

Taylor BS, Schultz N, Hieronymus H et al. (2010) Integrative genomic profiling of human prostate cancer. Cancer Cell 18: 11-22

Thomas L (1982) On immunosurveillance in human cancer. Yale J Biol Med 55: 329-333

Trachootham D, Alexandre J, Huang P (2009) Targeting cancer cells by ROS-mediated mechanisms: a radical therapeutic approach? Nature reviews Drug discovery 8: 579-591

Tung EK, Mak CK, Fatima S et al. (2011) Clinicopathological and prognostic significance of serum and tissue Dickkopf-1 levels in human hepatocellular carcinoma. Liver Int 31: 1494-1504

Uva P, Aurisicchio L, Watters J et al. (2009) Comparative expression pathway analysis of human and canine mammary tumors. BMC Genomics 10: 135

Vivarelli S, Wagstaff L, Piddini E (2012) Cell wars: regulation of cell survival and proliferation by cell competition. Essays Biochem 53: 69-82

Wang HW, Joyce JA (2010) Alternative activation of tumor-associated macrophages by IL-4: priming for protumoral functions. Cell cycle 9: 4824-4835

Wang YY, Wu SX, Liu XY et al. (2003) Effects of c-fos antisense oligodeoxynucleotide on 5-HT-induced upregulation of preprodynorphin, preproenkephalin, and glutamic acid decarboxylase mRNA expression in cultured rat spinal dorsal horn neurons. Biochem Biophys Res Commun 309: 631-636

Wu H, Ding Z, Hu D et al. (2012) Central role of lactic acidosis in cancer cell resistance to glucose deprivation-induced cell death. J Pathol 227: 189-199

Xu K, Mao X, Mehta M et al. (2013) Elucidation of how cancer cells avoid acidosis through comparative transcriptomic data analysis. PloS one 8: e71177

Yao CK, Lin YQ, Ly CV et al. (2009) A synaptic vesicle-associated Ca2+ channel promotes endocytosis and couples exocytosis to endocytosis. Cell 138: 947-960

Yusuf N, Nasti TH, Katiyar SK et al. (2008) Antagonistic roles of CD4+ and CD8+ T-cells in 7,12-dimethylbenz(a)anthracene cutaneous carcinogenesis. Cancer research 68: 3924-3930

Zambelli F, Prazzoli GM, Pesole G et al. (2012) Cscan: finding common regulators of a set of genes by using a collection of genome-wide ChIP-seq datasets. Nucleic Acids Res 40: W510-515

Ziosi M, Baena-Lopez LA, Grifoni D et al. (2010) dMyc functions downstream of Yorkie to promote the supercompetitive behavior of hippo pathway mutant cells. PLoS genetics 6: e1001140

Chapter 9
Cell Proliferation from Regulated to Deregulated States Via Epigenomic Responses

It was established in the previous chapters that cells, under chronic conditions of hypoxia and/or ROS accumulation, must evolve for survival to overcome the pressure created by continuous glycolytic-metabolite accumulation plus possibly other pressures. This leads to continuous synthesis and export of hyaluronic acids in the early stage of a cancer development, as observed in many cancers. The fragments of the hyaluronic acid chains released into the pericellular space immediately become signals for inflammation, cell-cycle activation, cell proliferation, cell survival and angiogenesis, all designed for tissue repair except that no tissue is injured here. These molecules are continuously generated as long as the hypoxic or ROS conditions persist, hence providing driving signals for tissue repair on a continuous basis. In contrast, when a tissue is indeed injured, the hyaluronic acid fragments are released from the damaged underlying ECM rather than from the cells directly, hence the signaling will not continue indefinitely.

As the neoplastic cells continue to evolve by changing their metabolism to adapt to the challenging microenvironment, they also continuously change their microenvironment as a side-product of their altered metabolism, making their environment increasingly more stressful and unfamiliar. In time, the environment may become so atypical and stressful that no condition-specific stress responses encoded in the genomes can overcome the conditions, hence putting these cells in a life-or-death situation. Recent studies on epigenomics suggest that cellular systems have a general stress-response mechanism to handle unfamiliar, persistent and severe stressors through epigenomic modifications, possibly as a last resort for survival. The increased utilization of the general stress-responses as observed in cancers reflects the challenging nature of the pressures imposed upon them. While the utilization of such a general stress-response mechanism aids the cells in overcoming the stress encountered, they also direct the cells onto evolutionary trajectories considerably less reversible than those created by condition-specific responses. In such a case the

© Springer Science+Business Media New York 2014
Y. Xu et al., *Cancer Bioinformatics*, DOI 10.1007/978-1-4939-1381-7_9

ensuing evolution becomes increasingly more irreversible and uncontrollable when genomic mutations are selected to replace certain cellular activities accomplishing the same functionalities, possibly driven by sustainability and energetic reasons, leading to increasingly more malignant cells.

9.1   Changing Microenvironments Caused by Evolving Neoplastic Cells

The development of neoplastic cells is the result of cells responding to the stressful microenvironment such as chronic hypoxia, increased ROS and persistent inflammation. Concurrently, the evolving cells also further change the micro- and intracellular environments as side-products of their altered metabolism, which may further drive the evolving cells to become increasingly more malignant, forming a vicious cycle. For example, it is known that cancer cells tend to create an acidic pericellular environment by releasing higher-than-normal quantities of lactic acid as a result of their altered glucose metabolism, initially induced by their hypoxic environment. As a result, the altered microenvironment induces further changes in the cellular metabolism as discussed in Chap. 8. The same can be said about other microenvironmental changes, such as hypoxia-induced angiogenesis leading to increased ROS level in neoplastic cells. Here a systematic examination is presented regarding how the microenvironment changes as a cancer evolves, taking multiple factors into consideration such as (1) hypoxia, (2) ROS, (3) the composition of the pericellular matrix, and (4) the composition of local stromal cells. We will also discuss how cells respond to such changes by altering the activities of various metabolic and other activities, leading to further altered microenvironments. From these analyses, it can be seen that it is the active interplay and co-adaptation between the neoplastic cells and their microenvironments directed by instructions encoded in the genomes that drive the interactions from the originally harmonic to the increasingly more antagonistic and stressful during their co-evolution. The example given in Chap. 5 on the co-adaptation and co-evolution between *Pseudomonas fluorescens* and its viral parasite serves a good framework of thinking when considering the co-evolving and antagonistic relationships between the cells and the microenvironment.

9.1.1 Hypoxia in a Cancer Microenvironment

Previous studies have established a strong correlation between the cellular O_2 level and the survival rate of cancer patients (Vaupel 2008). Specifically, the level of hypoxia plays a vital role in defining the malignancy of a cancer. Expression data of tissue samples of multiple cancer types are examined here in terms of how their hypoxic level changes as a cancer progresses. In Fig. 9.1, the average expression level of the *HIF1*α gene, a most widely used marker gene for hypoxia, was

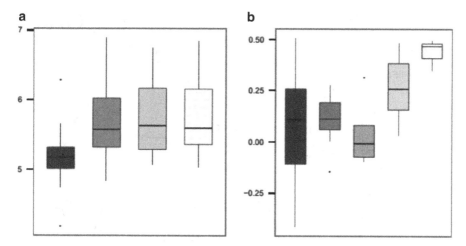

Fig. 9.1 Expression levels of *HIF1α* normalized with respect to the internal control for each dataset across stages 1 through 4. (**a**) Gastric cancer (dataset GSE13195 in GEO) and (**b**) melanoma (dataset GSE12391 in GEO). The detailed information about the two datasets is given in Appendix. The y- and x-axes give the average gene-expression level and the stage, respectively, with the *leftmost* denoting normal tissue and the following three representing stages 2 through 4 for (**a**) and stages 1 through 4, namely, common melanocytic nevus, dysplastic nevus, radial growth phase melanoma, and vertical growth phase for (**b**). For each boxplot, *three horizontal lines* are shown, the 75 percentile line in the *top*, the *middle thick line* inside a *box* for the mean, and the 25 percentile line in the *bottom*

calculated and plotted across multiple samples for each of the four stages of melanoma and gastric cancer, respectively. These two cancer types were selected since the former represents one of the most aggressive cancers and the latter is a relatively slow growing cancer. One can see from Fig. 9.1 that the hypoxia level generally increases from the early stage to stage 4 for both cancer types, while the detailed increasing patterns are different between the two: gastric cancer exhibits an increase from stage 1 to stage 2, but then remains flat after while melanoma shows a steady increase over the first three stages, but with a substantial jump from stage 3 (radial growth stage) to stage 4 (radial vertical growth stage), which is consistent with the general knowledge about the growth pattern of this cancer type.

Generally cells will adjust their O_2 consumption by altering their metabolism, redox homeostasis and other O_2 consuming processes, in response to reducing oxygen levels. As a result, numerous pathways will alter their activity levels. These changes are coordinated by a few key pathways such as the unfolded protein response (*UPR*), *mTOR* signaling and transcription regulation by the *HIF* genes. A number of fundamental processes are regulated by *UPR* and *mTOR* signaling, such as (1) translation attenuation, (2) cell cycle arrest in the G_1 phase, (3) induction of ER stress, (4) production of proteins involved in chaperon function and protein folding, (5) activation of *JNK* and *TRAF2* (tumor necrosis factor receptor-associated factor 2) and (6) apoptosis inhibition. In addition, a large number of pathways are regulated by the

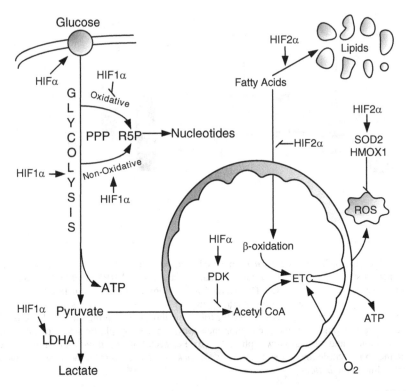

Fig. 9.2 Roles of the *HIF* genes in central metabolism (adapted from (Majmundar et al. 2010))

HIF genes as discussed in Chaps. 5 and 6. As an example, Fig. 9.2 shows a few pathways in central metabolism that are directly regulated by the two *HIF* genes. Some of the induced responses will make the tumor environment more stressful and unfamiliar to the underlying cells such as the increased acidity in the pericellular environment as a direct result of the increased activity level of glycolytic fermentation caused by increased hypoxia. Another example is the production of hyaluronic acid caused ultimately by hypoxia, which results in cell proliferation.

9.1.2 Intermittent Hypoxia

Intermittent hypoxia occurs when the cellular O_2 level alternates between normoxia and hypoxia, which takes place repeatedly throughout the development of a cancer when the tumor goes through cycles of re-oxygenation due to repeated tumor angiogenesis. Studies on responses to chronic intermittent hypoxia have shown that it can induce ROS production (Peng et al. 2006). Intermittent hypoxia in cancer may trigger a variety of processes by way of increased ROS levels. A recent study revealed that intermittent hypoxia can enhance stem-like characteristics and suppress differentiation propensities in neuroblastoma (Bhaskara et al. 2012). Specifically, the study found that stem cell markers *CD133* and *OCT4* are up-regulated in intermittent hypoxia-conditioned cells compared with cells grown under normoxia, suggesting that intermittent hypoxia may play a selective role for more malignant cancer cells. Another study showed that metastasis-associated genes were significantly up-regulated in tumors of intermittent hypoxia-conditioned mice (Chaudary and Hill 2009).

9.1.3 Changes in Cellular ROS Level

Cellular ROS is predominantly produced by the NADPH oxidase complexes. Previous studies have found that cancer cells tend to have increased ROS levels compared to normal cells, possibly due to the combination of increased metabolic activities, re-oxygenation and mitochondrial malfunction. The current knowledge is that moderately increased ROS is beneficial to the growth of cancer cells while substantially elevated ROS may drive cancers to metastasize (Pani et al. 2010) or lead to cell death (Cai and Jones 1998).

To elucidate the detailed functional roles played by ROS throughout a cancer development, it is important as a starting point to examine how the ROS level changes as a cancer progresses from the early to advanced stages. Also it is important to scrutinize if cancers with distinct clinical phenotypes, such as fast *versus* slow growing cancers, may correlate with their cellular ROS levels. While it is not trivial to directly measure the ROS level *in situ*, one can possibly estimate the ROS level through analyses of the expression levels of genes that are known to respond to such changes, one being *SOD2* (superoxide dismutase 2), whose expression level has been used as a marker for the cellular ROS level (Zelko et al. 2002). Figure 9.3 shows changes in *SOD2* expression levels with the progression of gastric cancer and melanoma, respectively, on the same set of transcriptomic data used for Fig. 9.1. It can be seen from the figure that the ROS level increases from the early stage to stage 4 in both cancer types.

When the ROS level exceeds the antioxidant capacities, the cells become *oxidatively stressed*. A number of biological processes are known to respond to the ROS increase to protect the cells. Specifically, ROS can regulate directly or indirectly the activities of some important proteins such as those involved in *GPR* signaling, apoptosis, angiogenesis, immune response and general stress response. Specifically this list includes *ATM, ERKs, HSF1, JAK, JNKs, NFκB, PI3K, PKC* (protein kinase C), *PLCγ1* (phospholipase C-γ1) and *STAT*, indicating the global impact of ROS-induced stress, many of which will lead to changes in a range of metabolic activities, hence further altering the cellular and extracellular environments. To find out all the pathways responsive to increased ROS, one can carry out statistical correlation analysis to identify genes whose expression patterns exhibit strong correlations with changes in ROS levels (e.g., the one shown in Fig. 9.3), and carry out pathway enrichment analysis among such genes.

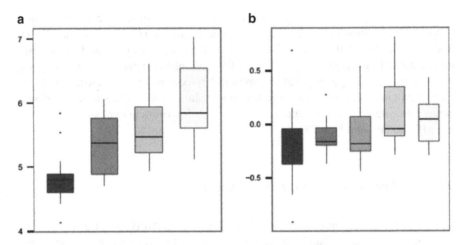

Fig. 9.3 Expression levels of *SOD2* normalized with respect to the internal control for each dataset across stages 1 through 4. (**a**) Gastric cancer and (**b**) melanoma. Other definitions are the same as in Fig. 9.1

9.1.4   Compositional Changes in ECM

Similar analyses to the above can be carried out on ECM composition changes and the corresponding responses. The ECM maintains tissue integrity by providing cell cohesion and organizing the constituent cells into the specific shape of a tissue. The mechanical properties of ECM have essential roles in cell differentiation, proliferation, adhesion, migration and apoptosis (Guilak et al. 2009). As discussed in Chap. 8, the relative concentrations of collagen, laminin and elastin are among the key determinants of the mechanical properties of a matrix. Specifically, collagen provides the tensile strength, hence the resistance to plastic deformation (Buehler 2006), while elastin defines the extensibility and the reversible recoil (Muiznieks and Keeley 2010). In addition, laminins, unlike the fiber-forming collagens and elastins, form network-like structures to provide resistance to tensile forces. The rich set of glycoproteins in ECM such as fibronectins, proteoglycans and glycosaminoglycans like hyaluronic acid also contribute to the mechanical properties of the ECM. An examination of the expression levels of ECM-component encoding genes can provide useful information about the composition of an ECM and changes in the composition as a cancer progresses when the examination is carried out over cancer samples at different stages.

To demonstrate the feasibility of obtaining such information, a number of genes whose proteins are known to contribute to the basic components of the ECM were examined here in terms of their expressions. The goal was to determine how the expression patterns of these genes change as a cancer advances, thus providing information, albeit indirect, of changes in the relevant component concentrations in the ECM. Figure 9.4 shows expression changes of three such genes in gastric cancer and melanoma, namely *COL4A2*, *HAS2* and *FN1*, where *COL4A* is a gene responsible

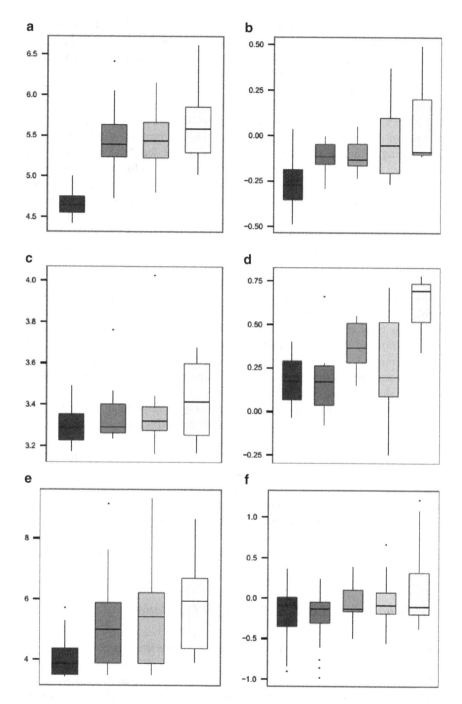

Fig. 9.4 Expression patterns of (**a** and **b**) *COL4A2*, (**c** and **d**) *HAS2* and (**e** and **f**) *FN1* genes, across stages 1 through 4. (**a**, **c**, and **e**) Gastric and (**b**, **d**, and **f**) melanoma. Other definitions are the same as in Fig. 9.1

for one form of collagen production, *HAS2* for hyaluronic acid and *FN1* for fibronectins. Overall, these genes are up-regulated in cancer *versus* normal tissues and generally increase as a cancer progresses, although the detailed up-regulation patterns are different between the two cancer types.

If one is interested in predicting the physical properties of a matrix from the expression data of the ECM genes, it is first necessary to carry out association analyses between the expression levels and the matching data collected on the physical properties of the matrix. Clearly, measurement of such properties *in situ* represents a challenging technical problem and is probably not feasible on a large scale. Fortunately, computational approaches can provide useful information here. Knowing that very little data are currently available on the physical properties of ECMs, one can directly conduct association analyses between clinical phenotype data and expression data of EMC-relevant genes as follows. One can select a few phenotypic characteristics such as fast *versus* slow growing or easy-to-metastasize *versus* difficult-to-metastasize cancers, and then identify ECM-related gene-expression patterns that are distinctly associated with each phenotypic characteristic in a similar fashion to the typing or staging prediction discussed in Chap. 3. Then the observed relationships between gene-expression patterns and phenotypic characteristics can be applied to the limited physical property data in order to establish associations between gene-expression patterns and ECM physical properties. For example, knowing that a stiffer matrix tends to make a cancer grow faster (Wells 2008), one may be able to establish links between the stiffness of a matrix and gene-expression patterns via the observed relationships between expression patterns and fast *versus* slow growing cancers.

A number of biological processes are known to respond to changes in the mechanical properties of the ECM. For example, the stiffness of a matrix modulates a diverse range of cellular/tissue activities such as focal adhesion, cytoskeletal assembly, cell-cell assembly, migration, cell proliferation, cell differentiation, tissue development, regeneration and repair (Mason et al. 2012). Computational analyses of *omic* data should be able to identify all, at least most of the biological processes that respond to changes in the physical property of an ECM through association analyses of gene-expression patterns, thus leading to a more comprehensive understanding of how changes in ECM invoke alterations in cellular metabolism and changes in the interplay between the neoplastic cells and the microenvironment.

9.1.5   Changes in Local Stromal-Cell Population

As discussed in Chap. 6, when a tissue is injured, various immune cells are recruited from circulation, leading to changes in the local stromal cell population and facilitating tissue repair. It has been well established that different types of injuries will lead to different sets of tissue-repair signals, resulting in distinct stromal cell populations.

The subpopulation sizes of different cell types can thus provide information about the nature of an injury. Since cancer is being treated like an injured tissue (see Chap. 6), knowing the subpopulation sizes of different immune and stromal cells can provide useful information about the main facilitators of a cancer's evolution and possibly their relationships with the evolutionary trajectories. That is, these cell population sizes may reflect, at least to some extent, nature of the underlying cancer.

As discussed earlier and detailed in Chap. 2, transcriptomic data analyses can be used to estimate the relative subpopulation sizes of different stromal cells and how the subpopulation sizes change as cancer advances. It is worth noting that some normalization is needed when using such data to ensure that the observed expression levels of cell type-specific genes across different samples can be compared directly with each other. Such normalization can help to overcome the inherent problem of different tissue samples being processed with varying methods by separate labs, leading to the possibility that samples have different types of systematic errors introduced in terms of the (relative) subpopulation sizes of distinct cell types. Only with such normalization can one assume that the relative population sizes of different cell types in the resected samples are the same as in the tissues *in situ*. Then one can argue that the average expression level of genes uniquely expressed in specific cell types provide a reasonable estimate of the relative subpopulation sizes of the relevant cell types.

To demonstrate that this is feasible, a number of genes known to be uniquely expressed in specific stromal cell types, such as *IL2RA* and *SELL* (*L-selectin*, also known *CD62L*) in T-cells and *CD68* in macrophages, were examined. Figure 9.5 shows the expression level changes of these genes as a cancer progresses in both gastric cancer and melanoma. From the figure, one should be able to estimate changes in the (relative) subpopulation size of a specific cell type based on the average expression changes of genes uniquely expressed across multiple samples in the cell type. To accurately estimate the subpopulation-size changes, say as needed for in-depth analyses such as a determination of which biological processes are triggered by such changes, it may be necessary to include all the genes uniquely expressed in a specific cell type. A reliable way to accomplish this may prove to be a very useful tool for the general community that does cancer transcriptomic data analyses for functional studies of cancer.

The data presented in this section indicate that the cancer cellular activities and microenvironment become increasingly different from the normal ones. A natural question is: *What happens when the environment-induced stresses become sufficiently unfamiliar and challenging to the underlying cells that none of the condition-specific stress response systems, as outlined in this section, can overcome the stress?* If not responded properly, the stressful conditions may kill the cells. It is likely that this is the point where epigenomic responses are being heavily used, as observed from transcriptomic data of different cancers, a topic that is addressed in the following section.

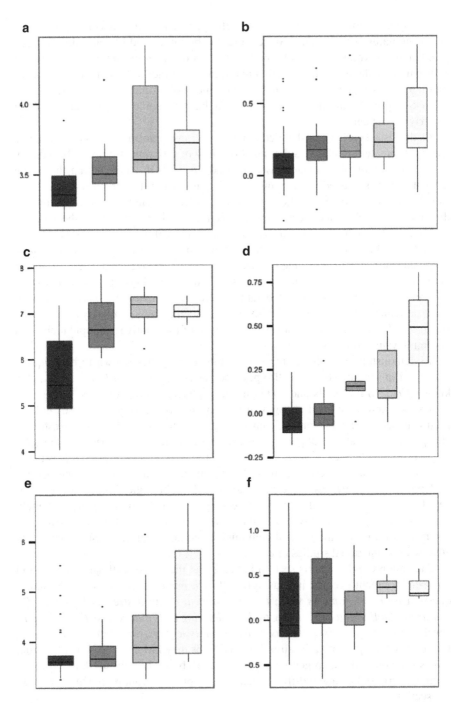

Fig. 9.5 Expression levels of (**a** and **b**) *IL2RA*, (**c** and **d**) *CD68*, and (**e** and **f**) *SELL*, each normalized with respect to the internal control for each dataset, across stages 1 through 4. (**a, c** and **e**) Gastric and (**b, d** and **f**) melanoma. Other definitions are the same as in Fig. 9.1

9.2   Epigenetic Response: A General Stress-Response System to Unfamiliar and Persistent Stresses

The epigenome of a cell refers to the collection of all the chemical modifications to the genomic DNA and the associated histones that impact on DNA-protein interactions, folding, packing and nucleosome positioning, thus facilitating or inhibiting various DNA-related functions such as transcription or DNA replication (see Chaps. 1 and 2 for details). Such changes can be passed on to the next generation during cell division. Among many possible epigenetic modifications, two have been extensively studied, namely DNA methylation and histone modification. Genes with methylated CpG islands are transcriptionally repressed or even inhibited, depending on the level of methylation. The interaction between a histone and DNA is formed by electrostatic attraction between the positive charges on the histone surface and the negative charges on DNA. Hence, modifications on histone proteins such as methylation or ubiquitination may change the charges of the surface residues. This in turn can alter the conformation and hence the transcriptional accessibility of a folded DNA and ultimately enhance or repress the expression of the relevant gene.

As an emerging area of cancer research, numerous studies have appeared on cancer epigenomics, resulting in the identification of a number of general characteristics of cancer epigenomes, as well as cancer type-specific epigenomic modifications (Jones and Laird 1999; Feinberg and Tycko 2004). For example, it has been reported that cancer genomes tend to have a global reduction in DNA methylation, while promoter regions of protein-encoding genes generally have increased methylation compared with the matching noncancerous genomes (Das and Singal 2004). Also, a global reduction in mono-acetylated and tri-methylated histone *H4* has been observed in the epigenomes of numerous cancer types (Fraga et al. 2005). To a lesser degree, over-production of histone methyltransferases for methylation of the K4 or K27 residue of histone *H3* has been found across multiple cancer types (Yoo and Hennighausen 2012; Yang et al. 2013). In addition, specific epigenomic changes have been consistently observed in tissue samples of specific cancer types. For example, hyper-methylation of the CpG island in the promoter region of the *GSTP1* (glutathione *S*-transferase pi 1) gene was observed in 90 % of prostate cancers (Nakayama et al. 2004), and increased methylations have been consistently observed in a number of nuclear genes such as *TERT* (telomerase reverse transcriptase), *DAPK1* (death-associated protein kinase 1), *RARβ* (retinoic acid receptor β), *MAL* (myelin and lymphocyte protein) and *CADM1* (cell adhesion molecule 1) in cervical cancer genomes.

Here the goal is to understand how the overall level of epigenomic activities differs between cancer and matching normal tissues, as well as how this level changes as a cancer progresses. To accomplish the goal, the total expression levels of genes involved in epigenomic activities were compared, and the expression levels of a comprehensive set of genes involved in DNA methylation and histone modification,

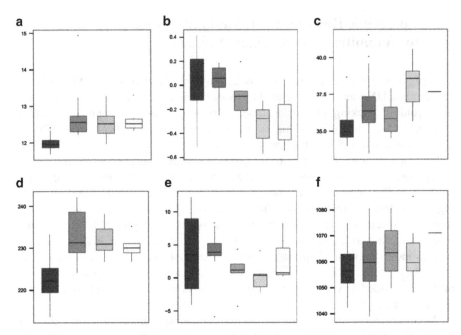

Fig. 9.6 Expression levels of histone-modification related enzymes (**a**, **b** and **c**) and DNA methylation enzymes (**d**, **e** and **f**). Three cancer types are considered: gastric cancer (**a** and **d**) (GEO:GSE13195), melanoma (**b** and **e**) (GEO:GSE12391), and lung cancer (**c** and **f**) (GEO:GSE19804). Detailed information on the three datasets is given in the Appendix. The y-axis denotes the average gene-expression level axis, and the x-axis represents the cancer axis, with the *leftmost* for normal tissue and the following being stages 2 through 4 for (**a** and **d**) and stages 1 through 4 (or common melanocytic nevus, dysplastic nevus, radial growth phase melanoma, and vertical growth phase) for (**b** and **e**; **c** and **f**). Other definitions are the same as in Fig. 9.1

respectively, were examined. The genes investigated include: three DNA methylation enzymes and 65 enzymes involved in histone modifications, namely (a) *DNMT1, DNMT3A* and *DNMT3B* for DNA methylation, and (b) *HDAC1, HDAC2, HDAC3, HDAC4, HDAC5, HDAC6, HDAC7, HDAC8, HDAC9, HDAC10, HDAC11, KAT2B, KAT2A, KAT7, KAT5, KAT1, KAT6B, KAT8, AKT6A, EP300, KMT2A, SETD1A, SETD2, SETD7, SETD8, EHMT2, EHMT1, SUV39H1, SUV39H2, SETDB1, DOT1L, SETMAR, CARM1, KMT2C, KMT2D, SETD2, KMT2B, NSD1, ASH1L, WHSC1L1, SMYD3, WHSC1, EZH1, KMT2E, SUV420H1, SETD3, SUV420H2, SETD1B, PRDM9, SETDB2, SMYD1, KMT2A, EZH2, KDM4A, KDM4B, KDM5D, KDM6A, JHDM1D, KDM5A, KDM5B, KDM5C, KDM4D, KDM4C, PHF2, KDM6B* and *PHF8* for histone modification. For each of these two sets of enzymes, the sum of the expression values of each gene is calculated over a set of samples for each of three cancer types: gastric cancer, melanoma and lung cancer. The results are plotted in Fig. 9.6. The detailed information of the datasets used here is given in the Appendix.

From the figure, one can see that gastric and lung cancers both have increased levels of epigenomic activities over the matching normal tissues, along with increasing trends in the utilization of epigenomic modification genes as a cancer advances. This pattern is shared by numerous other solid cancer types. Melanoma has a complex expression pattern in its epigenomic modification genes, showing that epigenomic activities have an overall decreasing trend for both DNA methylation and histone modification. This clearly represents an interesting case and warrants further investigation to understand the possible reason for the reduced epigenomic activities in the advanced stage of melanoma and the possible implications to the explosive growth of such cancer.

9.2.1   Towards Understanding Cancer Epigenomics

While some distinct patterns of epigenomic modifications have been observed in cancer *versus* normal cells, very little is understood as to why cancers in general tend to have increased epigenomic activities as revealed in Fig. 9.6. An understanding of this issue could potentially provide fundamental and novel insights into cancer development and possibly tissue development in general. Some proposals have been made regarding the possible reasons for this observation. One suggestion is that when a gene or a group of genes needs to be silenced for an extended period, such as tumor suppressor genes during the development of cancer, epigenomic modifications may be selected to provide the stability for gene silencing (Sharma et al. 2010). Another proposal argues that it is the "accumulated stress" (defined as the activation duration of stress response pathways) that triggers short-term or long-term deregulation of the cancer epigenome, leading to changes in repression states of stress-response genes on a global scale (Vojta and Zoldos 2013). Specifically, the model suggests that as the environmental stresses intensify and become constant, cancer cells will "traverse their adaptive response continuum" to search for a long-term and effective response.

We propose here another possible cause for the increased utilization of epigenomic activities. Specifically, the epigenome encodes a general stress-response mechanism possibly as the last resort protection of severely stressed cells; the mechanism is activated when all the applicable condition-specific stress-responses fail and the stress level goes beyond the tolerance capacity of the cells. This proposal is based on the following three considerations.

1. It has been proposed previously that the epigenome emerged originally as a defense mechanism for genome-integrity protection (Johnson 2007), which has been accepted by some authors (Johnson and Tricker 2010). This defense

mechanism may be triggered by genome integrity-related stress sensors or by some general stress sensors which may be triggered by genome integrity-related stress.

2. A major side-product of the rapidly evolving cancer cells is their increasingly deteriorating cellular and microenvironment, largely due to the altered cellular metabolism, reflected by the increased proliferation rates as a cancer advances (recall the antagonist co-adaption example of *Pseudomonas fluorescens* and its viral parasite phage in Chap. 5). Published studies support this view since it has been well established that higher hypoxia levels tend to lead to more malignant cancers (Vaupel 2008). Consequently, the issue becomes that which was asked at the end of Sect. 9.1: *What happens to the cells if all the applicable condition-specific stress responses fail to overcome the stress?* It has been hypothesized that a yet to-be-understood general stress-response mechanism at the epigenomic level will be triggered, which allows the cells to search for effective stress-responses that are encoded in the genome but not designed for the current stress. A few authors have suggested similar ideas (Vojta and Zoldos 2013), where it was referred to as *epigenomic deregulation* or *relaxation of epigenetic regulation* as has been used in the plant epigenomic literature (Madlung and Comai 2004). While such a general stress-response system is speculated to exist, no proposals have been made in terms of its detailed trigger and effector mechanisms in the literature. Interestingly, a recent study on fly larvae development under an unfamiliar and life-or-death stress may help clarify this fundamental issue, not only to cancer biology but also to developmental biology in general (Stern et al. 2012).

The goal of the study (Stern et al. 2012) was to demonstrate how fly larvae cells, utilizing a more basic and probably more primitive mechanism, handle severe and persistent stress that is unfamiliar to the encoded condition-specific stress-response system. Specifically, larvae tissues were treated with the toxin G418 at concentrations that are lethal to the organism. A toxin-resistance gene had been genetically inserted into the larvae genome with a randomly selected developmental promoter that is unrelated to the toxin, while the protein encoded by the gene is capable of pumping out the toxin when it is activated. Systematic tests found that the wild-type larvae without the inserted gene were all killed by the toxin. In contrast, eight out of nine genetically engineered larvae, each having a distinct developmental promoter, survived and developed into adulthood but with some developmental delays. This finding indicates that a general survival mechanism was able to overcome the toxin, which evidently located and activated the inserted gene.

Interestingly, this capability is heritable for at least four generations, and the offspring of the first generation of genetically engineered larvae develop into adulthood without any developmental delay when treated with the same toxin G418. These data strongly suggest that an epigenetic modification was made, giving rise to a new and heritable survival capability that has integrated the added toxin resistance gene and the randomly selected promoter as an encoded response to the toxin in the next generation. Further investigation by these authors revealed that there is a general stress-response mechanism in flies that can broaden the range of activation of all developmental promoters. With the broadened activation range, the genetically engineered larvae have managed to activate their randomly selected promoters, and hence activated the toxin-resistance gene and gave rise to the observed survival of the larvae. This study also found that the broadened range of activation is achieved through lowering the expression of the *PcG* (polycomb group) genes that are known to repress many developmental regulators epigenetically. The authors suggest that this broadening mechanism of developmental regulators may be a general mechanism for coping with unfamiliar stressors. Since this mechanism is accomplished epigenetically, this gained function is inherited for a number of generations as demonstrated in the study.

3. To determine if a similar mechanism may exist in humans, the expression data of the human *Polycomb* genes, which are known to have key roles in many aspects of organism development, were examined in cancer *versus* normal tissues. Figure 9.7 shows the summarized expression-level changes of these genes with the cancer progression of the same three cancers used in Fig. 9.6. From the figure, it can be seen that all three cancers have increased *Polycomb* expression over the matching normal tissues, suggesting the possibility that *Polycomb* may potentially be involved in the regulation of large-scale epigenomic activities.

Based on the above discussion and analyses, one may speculate that as neoplastic cells (except possibly for melanoma) continue to evolve for survival, their microenvironmental stress becomes increasingly more challenging and unfamiliar for their condition-specific stress-response systems to overcome. The result is that the stress level goes beyond certain thresholds, which triggers a currently unknown general stress-response mechanism encoded at the epigenomic level, possibly as the last resort to save the cell. It is reasonable to speculate that a general stress-response mechanism similar to that in flies may exist in humans as hinted by Fig. 9.7, which may be the main reason for the observed increased utilization of epigenomic activities in cancer *versus* normal cells. As suggested by the above study on flies, this mechanism may consist of trial-and-error search strategies, and hence take a longer time to respond, as was observed. Knowing that melanoma is a very fast growing cancer (at the vertical growth stage), it is likely that the cancer may have used a different system to overcome the severe stress issue, as the above mechanism may require too much time to be consistent with the observed behavior of melanoma development.

Overall, this may represent another problem that could benefit significantly from computational analyses of *omic* data. In this case, association analyses could potentially identify the key relevant players and potential relationships in this still enigmatic but fundamentally important mechanism in cancer.

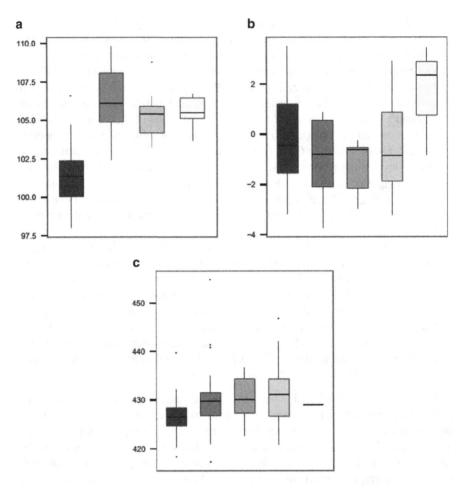

Fig. 9.7 Expression levels of *Polycomb* genes. Three cancer types are considered, namely gastric cancer, melanoma, and lung cancer, from *left* to *right*. For each figure, the sum of the expression values of each of the following 24 *Polycomb* genes: *EZH2, EZH1, EED, SUZ12, RBBP4, RBBP7, RING1, RNF2, CBX2, CBX4, CBX6, CBX7, CBX8, PHC1, PHC2, PHC3, PCGF1, PCGF2, PCGF3, BMI1, PCGF5, PCGF6, ZNF134, SCMH1*, which form two large complexes (Cavalli-lab 2014), over all the samples for each of the three cancer types (see Fig. 9.6 for the detailed data information). Other definitions are the same as in Fig. 9.6

While this to-be-elucidated mechanism facilitates the survival of the severely stressed cells by finding novel combinations of stress-responses to overcome the current stress, it also makes the changes substantially less reversible than those accomplished through the condition-specific stress responses generally induced through transcriptional regulation, hence moving further away from the normal behavior of healthy human cells. In addition, recent studies have revealed that epigenomic modifications have a strong connection to genome instability (Suzuki and Bird 2008).

This may represent another reason for the observed increase in genomic mutations in cancer in general, in addition to the increased utilization of emergency DNA repair under persistent hypoxic conditions and increased proliferation rates as cancer progresses (see discussion in Chap. 4).

9.3 Genome Instability and Cancer Evolution

Genome instability is considered as one of the cancer hallmarks (Hanahan and Weinberg 2011), as discussed in Chaps. 1 and 4. It typically refers to the increased genomic structural changes and copy-number variations in the genome with the progression of a cancer. Two general types of genomic instabilities have been observed across the genomes of virtually all cancer types, with *chromosomal instability* (CIN) as the dominating one and *microsatellite instability* (MSI) as the distant second. While these genome instabilities are known to be associated with cancer genomes, their causes are not well understood (Negrini et al. 2010), even though multiple proposals have been made as discussed earlier.

One suggestion was that about 60,000 mutations take place on average in the genome of a healthy human cell per day, caused by endogenous factors such as ROS and other reactive metabolites, the vast majority of which are single-point mutations (Bernstein et al. 2013). If a neoplastic cell loses its ability for accurate repair of DNA damage, such mutations will accumulate rapidly over time. In addition, ROS, which tends to be elevated in cancer cells, is known to be responsible for many of the single-point mutations in cancer genomes. Clearly, these factors cannot explain the genomic instabilities defined as above. A few proposals about the possible causes for cancer genome instabilities have been made such as: (1) being induced by the activation of oncogenes and inactivation of tumor suppressor genes (Halazonetis et al. 2008); (2) related to epigenomic modifications (Qu et al. 1999); and (3) due to aberrant behavior of DNA repair systems associated with cancer-promoting microenvironments (Wimberly et al. 2013). As can be seen from the following, there is evidence for each of these possibilities; hence they may apply under different conditions, and possibly are all correct. However, it remains unknown how these different mechanisms relate to each other in producing the genome instability patterns observed in individual cancer types—recall from Chap. 4 that different cancer types may have distinct genome alteration patterns. We expect that careful statistical analyses of cancer genomic sequences, along with the matching transcriptomic data, could potentially reveal useful information and insights about this issue. Moreover, such an analysis could possibly lead to mutation models that can explain the distinct mutation patterns observed in cancer genomes of different types.

9.3.1   Genome Instability Due to Compromised DNA Repair Machinery

There are multiple models on how compromised DNA repair machinery is responsible for the large number of mutations observed in cancer genomes such as: (a) those that attribute structural alterations to loss-of-function mutations in DNA repair machinery or in mitotic checkpoint genes (Negrini et al. 2010), and (b) those that suggest such mutations are the result of inaccurate DNA repair under hypoxic conditions. The latter relies on microhomology rather than the normal 200-base pair homology, as discussed in Chap. 2, when using the sister chromosome as the template to repair damaged DNA such as single-point mutations and double-strand DNA breaks, possibly caused by rapid proliferation and lack of a sufficient supply of nucleotides.

9.3.2   Epigenomic Modification and Genome Instability

The earliest discovery of the association between cancer and DNA hypo-methylation can be traced back to the 1980s when colon cancers were found to have a substantially reduced methylation content on a genome scale (Goelz et al. 1985). In the past decade, a number of studies have detected strong correlations between changes in methylation levels and genome instability.

One early study found that decreasing the level of DNA methylation on a genomic scale by activating methylation inhibitors can lead to the instability of the CTG·CAG repeat in the genome, specifically having increased the run of the tri-nucleotide repeat (Gorbunova et al. 2004), a hallmark of a number of human genetic diseases (Ashley and Warren 1995; Mitas 1997). A few other studies have found that hypo-methylation tends to be associated with increased genomic instability (Chen et al. 1998) in general, including: (a) increased centromere instability correlates with hypo-methylated centromeres (Ehrlich 2002); (b) deficiency in methylation enzymes, specifically *DNMT1*, has been found to co-occur with chromosomal instability; and (c) DNA hypo-methylation precedes extensive genomic damage in gastrointestinal cancer (Suzuki et al. 2006). A more recent study suggests that it is the activity of DNA demethylation rather than the hypo-methylation state of a genome that correlates with genome instability, an observation made on colon cancer (Rodriguez et al. 2006). Some studies have demonstrated causal relationships between changes in methylation and genetic instability for specific cases (Rizwana and Hahn 1999; Robertson and Jones 2000; Rodriguez et al. 2006).

Intuitively the above observations seem reasonable since hypo-methylation suggests the possibility of increased transcription, which requires unwinding the double-stranded DNA into single-stranded DNA, hence increasing the chance of DNA breaks and the need for DNA repair. In addition, there is a deep connection between DNA repair and methylation mechanisms (Schar and Fritsch 2011).

Again, computational analyses could have a key role in identifying association relationships among the key players in DNA usage, DNA repair and methylation utilization.

9.3.3 Genome Instability Relevant to Aberrant Behavior of Proto-oncogenes and Tumor Suppressor Genes

A recent model suggests that genome instability is largely induced by double-stranded DNA breaks generated as a result of oncogene-induced rapid cell replication (Halazonetis et al. 2008). The model speculates that rapid DNA replication induced by oncogenes may result in stalling and collapse of DNA replication forks and hence double-stranded DNA breaks. This is attributed to the fact that nucleotide production cannot maintain pace with the cell replication rates, as often observed in cancer. This, we speculate, will take place only after oncogenes start to drive the proliferation process (instead of hyaluronic acid fragments) once a cancer reaches certain developmental stages, based on our discussion in Chaps. 5 and 6. In conjunction with the impaired DNA repair systems, this may ultimately lead to the observed genome instability. However, this model has been questioned because it may not represent the majority of the oncogene-induced genome instabilities observed in cancer tissues (Corcos 2012).

A related but improved model suggests that it is the unbalanced DNA replication, resulting from a combination of growth signaling, rapid replication driven by oncogenes and the inactivation of certain tumor-suppressor genes, that leads to genome instability (Corcos 2012). Specifically, during cancer cell proliferation, the cell cycle may enter a G_1 quiescent state due to some suboptimal growth conditions such as DNA damage caused by ROS. When the *MYC* gene is overexpressed, the cell cycle will move forward but is arrested at the G_2 phase (Felsher et al. 2000), indicating that *MYC* is able to overcome the G_1/S checkpoint under suboptimal growth conditions but not the G_2 checkpoint. The aberrant activation of other oncogenes and the inactivation of some tumor suppressor genes may also lead to arrests in the G_1 or G_2 checkpoint in general. For example, *RB* deficiency can lead to a G_2 arrest. When the loss of the G_2 checkpoint occurs due to additional mutations, the cell cycle may finish, but with an extended or shortened S phase, depending on specific cellar conditions. A direct result of such inhomogeneous replications, which may take place frequently in cancer cells, is that some portions of the genomic DNA can be under-generated while other portions might be over-generated. Such inhomogeneous abundances of different segments of the DNA will clearly lead to genome instabilities. This interesting aspect of the model is that it requires only high proliferation rates but not a compromised DNA repair system. Hence it represents a fundamentally different cause, possibly a more realistic one, of DNA instability compared to the other proposals.

Based on the above discussion, one can possibly assemble a simple model for genome instabilities as observed in cancer as follows. Endogenous factors in normal and in cells undergoing transformation, such as increased ROS, may lead to the initial single-point mutations. It is worth emphasizing that all the structural alterations and copy-number changes are not produced by such mechanisms, instead they are the results of compromised DNA repair machinery under persistent abnormal conditions such as hypoxia (Hastings et al. 2009). As the intracellular or pericellular stresses become increasingly more unfamiliar and persistent, epigenomic responses will be utilized more frequently to handle the more challenging stresses. As a side-effect, the increased epigenomic activities may generate more genomic alterations as discussed earlier. The relationship between epigenomic activities, particularly methylation, and genome instability could be quite complex and possibly fundamentally important to the transformation process of proliferating cells to malignant cells. One reason is that methylation seems to be connected with the DNA repair machinery at a fundamental level, which involves DNA replication when doing template-based repair.

From studies on meiotic replications, methylation and replication are closely associated with each other. This could prove useful in guiding computational studies on the relationships between methylation and short-range replications resulting from DNA repair in cancer cells. To date, all the genomic instabilities are not relevant to the increased replication rate. As cancer starts to be driven by oncogenes such as *MYC*, which may lead to inadequacies in supplying nucleotides needed for the rapid DNA replication and hence give rise to double-strand DNA breaks, the genome instability may move to a different complexity level. Clearly during these processes, when DNA repair genes or cell-cycle checkpoint genes become compromised, e.g., via mutations, it will add to the complexity. To ascertain how such a simple model may apply to various cancers, one may need to build dynamic models that integrate models developed for individual mutation-generation mechanisms along with the triggering conditions. These can be estimated based on the available genomic, epigenomic and transcriptomic data for cancer tissue samples collected at different developmental stages, followed by an assessment of how well such models can explain the observed mutation patterns for individual cancer types.

9.4   Concluding Remarks

The complexity in studying cancer may arise largely from the challenging problem in elucidating the impact of the evolving microenvironment on the underlying cells. Initially the impact arising from either hypoxia or increased ROS is quite simple, i.e., leading to increased activity of glycolytic fermentation. As the cells continue to evolve through adaptation, their metabolism changes and this further alters their microenvironment, forming a self-propelled vicious cycle and driving the underlying cells to become increasingly more malignant. In this chapter, a number of micro-environmental factors were considered with the finding that the co-adaptation and antagonist coevolution ultimately drive the environmental conditions to become sufficiently stressful that none of the condition-specific stress responses can overcome them. Consequently, this will lead to the frequent activation of general stress-response mechanisms at the epigenomic level. Two consequences are probably the direct result of the increased utilization of epigenomic activities: (a) increased irreversibility of the evolutionary trajectories by the neoplastic cells; and (b) increased genomic instability due to the fundamental link between the two, possibly a pathway encoded in human genomes for increasing the chances in creating genomic mutations relevant to specific functions as we hypothesize. From the discussion presented here, one may get the tantalizing impression that epigenomic responses, which can be altered by the changing environment, may have a more pivotal role than genomic changes in cancer development, consistent with many recent studies. Currently there are no epigenomic landmarks for measuring the return-ability or the potential for metastasis of neoplastic cells in their journey for survival. Again, computational *omic* data analyses, in an integrated fashion, may prove to be a key in addressing these important questions.

Appendix

Table 9.1

Data set	Cancer types	Sample size	Platform
GSE13195	Gastric cancer	49	GPL5175
GSE12391	Melanoma	41	GPL1708
GSE19804	Lung cancer	120	GPL570

References

Ashley Jr CT, Warren ST (1995) Trinucleotide repeat expansion and human disease. Annual review of genetics 29: 703-728

Bernstein C, Prasad AR, Nfonsam V et al. (2013) DNA Damage, DNA Repair and Cancer. New Research Directions in DNA Repair.

Bhaskara VK, Mohanam I, Rao JS et al. (2012) Intermittent hypoxia regulates stem-like characteristics and differentiation of neuroblastoma cells. PloS one 7: e30905

Buehler MJ (2006) Nature designs tough collagen: Explaining the nanostructure of collagen fibrils. Proceedings of the National Academy of Sciences 103: 12285-12290

Cai J, Jones DP (1998) Superoxide in Apoptosis MITOCHONDRIAL GENERATION TRIGGERED BY CYTOCHROMEc LOSS. Journal of Biological Chemistry 273: 11401-11404

Cavalli-lab (2014) The Polycomb and Trithorax page.

Chaudary N, Hill RP (2009) Increased expression of metastasis-related genes in hypoxic cells sorted from cervical and lymph nodal xenograft tumors. Laboratory investigation; a journal of technical methods and pathology 89: 587-596

Chen RZ, Pettersson U, Beard C et al. (1998) DNA hypomethylation leads to elevated mutation rates. Nature 395: 89-93

Corcos D (2012) Unbalanced replication as a major source of genetic instability in cancer cells. Am J Blood Res 2: 160-169

Das PM, Singal R (2004) DNA methylation and cancer. J Clin Oncol 22: 4632-4642

Ehrlich M (2002) DNA hypomethylation, cancer, the immunodeficiency, centromeric region instability, facial anomalies syndrome and chromosomal rearrangements. J Nutr 132: 2424S-2429S

Feinberg AP, Tycko B (2004) The history of cancer epigenetics. Nature Reviews Cancer 4: 143-153

Felsher DW, Zetterberg A, Zhu J et al. (2000) Overexpression of MYC causes p53-dependent G2 arrest of normal fibroblasts. Proc Natl Acad Sci U S A 97: 10544-10548

Fraga MF, Ballestar E, Villar-Garea A et al. (2005) Loss of acetylation at Lys16 and trimethylation at Lys20 of histone H4 is a common hallmark of human cancer. Nat Genet 37: 391-400

Goelz SE, Vogelstein B, Hamilton SR et al. (1985) Hypomethylation of DNA from benign and malignant human colon neoplasms. Science 228: 187-190

Gorbunova V, Seluanov A, Mittelman D et al. (2004) Genome-wide demethylation destabilizes CTG.CAG trinucleotide repeats in mammalian cells. Hum Mol Genet 13: 2979-2989

Guilak F, Cohen DM, Estes BT et al. (2009) Control of stem cell fate by physical interactions with the extracellular matrix. Cell Stem Cell 5: 17-26

Halazonetis TD, Gorgoulis VG, Bartek J (2008) An oncogene-induced DNA damage model for cancer development. Science 319: 1352-1355

Hanahan D, Weinberg RA (2011) Hallmarks of cancer: the next generation. Cell 144: 646-674

Hastings P, Lupski JR, Rosenberg SM et al. (2009) Mechanisms of change in gene copy number. Nature Reviews Genetics 10: 551-564

Johnson LJ (2007) The Genome Strikes Back: The Evolutionary Importance of Defence Against Mobile Elements. Evol Biol 34: 121-129

Johnson LJ, Tricker PJ (2010) Epigenomic plasticity within populations: its evolutionary significance and potential. Heredity 105: 113-121

Jones PA, Laird PW (1999) Cancer-epigenetics comes of age. Nature genetics 21: 163-167

Madlung A, Comai L (2004) The effect of stress on genome regulation and structure. Annals of botany 94: 481-495

Majmundar AJ, Wong WJ, Simon MC (2010) Hypoxia-inducible factors and the response to hypoxic stress. Molecular cell 40: 294-309

Mason B, Califano J, Reinhart-King C (2012) Matrix Stiffness: A Regulator of Cellular Behavior and Tissue Formation. In: Bhatia SK (ed) Engineering Biomaterials for Regenerative Medicine. Springer New York, pp 19-37

Mitas M (1997) Trinucleotide repeats associated with human disease. Nucleic Acids Research 25: 2245-2253

Muiznieks LD, Keeley FW (2010) Proline periodicity modulates the self-assembly properties of elastin-like polypeptides. J Biol Chem 285: 39779-39789

Nakayama M, Gonzalgo ML, Yegnasubramanian S et al. (2004) GSTP1 CpG island hypermethylation as a molecular biomarker for prostate cancer. Journal of Cellular Biochemistry 91: 540-552

Negrini S, Gorgoulis VG, Halazonetis TD (2010) Genomic instability–an evolving hallmark of cancer. Nat Rev Mol Cell Biol 11: 220-228

Pani G, Galeotti T, Chiarugi P (2010) Metastasis: cancer cell's escape from oxidative stress. Cancer metastasis reviews 29: 351-378

Peng YJ, Yuan G, Ramakrishnan D et al. (2006) Heterozygous HIF-1alpha deficiency impairs carotid body-mediated systemic responses and reactive oxygen species generation in mice exposed to intermittent hypoxia. J Physiol 577: 705-716

Qu G-z, Grundy PE, Narayan A et al. (1999) Frequent Hypomethylation in Wilms Tumors of Pericentromeric DNA in Chromosomes 1 and 16. Cancer Genetics and Cytogenetics 109: 34-39

Rizwana R, Hahn PJ (1999) CpG methylation reduces genomic instability. J Cell Sci 112 (Pt 24): 4513-4519

Robertson KD, Jones PA (2000) DNA methylation: past, present and future directions. Carcinogenesis 21: 461-467

Rodriguez J, Frigola J, Vendrell E et al. (2006) Chromosomal instability correlates with genome-wide DNA demethylation in human primary colorectal cancers. Cancer Res 66: 8462-9468

Schar P, Fritsch O (2011) DNA repair and the control of DNA methylation. Progress in drug research Fortschritte der Arzneimittelforschung Progres des recherches pharmaceutiques 67: 51-68

Sharma S, Kelly TK, Jones PA (2010) Epigenetics in cancer. Carcinogenesis 31: 27-36

Stern S, Fridmann-Sirkis Y, Braun E et al. (2012) Epigenetically heritable alteration of fly development in response to toxic challenge. Cell Rep 1: 528-542

Suzuki K, Suzuki I, Leodolter A et al. (2006) Global DNA demethylation in gastrointestinal cancer is age dependent and precedes genomic damage. Cancer Cell 9: 199-207

Suzuki MM, Bird A (2008) DNA methylation landscapes: provocative insights from epigenomics. Nature reviews Genetics 9: 465-476

Vaupel P (2008) Hypoxia and aggressive tumor phenotype: implications for therapy and prognosis. The oncologist 13 Suppl 3: 21-26

Vojta A, Zoldos V (2013) Adaptation or malignant transformation: the two faces of epigenetically mediated response to stress. BioMed research international 2013: 954060

Wells RG (2008) The role of matrix stiffness in regulating cell behavior. Hepatology 47: 1394-1400

Wimberly H, Shee C, Thornton PC et al. (2013) R-loops and nicks initiate DNA breakage and genome instability in non-growing Escherichia coli. Nature communications 4: 2115

Yang YJ, Song TY, Park J et al. (2013) Menin mediates epigenetic regulation via histone H3 lysine 9 methylation. Cell Death Dis 4: e583

Yoo KH, Hennighausen L (2012) EZH2 methyltransferase and H3K27 methylation in breast cancer. Int J Biol Sci 8: 59-65

Zelko IN, Mariani TJ, Folz RJ (2002) Superoxide dismutase multigene family: a comparison of the CuZn-SOD (SOD1), Mn-SOD (SOD2), and EC-SOD (SOD3) gene structures, evolution, and expression. Free Radical Biology and Medicine 33: 337-349

Chapter 10
Understanding Cancer Invasion and Metastasis

Cancer is a deadly disease in large part because, if not stopped, will generally evolve to the metastatic stage, i.e., cancer cells spread from the primary site to new locations (generally different organs) through blood circulation or the lymphatic system. For largely unknown reasons, metastatic cancers tend to exhibit distinct growth patterns from its primary cancer counterpart, growing substantially faster and metastasizing more easily. Recent statistics show that metastatic cancer is responsible for approximately 90 % of all cancer-related mortalities. While it is known to be the deadliest stage of a cancer, the current understanding of the biology of metastatic cancer is rather limited. Some of the very basic questions such as: *what drives a primary cancer to metastasize*; *why some cancers tend to metastasize more easily than the other cancers, e.g., melanoma versus basal cell carcinoma*; and *why metastatic cancers tend to grow much faster than the corresponding primary cancer*, still have no clear answers. This may be the result of: (1) the true challenging nature of these questions, and (2) the lack of adequate investment and hence efforts into metastatic cancer research. This unfortunate reality is probably due to the general belief in the field that little can be done once a cancer has metastasized.

In this and the following chapter, we present the current knowledge about the potential drivers of metastasis, the key mechanisms in executing cancer metastasis and our recent understanding about the biology of metastatic cancers in their new microenvironment. As in the previous chapters, cancer evolution is viewed as a process for the diseased cells to escape from the deadly pressures imposed on them from their microenvironment. As part of their adaption to the challenges, the altered metabolism of the cells may be responsible to a significant degree for their increasingly more challenging microenvironment, this following the initial pressure caused by the accumulation of glucose metabolites due to chronic hypoxia and/or ROS accumulation (see Chap. 5).

© Springer Science+Business Media New York 2014
Y. Xu et al., *Cancer Bioinformatics*, DOI 10.1007/978-1-4939-1381-7_10

10.1   Local Invasion by Cancer Cells

The first step in cancer metastasis is tumor invasion, i.e., cancer cells breach their basement membrane (a type of ECM) and enter the stromal compartment, where stromal cells (fibroblasts and pericytes), immune cells and blood capillaries reside, as introduced in Chap. 1 and depicted in Fig. 10.1. To understand the process of tumor invasion, one needs to first understand how epithelial cells, from which most of the solid tumors evolve, are organized to facilitate their division and inhibition of division when needed.

10.1.1   Tumor Invasion and the Roles Played by Hyaluronic Acid

Epithelial cells are arranged adjacent to each other, much like a sheet, on top of the basement membrane, which is a knitted network consisting of collagen and hyaluronic acid fibrils and multiple types of linker proteins such as fibronectins, elastins and laminins (Hay 1981), as discussed in Chap. 1 and a few other chapters. It is known that cell-cell contacts inhibit cell division, a phenomenon referred to as *contact inhibition* under physiological conditions, and their anchorage to the basement membrane is generally required before they can divide. Structurally, cell-cell adhesion is provided by *adherens junctions*, one of the three types of intercellular junctions connecting two neighboring cells while the other two types, *tight* and

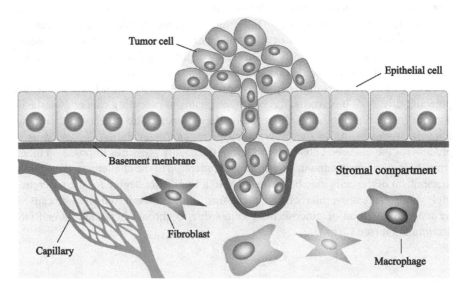

Fig. 10.1 A schematic of epithelial cells located above a basement membrane and associated stromal compartment, along with developing neoplastic cells

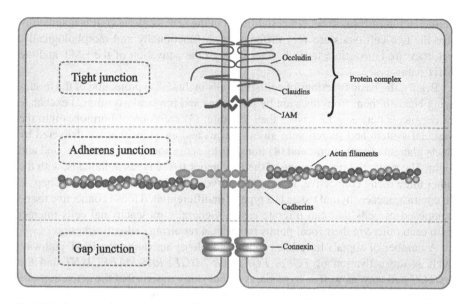

Fig. 10.2 A schematic of three types of junctions connecting two neighboring cells with adherens junctions providing the actual binding between the cells, each represented as a *rectangular box*

gap junctions, serve mainly as communication channels to allow molecules, including nutrients and signals, to pass between cells. An adherens junction consists of cadherin and a number of cytoplasmic proteins such as actin and catenin bound to cadherin, providing the actual intercellular adhesion as shown in Fig. 10.2. The cadherin protein in epithelial cells, specifically *E-cadherin*, is constantly regenerated with a 5-h half-life on the cell surface. As recently reported, reduced expression of this protein allows cells to migrate (Chen et al. 1997). A possible mechanism for this may be that repression of E-cadherin, in conjunction with other factors, can lead to the activation of the EMT (epithelial-mesenchymal transition) pathway since the repression of E-cadherin is crucial to the EMT activation as reported in a recent study (Lee et al. 2006). It has been demonstrated that *SNAIL* is a key regulator in repression of E-cadherin (Peinado et al. 2004; Montserrat et al. 2011). Interestingly, high molecular-weight hyaluronic acid has been reported to have a key role in the regulation of *SNAIL* (Craig et al. 2009), strongly suggesting its roles in repression of E-cadherin, as well as in activation of EMT as discussed in Chap. 6. [*N.B. A mesenchymal cell is a type of stem cell that can differentiate to different cell types and move between different locations.*]

We first briefly introduce the EMT pathway and its associated functions. The EMT pathway is involved in organ formation during embryogenesis. Under physiological conditions, its activation facilitates invasion of the endometrium and placenta placement to enable nutrient and gas exchange to the embryo. Cancer cells have apparently adapted to utilizing this pathway to facilitate their migration and then its reverse pathway, MET, to convert the migrated cells back to the original epithelial tumor form to become established in the new location(s). It is noteworthy that the two cell types are very different, both functionally and morphologically; yet, they are convertible to each other through the activation of the EMT and the MET pathways.

Briefly, the basic functions of epithelial cells include: (1) protection of the tissues lying beneath them from invasion by pathogens and physical assault; (2) exchange of chemicals between the tissues they separate; (3) secretion of hormones into the vascular system and secretion of sweat, mucus and enzymes that are delivered by ducts glandular epithelium; and (4) transferring sensation such as smell, sound and sight. The epithelial tissue is one of the four major tissue types in humans, with the other three being connective, muscle and nervous tissues as introduced in Chap. 1. In contrast, mesenchymal tissue is a type of undifferentiated loose connective tissue composed of cells that can migrate easily. Generally mesenchymal cells interact with each other via their focal points rather than requiring cell-cell adhesion.

A number of signals have been found capable of activating the EMT pathway, such as the activation of *TGFβ, FGF, EGF, HGF, RAS-MAPK, WNT* and the *NOTCH* pathway, as well as hypoxia. A recent study reports that the activation of a specific isoform of *CD44* (see Chap. 6), namely *CD44s*, is a necessary condition for the activation of EMT. By integrating the information above and the discussion in Sect. 6.2, one can speculate that the excess production of hyaluronic acid may be a key initiator for abolishing cell-cell adhesion through a sequence of events comprising activation of *SNAIL*, repression of E-cadherin, mechanical stretches induced by hyaluronic acid and interactions between *CD44s* and hyaluronic acid, leading to the disconnection between two cells at the end. Potentially a well-designed computational analysis of transcriptomic data and statistical inference could lead to a detailed model of how cells lose their cell-cell adhesion in specific cancer types.

In addition to cell-cell adhesion, the adhesion between epithelial cells and basement membrane is provided by interactions between integrins on the cell surface and fibronectins of the ECM. While the current knowledge of the regulation of the interactions between such proteins is not complete, it has been observed that the spatial distance between the two is one key regulating factor, namely mechanical forces that stretch the connection can lead to their separation (Li et al. 2008; Schwartz 2010).

Furthermore, cancer cells also need to breach the basement membrane in order to migrate. This is accomplished through assembly and activation of a large complex structure named *invadopodium*, which consists of a dense actin core surrounded by actin-assembly proteins, membrane trafficking proteins, signaling proteins and transmembrane proteinases (Hagedorn and Sherwood 2011; Hagedorn et al. 2013). When activated, invadopodia create tunnels in the basement membrane for delivery of *MMPs* to the desired locations to degrade the membrane, which will be followed by tumor growth into the newly created space as depicted in Fig. 10.3. The current understanding is that the assembly of invadopodia is regulated by pericellular accumulation of excess hyaluronic acid and its interactions with *CD44* and *PKC* (Artym et al. 2006; Hill et al. 2006; Montgomery et al. 2012).

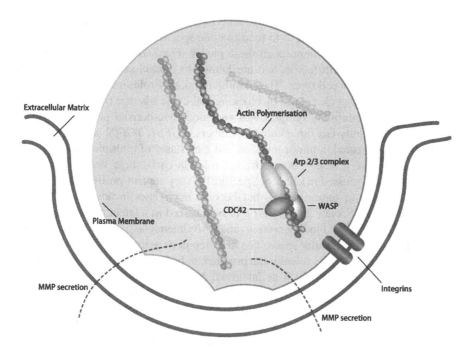

Fig. 10.3 A schematic of an invadopodium complex in action to break an extracellular matrix, where the *ARP2-3* protein complex has a key role in the regulation of the actin cytoskeleton; *CDC42* is involved in the regulation of cell cycle; and *WASP* is related to the Wiskott–Aldrich Syndrome

The stiffness of a basement membrane has been found to have a key regulatory role in promoting the activity of invadopodia (Alexander et al. 2008; Parekh et al. 2011), in addition to its role in stimulating cell proliferation as discussed in Chap. 8. As reviewed in the earlier chapters, the stiffness of a basement membrane is mainly determined by the relative concentrations of collagen, elastin and laminin (Alberts et al. 2002; Owen and Shoichet 2010). Interestingly, hyaluronic acid fragments may have an important role in determining the relative concentrations of these macro-molecules, hence the stiffness of the matrix. Specifically, hyaluronic acid fragments are able to up-regulate collagen-encoding gene (Chung et al. 2009) and enhance the synthesis of matrix elastin (Kothapalli and Ramamurthi 2009; Kothapalli et al. 2009). In addition, hyaluronic acid has also been linked to the production of laminin in various diseases such as cirrhosis (Lindqvist 1997). Overall, increased stiffness results in an increased concentration of invadopodia, as well as increased activities by invadopodia via the myosin II-*FAK/CAS* (Crk-associated substrate) pathway (Alexander et al. 2008), which could be triggered by the increased production of hyaluronic acid, ultimately induced by increased ROS levels as discussed later in this chapter.

To ascertain how these distinct components may functionally cooperate to initiate the metastatic process, one needs to focus on one specific protein family, *TGFβ*, as it may be the thread that connects all these pieces. The *TGFβ* proteins are a well-studied family of growth factors and are known to control cell proliferation and differentiation in most cell types. They exhibit regulatory roles in: (1) the activation of apoptosis; (2) cell cycle control by blocking cell-cycle advance from the G_1 to the S phase; and (3) inhibiting lymphocytes and monocyte-derived phagocytes from formation. The family has three known members, *TGFβ1*, *TGFβ2* and *TGFβ3*, all having been implicated in tumor invasion and metastases of multiple cancer types. Interestingly, these proteins serve different roles in early stage *versus* advanced stage cancers as discussed in Chap. 6. Specifically, they are anti-proliferative factors in the early stage of tumorigenesis but become oncogenes in advanced cancers (Prime et al. 2004; Seoane 2006). *TGFβ* is synthesized as a latent protein complex with *LTBP* (latent *TGFβ* binding protein) and *LAP* (latency-associated peptide), and secreted into the extracellular space. The first step in its activation is that of release from the complex. After its release, the protein can be activated by multiple factors under different conditions, such as integrins, *MMPs*, the tissue-injury responder protein *TSP1* (thrombospondin 1) (Rifkin and Sheppard 1999; Yu and Stamenkovic 2000) and even by changes in the ROS (Barcellos-Hoff and Dix 1996) and pH levels (Lyons et al. 1988). A particular mechanism is most relevant here, that of mechanochemical signaling through integrin-αvβ5 (Wipff et al. 2007), strongly suggesting that tissue growth may play a role in the activation of *TGFβ*.

To examine if hyaluronic acid may have a role in the breaching of the basement membrane, an analysis of transcriptomic data of three cancer types, namely brain, liver and lung, was conducted. The statistical analysis revealed that, when *TGFβ* is activated, key hyaluronic acid synthesis and export genes, e.g., HAS2 and ABCC5 (see Chap. 6), tend to be expressed as shown in Fig. 10.4, suggesting that *TGFβ* may be a regulator of hyaluronic acid synthesis. Interestingly, multiple studies have reported that *TGFβ* can indeed increase the synthesis of hyaluronic acid (Wang et al. 2005; Nataatmadja et al. 2006), hence providing strong supporting evidence to the hypothesis.

In addition to its role in promoting hyaluronic acid synthesis, *TGFβ* can also activate the EMT pathway through multiple mechanisms. One pathway involves the activation of *SMAD2-3*, which then forms a complex with *SMAD4*, together serving as a transcription factor to trigger the EMT pathway (Miyazawa et al. 2002; Derynck and Zhang 2003; ten Dijke and Hill 2004; Gui et al. 2012), where *SMADs* are a family of proteins that transduce extracellular signals from *TGFβ* ligands to the nucleus. Another pathway does not involve the *SMAD* proteins, instead through the activation of *ERK MAP* kinases, *RHO GTPases* and the *PI3K/AKT* pathway (Derynck and Zhang 2003; Xu et al. 2009; Zhang 2009). Overall, the current understanding is that *TGFβ* can activate both *SMAD* and non-*SMAD* pathways, which crosstalk with various signaling pathways to trigger EMT and possibly other pathways depending on the specific context.

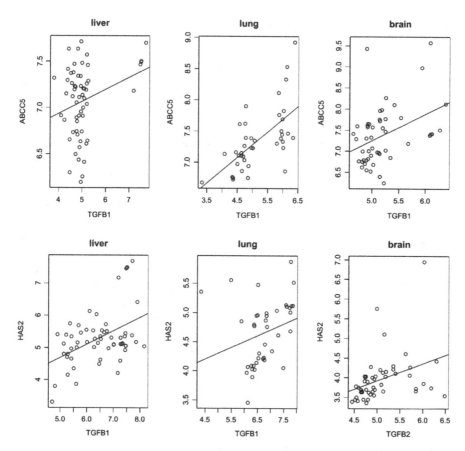

Fig. 10.4 Gene expression of *TGFβ versus* genes responsible for hyaluronic acid synthesis across multiple tissue samples of three types of advanced cancer, where *HAS2* and *ABCC5* are hyaluronic acid synthesis and export genes, respectively (data from the GEO database)

Joining the above information and discussion, one can postulate the following sequence of events that could lead to the breach of the basement membrane by a growing tumor. The continuous growth of a tumor, coupled with inflammation, may create mechanical forces that promote the activation of integrin-αvβ5, which in turn activates *TGFβ*, leading to the synthesis and export of hyaluronic acid. The increased production and export of hyaluronic acid can further increase the aforementioned mechanical forces, further increasing bioactive integrin-αvβ5 and *TGFβ*, which, in conjunction with the repressed E-cadherin due to hyaluronic acid, can ultimately lead to the activation of EMT, as well as increased mobility and invasiveness of the cancer cells. The actual breaching of the basement membrane is accomplished by *MMPs* delivered to the right locations through the assembly and activity of invadopodia, which seem to be initiated by the production and degradation of hyaluronic acid. Overall, hyaluronic acid exerts an essential role in making this sequence of events possible.

10.1.2   Interactions with Stromal Cells

The second key event during local invasion of tumor cells is that they enter the stromal compartment where they can interact directly with stromal cells, namely fibroblasts and pericytes, which are the supporting cells to the parenchymal cells in an organ. The local immune cells are also considered as stromal cells because of their supporting roles.

The main physiological functions of fibroblasts are to synthesize and export ECM proteins, glycoproteins and glycosaminoglycans, and to function in wound healing. Multiple diseases are closely related to excess production, deposition and contraction of the ECM, such as diabetic nephropathy, liver cirrhosis, arteriosclerosis and rheumatoid arthritis. The current understanding is that *TGFβ* can induce not only the synthesis by fibroblasts but also the contraction of an ECM. Specifically, it induces a specific form of fibronectin, *EDA* (ectodysplasin-A), which together with *TGFβ1* can trigger the enhancement of α-SMA (alpha-actin-2) and accelerate the contraction of fibroblasts (Ina et al. 2011), a major cue for activation of latent *TGFβ1* (Wipff et al. 2007). The information here complements the above model for mechanical force-induced activation of integrin-αvβ5, ultimately leading to the activation of EMT. Moreover, these two processes may interact, sending signals to each other and together accomplishing EMT activation. Well-designed analyses of transcriptomic data for various advanced stage cancers may lead to the establishment of a self-consistent model for the entire process of EMT activation in the microenvironment of advanced cancers. In addition to this role, cancer associated fibroblasts are also known to release a variety of proteases such as *MMPs*, hence further facilitating remodeling of the ECM needed by cancer invasion and metastasis.

Pericytes are contractile cells that surround the endothelial cells of capillaries. Their physiological function is to regulate capillary blood flow and clearance of cellular debris. Previous studies have discovered that pericytes serve as a gatekeeper in preventing cancer cells from spreading as it has been demonstrated that pericyte-deficiency in mice increases cancer metastasis (Xian et al. 2006). A report on diabetic retinopathy may provide a strong clue as to why cancer tissues tend to have decreased numbers of pericytes as has been observed. The study concludes that the activation of the angiopoietin-2 protein leads to a reduction in pericyte population (Hammes et al. 2004). Hence, one can infer that the increased expression of angiopoietin-2, triggered by the need for angiogenesis in a tumor microenvironment, results in a decrease in pericyte population, hence gradually losing its safeguard against metastasis. This model is well supported by the transcriptomic data collected on a large number of tissue samples of four cancer types. Specifically the pericyte concentration decreases in tumor samples as a cancer advances, as reflected by the decreased expressions of pericyte marker genes: *ACTA2* (actin, aortic smooth muscle), *CSPG4* (chondroitin sulfate proteoglycan 4), *ENPEP* (glutamyl aminopeptidase) and *ANPEP* (alanyl aminopeptidase), as well as by increased expression of *ANGPT2* (angiopoietin-2) as shown in Fig. 10.5. It can be expected that data mining and statistical inference on larger datasets in a more systematic manner could lead to the development of a detailed model for this hypothesis.

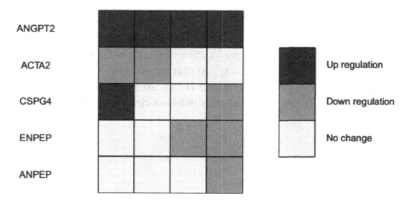

Fig. 10.5 Expression-level changes of five genes related to pericyte population in cancer *versus* control samples. Each *column* represents one cancer from *left* to *right*: renal cell carcinoma, leukemia, liver, and lung adenocarcinoma, and each *row* represents a gene. *Black* and *gray* denote increase and decrease in gene-expression levels, respectively, while *white* is for no change. The detailed dataset used here is given in Appendix

In addition, there seems to be a partner relationship between tumor cells and the tumor-associated stromal cells from an energy-metabolism perspective. As discussed in Chaps. 5 and 8, cancer cells produce high concentrations of lactic acid in their microenvironments, which is an incompletely used energy source. The lactate acid can be further used by the stromal cells, which do not necessarily use glycolysis for ATP generation as tumor cells do after its oxidation back to pyruvate. A proposal has been made with supporting data that the regenerated, excess pyruvate in stromal cells can also be released and reused subsequently by the cancer cells (Martinez-Outschoorn et al. 2011). This model is also partially supported by the observation that tumor-associated stromal cells tend to have low expression levels of their glucose transporters, indicating the possibility of low glucose uptake, thus enabling tumor cells greater access to glucose. In addition, epithelial cancer cells have been found capable of inducing the expression of metabolic genes in the neighboring fibroblasts and enhancing their output for the production of energy-rich metabolites (Pavlides et al. 2009; Migneco et al. 2010; Martinez-Outschoorn et al. 2011), hence making these two cell types adopt a close "host-parasite" relationship.

The results of interactions between cancer and immune cells can be very complex as different immune cells may have different roles, including both anti- and pro-cancer, at different stages of cancer development. In addition to the direct involvement of immune cells in tumorigenesis, they also serve as a selection process for those tumor cells that elude destruction by immune cells, as discussed in detail in Chap. 8.

10.2   Traveling Cancer Cells

While cancer cells can travel to distant locations through both blood circulation and the lymphatic system, clinical data suggest that the majority of cancer cells travel through the circulatory system (Wong and Hynes 2006; Eccles and Welch 2007). Hence, we focus on the fate of cancer cells in blood circulation in this section.

10.2.1   Intravasation

The first step for the invading cancer cells to enter circulation is to cross the endothelial cell barrier that forms the wall of capillaries. Innate immune cells are known to have essential roles in promoting cancer cell intravasation, analogous to their roles during cancer initiation as discussed in Chap. 7. Specifically, tumor-associated macrophages (TAMs) promote tumor angiogenesis through release of *VEGFs* to stimulate the formation of tumor blood vessels, which tend to be leaky compared to normal blood vessels, hence allowing cancer cells to enter the circulation more easily.

10.2.2   In Circulation

It is estimated that for each gram of tumor tissue accessible to blood circulation, about one million tumor cells actually escape into the circulation, and they typically remain there for just a few hours before they lodge on the inside wall of the blood vessels or are destroyed. Another estimate suggests that 1 out of 10,000 cancer cells in circulation can survive and ultimately settle in a distant location.

Circulating tumor cells (*CTCs*) need to overcome a number of challenges to survive in circulation, including mechanical forces and immune destruction. In addition, they also need to overcome a programmed cell death, called *anoikis*, which is self-induced when cells leave their original habitats and become anchorage-free (Douma et al. 2004; Gupta and Massague 2006). The mechanism(s) by which the CTCs avoid this programmed cell death is poorly understood, but the following information provides some hints about this. It is known that tumor cells can continue to thrive in an anchorage-independent manner through a signaling process mediated by cell-surface hyaluronic acid to avoid anoikis. However, it is not known whether the CTCs have any hyaluronic acid on their surfaces. While determination of whether this is the case can be done through metabolic analyses of cells using techniques such as mass spectrometry, no data are currently available in the public domain to give a direct answer to the question.

Recent findings suggest that protein *TRKB* (tyrosine receptor kinase B) may have an important role in rendering tumor cells anoikis-resistant (Kim et al. 2012), where *TRKB* is a growth factor that can induce cell survival and differentiation pathways, upon binding with and being activated by its cognate ligand such as *BDNF* (brain derived neurotrophic factor). Another study has reported that the hyaluronic acid tetrasaccharide can increase the expression of *BDNF* and *VGEF* in an *in vitro* experiment (Wang et al. 2012). Furthermore, a study on tissue regeneration has shown that high molecular-weight hyaluronic acid can serve as a scaffold for *BDNF* during tissue regeneration (Takeda et al. 2011). Together this information suggests the possibility that hyaluronic acid may be generated by CTCs, which trigger the expression of *BDNF*, possibly along with other factors, that in turn activates *TRKB* and provides the CTCs with anoikis-resistance. Clearly this possibility requires further experimental validation.

While in circulation, CTCs tend to aggregate into clusters with platelets (Cho et al. 2012a). Such a formation should give CTCs an advantage for their survival against the mechanical forces of the blood flow, the shear force and immune attack in circulation. Platelets seem to have an essential role in transporting the CTCs and maintaining their viability as multiple studies have demonstrated that platelet depletion, or even an inhibition of tumor cell-induced platelet aggregates, diminishes metastasis (Gasic et al. 1968; Gasic 1984; Amirkhosravi et al. 2003; Palumbo et al. 2005). While the detailed binding mode between CTCs and platelets has not been thoroughly elucidated, it has been proposed that the interaction is through binding of integrins on the cell surfaces of the two types to common fibronectin or collagen, or through binding of *PARs* (protease activated receptors) on the two cell surfaces to common thrombin (Gay and Felding-Habermann 2011). Note that integrins are a family of transmembrane receptors for providing linkages among cells or between cells and macromolecules in the ECM as discussed earlier.

An analysis of gene-expression data of breast cancer CTCs in the public domain (Molloy et al. 2012) revealed that genes encoding integrin-$\alpha 2b\beta 3$, integrin-$\alpha 2\beta 1$, *GPI* receptors (responsible for platelet adhesion), *ADP* (adipose) receptors and the "platelet aggregation plug formation" pathway are all up-regulated, hence providing strong evidence in support of the above proposal.

10.2.3 Extravasation

Tumor extravasation is the process through which the CTCs lodge to the inner wall of a blood vessel of a distant organ and then penetrate the wall to settle in the stromal compartment of the organ. Little is known about the mechanism of tumor extravasation, but it has been speculated that it is probably similar to that of leukocyte extravasation into inflammatory tissues during immune responses (Strell and Entschladen 2008).

Briefly, the extravasation process of leukocytes, such as T-cells, natural killer cells, neutrophil granulocytes and monocytes, consists of the following steps: (1) *rolling*: the vascular endothelial cells recruit leukocytes through the protein selectin on cell surfaces, which bind with selectin ligands such as *SELPLG* (p-selectin glycoprotein ligand-1, also known as *CD162*) on the surfaces of leukocytes, forming loose interactions, where selectins are a family of cell-adhesion glycoproteins. Because the interactions tend to be relatively loose, the recruited leukocytes have rolling motions in response to the blood flow, hence the name; (2) *adhesion*: integrins are activated on both leukocytes and endothelial cells during the rolling step, and their binding gives rise to tight adhesion between the cells; and (3) *trans-migration*: leukocytes transmigrate through the endothelium without irreversibly impairing its integrity as they tend to move through the endothelial monolayer between the endothelial cells (Hofbauer et al. 1999).

It has been speculated that while cancer cells may use a similar mechanism for extravasation, they may use a different set of selectin ligands in different metastasis types to accomplish the initial loose binding with endothelial cells, such as *HCELL* (an E- and L-selectin ligand), *CD44*, *ELAM1* (E-selectin ligand-1, also known as *CD62E*) for bone metastasis and *CEA* (carcinoembryonic antigen) for colon metastasis (Dimitroff et al. 2005; Strell and Entschladen 2008; Thomas et al. 2008; Dallas et al. 2012; Hiraga et al. 2013). The adhesion between cancer and endothelial cells may be accomplished via binding between integrins as in the case of leukocytes discussed above, but possibly by subgroups different from those used by leukocytes, specifically the β2 subgroup. A recent study has observed that the α4 subgroup of integrins is used in some cancers (Okahara et al. 1994; Garofalo et al. 1995; Bendas and Borsig 2012). For the transmigration step, cancer cells may have evolved a strategy different from the one used by leukocytes as they tend to be highly destructive by damaging the endothelium, possibly because of their substantially larger sizes compared to those of leukocytes and no restraint being placed on them for not impairing the integrity of endothelium. No specific genes have been implicated for this, but it is expected that computational analyses of transcriptomic data on multiple metastasis types may lead to candidate genes.

The CTCs that reached the new locations are referred to as *DTCs* (disseminated CTCs) in the literature, which specifically refer to the direct progeny of the primary cancer rather than highly transformed metastatic cancer. Publicly available gene-expression data of *DTCs* originating from prostate cancer have been analyzed. It was found that the following gene groups are up-regulated in DTCs in comparison with the corresponding CTCs: cell cycle related genes such as the G_1-phase and S-phase genes, actin-cytoskeleton remodeling genes, *WNT*-signaling pathways, DNA synthesis, glucose metabolism, steroid metabolism and sphingolipid metabolism. These data strongly suggest the following: (1) these DTCs are in a state of proliferation; and (2) these cells are under oxidative stress.

For (1), it is worth noting that a DTC population tends to remain stable for months or even years in their new habitats, hence possible dormancy of these cells having been suggested (Meadows 2005; Wang et al. 2013). Clearly this hypothesis was not supported by the above results from data analysis. One possible explanation for these two pieces of seemingly conflicting information is that DTCs may initially be in a proliferation state but gradually stop proliferation to remain in a growth-arrest state in the cell cycle, possibly triggered by their incompatibility with the new microenvironment. Such incompatibility may include: (1) attacks from the local immune cells that have not become associated with the cancer cells; (2) limitation in the blood supply, which is designed only to support the local normal cells before the establishment of tumor angiogenesis; (3) toxic effect by the increased O_2 level in comparison with their previous habitats (see Chap. 11 for details); and (4) altered ROS and pH levels, both of which will be quite different from their original sites, hence possibly causing substantial changes in the cellular metabolism of the DTCs if not killed by the altered ROS and pH. Another possibility is that the majority of the proliferated cells from the DTCs may be destroyed due to their incompatibility with the new microenvironment (see Chap. 11 for a detailed discussion). Therefore, the observed population stability of the DTCs may represent a dynamic equilibrium, rather than no proliferative activities. Potentially both possibilities may be true for some metastatic cancers.

For (2), the possibility is an interesting one that has not received much attention in the cancer literature. A detailed analysis of the DTC transcriptomic data revealed that a number of *CYP* genes, all encoding anti-oxidant P450 enzymes, are up-regulated along with increased sphingolipid metabolism *versus* those in CTCs. Similar expression patterns are also observed between metastatic cancer and the matching primary cancer tissues (see Chap. 11 for details). These data strongly suggest that the DTCs are under increased oxidative stress and the plasma membranes are damaged. One possible explanation is: the DTCs have just migrated from a highly hypoxic condition where they had been residing for an extended period of time, possibly up to 10 or 15 years, and the cancer cells may have evolved to become nearly anaerobic. When they are suddenly exposed to an oxygen-rich condition, their cellular responses to the new stress induced by the increased O_2 level include the activation of *CYPs* and other anti-oxidant genes. In addition, the plasma membrane damage, suggested by the increased sphingolipid metabolism, may be the result of increased and continuous lipid peroxidation produced by the increased O_2 level as discussed in Chap. 11, where the implications of these observations are also offered.

Overall, one can see that cancer cells leave their primary bases and travel through the circulatory system, with help from multiple local environmental factors such as hyaluronic acid and stromal cells. This journey occurs in various steps, including disconnection from the original site and other cells, protection while in circulation and establishment in distant organs. The multi-faceted roles played by hyaluronic acid highlight this class of molecules as probably a most important facilitator throughout the whole process of cancer development (see Chap. 6). The relatively simple data analyses as done here have revealed interesting and previously unknown information about the activities of stromal cells and DTCs as they leave their base and seek establishment in the new location(s). The following section discusses how the DTCs survive the new environment.

10.3 Adaptation to the New Microenvironment

As mentioned earlier, it typically takes just a few hours for the CTCs to adhere to the endothelial cells along the blood vessels after leaving their primary sites, but they may remain dormant for weeks, months or even years before they begin to actively proliferate (Meng et al. 2004; Alix-Panabieres et al. 2008). During this period, these cells must overcome a number of obstacles to remain viable and to retain the ability to achieve reactivation.

The questions we are interested in understanding here are: (a) *What challenges must the metastatic tumor cells overcome in their new microenvironment*; (b) *What changes must these cells make in their microenvironment before they begin proliferating again*; and (c) *What determines the rate of proliferation of a metastatic cancer, knowing that some metastatic cancers grow substantially faster than others*?

To address these questions, it is prudent to first review a hypothesis proposed by British surgeon Stephen Paget over 100 years ago when he observed that breast-cancer patients tended to develop secondary cancer in their livers. Since then, it has been widely observed that cancers from different origins have propensities to different destinations. At the heart of the Paget hypothesis, different organs, in terms of their microenvironment, may have different levels of compatibility with specific metastasizing cells, the so-called "seed and soil" hypothesis (Fidler 2003; Fidler and Poste 2008). The hypothesis has recently regained some momentum and is being considered as a good model for distant metastasis because of the finding that the expression patterns of genes involved in mediating the metastasis of breast cancer to bone are rather different from those that direct the metastasis to lung (Langley and Fidler 2011). While validating the "seed and soil" hypothesis experimentally may prove to be tricky, it can potentially be computationally validated (or rejected, refined) through comparative transcriptomic data analyses of primary *versus* matching metastatic cancer samples across multiple cancer types, particularly samples with the same type of primary cancer that has metastasized to different organs. This can be accomplished by checking if primary cancers that metastasized to different organs tend to share similar expression patterns of some to-be-identified genes among those that spread to the same organs, which are not shared by those metastasized to different organs. Similar analyses can be carried out on cancers that have metastasized to the same organ but from different origins.

Interestingly, if one compares the microenvironments of wherever the primary cancer is located and wherever it may spread to, the difference between the old microenvironment and the new one is substantial and multi-faceted. The low compatibility between the metastatic cells and their new environment can make the survival of the new settlers very challenging. One key piece of information that the suggested analyses above could potentially reveal is: which aspects in the cell-microenvironment compatibility are the most essential factors in determining if DTCs can remain viable and develop in a specific new location? An answer to this question could potentially have a profound impact on our understanding of metastatic cancer and identifying possible ways to slow their growth. In the following, we discuss the adaptations the DTCs must make in order to survive in their new environment.

10.3.1 Challenges to Cancer Cells in the New Microenvironment

A key challenge for the arriving metastatic cancer cells is survival in the new microenvironment. A fundamental difference between the new and the old one is that the original environment is a tumor environment that is pro-cancer growth while the new one is a normal microenvironment in a healthy organ, which is anti-cancer. This difference could be substantial. Using pH as an example, the pH level in the new microenvironment will be higher than that of the old one as discussed in Chap. 8; furthermore, the environment is not lactate-rich as in the original one. Hence, T-cells in the new environment may be much more aggressive against the new settlers, compared to the T-cells in their old habitats where they have become less active against cancer cells due to the lactate-rich environment. More generally, the original microenvironment offers a variety of pro-cancer signals, such as anti-apoptotic, angiogenesis, cell survival and proliferation signals, but these will not be available in the new location. Similarly, their supporting stromal cells, such as TAMs, will not be available upon their arrival. Another key challenge is that, unlike the highly hypoxic environment typically associated with primary cancers, the new environment is rich in oxygen. Having the tumor cells in the new environment is analogous to putting anaerobic cells in an aerobic environment; it may either kill the new comers or yield substantial changes, for example through regulation or selection of specific mutations, in their cellular states to protect them against the toxic oxygen. A similar argument can be made about the need to adapt to the local redox states by the new settlers. Further discussion along this line is given in Chap. 11.

Another key factor that may affect the viability of the arriving tumor cells is their interactions with the ECM. As discussed in Chaps. 5 and 6, such interactions play essential roles in both the transformation and viability of the tumor cells in their original habitat. One obvious difference now is that the old ECM is pro-growth, and hence possibly very stiff, while the new one is clearly not. This may be one of the reasons for the relatively slow growth during the early stage of the new comers. Other than this, very little is currently known about the differences between the physicochemical properties of the new *versus* the original ECM. In addition, cell-cell competition will be a key factor that may affect the fate of the new settlers (as discussed in Chap. 8), knowing that they continue to proliferate as revealed by our analyses of gene-expression data of DTCs and shown in Fig. 10.6. One possibility could be that the long dormancy time may represent the preparation time needed by a metastatic cancer to select the fittest cells for the new environment through rounds of proliferation and cell-cell competition (see Chap. 8 for the details). It is worth noting that cell-cell competition does not change the overall biomass of a tumor but only serves as a selection process for more robust cells in the new environment, as discussed in Chap. 8. Overall, one can imagine that, for the cancer cells to become established and thrive in the new environment, they must undergo substantial adaptations to make the cell population strong enough to survive the new environment.

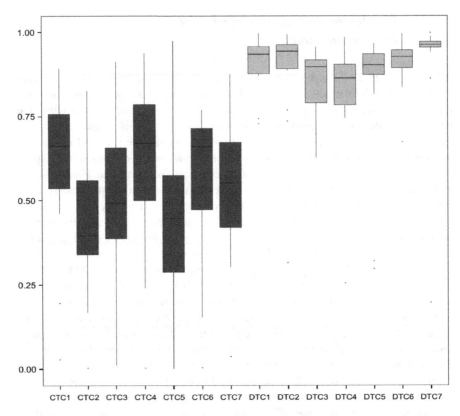

Fig. 10.6 The rankings of 13 cell-cycle related genes, namely *CCNA2* (cyclin A2), *CCNB1*, *CCNB2*, *CCND2*, *CCNE1*, *CCNF*, *CDH1* (cadherin 1, type 1, E-cadherin), *E2F1*, *MCM2* (mini-chromosome maintenance complex component 2), *MCM3*, *MCM4*, *MCM5*, *MCM6* among all N human genes in terms of gene expression levels, where N is set at 20,000 here. For each of the seven CTC samples (on the *left*) and the seven DTC samples (on the *right*), a gene's normalized expression rank is calculated as: $(N-\text{rank of the gene's expression})/N$. Each *dot* in a box plot is the normalized rank for 1 of the 13 above genes in a specific sample

10.3.2   Changing the Microenvironment

One proposed mechanism by which the new settlers alter their new environment to enhance their chance for survival is through the release of exosomes. *Exosomes* are derived from cancer cellular endosomes through a process termed *inward budding*, where cytoplasmic RNA molecules and functional proteins are encapsulated into exosomes and then secreted through a process driven by *RAB* (Rab escort protein 1) *GTPases* (Hsu et al. 2010; Ostrowski et al. 2010). The tetraspanin–integrin complex enables the binding of exosomes to the target cells that express adhesion molecules such as *ICAM1* (intercellular adhesion molecule 1) on the cell surface. Such adhesion molecules can be activated by pro-inflammatory signals. Cancer cells *in situ* have been found to exchange proteins endowed with oncogenic activities with each other through exosome-mediated transfer for their survival (Kahlert and Kalluri 2013). A number of studies have reported cases where cancer cells change their environment by releasing exosomes into the extracellular space. For example, exosomes from breast cancer cells have been found to convert mesenchymal cells to myofibroblasts via a *SMAD*-mediated pathway. Myofibroblasts, a less differentiated form of fibroblasts, are a key source of matrix-remodeling proteins within the tumor microenvironment and participate in tumor angiogenesis (Vong and Kalluri 2011; Cho et al. 2012b). Another example is that melanoma-derived exosomes enhance the lung endothelial permeability and increase lung metastases in mice (Peinado et al. 2012).

10.3.3   *From Proliferation to Dormancy*

Generally, DTCs enter a period of dormancy after becoming lodged in their new locations. This could result from the limited availability of blood supply (referred to as *angiogenic dormancy*), immune surveillance and attack on fast growing cells, or by cellular quiescence triggered by the incompatibility with the new environment, where growth may be arrested in the G_0-G_1 phase of the cell cycle. The duration of dormant time varies substantially across different cancers, even cancers of the same type. One observation is that the less differentiated (i.e., more stem-cell like) tumor cells tend to have shorter dormancy times and become more aggressive in their renewal to the proliferative phase (Aguirre-Ghiso 2007; Wikman et al. 2008). The overall level of understanding of cancer dormancy is quite limited at this point, partially due to the reality that very limited experimental data on such cells have been collected, possibly due to the challenging nature in identifying these cells *in vivo*.

One proposed tumor-dormancy model is that after the DTCs arrive and adhere to the local cells, their metastasis suppressor genes may regulate cell dormancy in response to the stresses invoked by the new environment, which will protect them from detection by the immune system (Horak et al. 2008). Another study has suggested that interactions between the arriving tumor cells and the local ECM may play a key role in sending the tumor cells into dormancy. Specifically, the study found that melanoma cells are growth arrested at the G_1/S checkpoint when they are in contact with polymerized fibrillar collagen. In comparison, alteration of the collagen formation to that of denatured collagen activates the cell cycle and moves to the S phase (Hansen and Albrecht 1999). This may represent another problem that can benefit from comparative statistical analyses of transcriptomic data across multiple types of early metastatic cancers, which could lead to discoveries regarding the validity of the model or whether it is true for only certain types of metastatic cancers.

A number of genes have been implicated in executing the growth inhibition of metastatic tumor cells, hence referred to as *metastasis suppressor* genes. Such genes have the ability to prevent proliferation as the tumor cells are becoming established in their new environment by inducing dormancy or apoptosis. *KISS1* (kisspeptin) is one such gene.

The Kisspeptins, processed by *KISS1*, can bind with *GPR54* and possibly regulate cellular cytoskeletal reorganization to block cell proliferation through the induction of dormancy (Nash et al. 2006; Paez et al. 2012). *KAI1* (R2 leukocyte antigen) is another metastasis suppressor gene. Its encoded protein can form a complex with integrins, and together they inhibit cell proliferation through induction of tumor-cell growth arrest. Other known metastasis suppressor genes include *MKK4* (an activator of *MAPK*, *P38* and *JNK*), *BRMS1* (inhibitor of angiogenesis by suppressing the *NFκB* activity) and *CTGF* (a regulator of cell adhesion, proliferation and differentiation). A recent study suggests that some ECM components may have important roles in maintaining metastatic dormancy (Barkan et al. 2010).

One should be able to design and carry out computational analyses of relevant transcriptomic data of early metastatic cancers to validate, refine or reject this hypothesis. One thing is clear, however, during dormancy the cancer cells continue some of their cellular activities, including proliferation as discussed earlier. It can be hypothesized that some of these activities are related to a change(s) in their metabolic state for their adaptation to the new microenvironment, which can also be computationally validated, refined or rejected when transcriptomic data for metastatic cancers at the early stages are available.

10.3.4   Reactivation to Proliferation from Dormancy

The remodeling pathway of the ECM is believed to have a key role in the reactivation of cancer growth from dormancy. Specifically, it has been reported that dormant cancer cells have a distinct cytoskeletal organization, which has only transient adhesion to the ECM (Barkan et al. 2010). Changing the components of the ECM, such as an increase in the fibronectin composition and hence the structure as well as the physical properties of the matrix, can reactivate the dormant tumor cells. In addition, type I collagen has been found to exhibit reactivation roles of dormant tumor cells, suggesting that it may not be specific molecular types, but rather the shape and the physical properties, such as stiffness of the matrix, that can reactivate dormant cells.

A few studies have been published that focused on the detailed molecular mechanisms of the reactivation process. One such investigation on bone metastasis found that local inflammation increases the expression of the *VCAM1* (vascular cell adhesion molecule 1) protein in cancer cells in a bone microenvironment. When *VCAM1* sheds from the cell surface, the soluble *VCAM1* molecule attracts osteoclast progenitors to the cancer cells through binding with the cognate receptor integrin α4β1, leading to adhesion of the progenitors to each other and ultimately resulting in an increase in osteoclast activity and the escape of dormancy (Lu et al. 2011). One possible cause of the local inflammation could be the result of certain renewed activities in the dormant cancer cells, which triggers the immune response and also alters the ECM properties.

As discussed throughout this chapter, hypotheses like the above can be computationally examined by determining changes in the expression patterns of genes believed to be involved in the aforementioned processes. Then, an assessment of the correlations between the expression changes of these genes and those possibly linked to the reactivation of cancer cells from dormancy could lead to a firm validation or rejection. The key for doing such analyses is the availability of gene-expression data of metastatic cancer cells in dormancy *versus* such cells that are exiting dormancy, which are currently lacking. Complementing such studies, it should be possible in time to also access proteomic data to determine protein content directly, including post-translational events.

A study on the transition from quiescence to proliferation of metastatic breast cancer showed that it is the cytoskeletal reorganization with *F-actin* (a protein that can form a linear polymer microfilament, relevant to cell mobility and contraction) that leads to actin stress fiber formation and reactivation of proliferative growth (Barkan et al. 2008). This study also showed that the ECM and tetraspanins play critical roles in enabling cell survival, proliferation and cytoskeletal changes required for the switch from dormancy to proliferation and invasion. Clearly this hypothesis can also be examined computationally as discussed earlier.

10.4 Hyaluronic Acid Is a Key Facilitator of Metastasis

Like the roles played by hyaluronic acid in cancer initiation, this glucosaminoglycan seems to also serve as a major facilitator of cancer metastasis and the initial development after migration (see discussion in the previous sections). Intuitively this makes sense since human cells have evolved in such a way that the ECM, of which hyaluronic acid is a part, serves as the main signal source for cell survival, proliferation and mobility among other cellular state transitions as introduced in Chap. 1. In these capacities, hyaluronic acid continues to serve cancer cells by facilitating their migration and survival in their new environment(s).

10.4.1 Motility

A question of interest here is: *Are there thresholds of some conditions beyond which primary cancers start to metastasize*? Clearly, various pericellular or intracellular environmental factors can be considered, such as the level of hypoxia, oxidative stress, extracellular pH or anything that can potentially lead to the increased production, and hence export, of hyaluronic acid as there are multiple lines of evidence suggesting that this molecule is a key to initiate the metastasis process. Here some discussion is provided on the accumulation of ROS and its role in increased hyaluronic acid production.

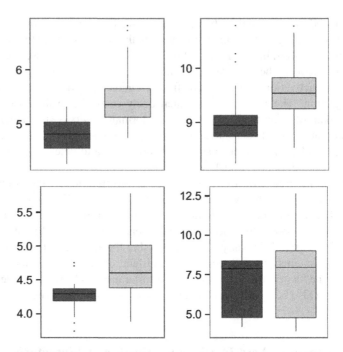

Fig. 10.7 Ranges of ROS level fluctuation, reflected by expression levels of ROS responsive genes, GSS (*top*) and GCLC (*bottom*) in normal (*dark gray*) *versus* cancer (*light gray*) in two datasets: GSE13195 consisting of 25 pairs of gastric cancer *versus* matching control tissues (the *panels* on the *left*) and GSE19804 consisting 60 pairs of lung cancer *versus* matching control tissues (the *panel* on the *right*)

It is known that ROSs such as the superoxide radical, hydrogen peroxide and hydroxyl radicals have an important role in cancer development. The general understanding has been that ROS tends to accumulate as a cancer evolves, leading to increased DNA damage (Waris and Ahsan 2006), faulty antioxidants, activation of cancer-related transcription factors such as *NFκB* and a gradual change in the cellular redox state (Gupta et al. 2012). A recent study even suggests that cancer metastasis is a cancer cell's escape from oxidative stress in their primary sites (Pani et al. 2010). Figure 10.7 shows a general trend of ROS levels in two cancer types as they progress, measured in terms of the expression levels of ROS stress-responsive genes, indicating that when the ROS levels fluctuate inside cells, their transient maximum ROS level tends to be higher in cancer *versus* matching controls.

Previous studies have reported that ROS can induce *TGFβ* (Barcellos-Hoff and Dix 1996; Jain et al. 2013), which can serve as an oncogene in advanced cancers and trigger the synthesis of hyaluronic acid by activating both hyaluronic acid synthases *HAS1* and *HAS2* (Liu and Gaston Pravia 2010). Actually it has been widely observed that advanced stage cancers tend to have increased hyaluronic acid production, which we posit is the result of the increased accumulation of ROS. In addition, it has been well established that an elevated level of hyaluronic acid increases the motility of tumor cells, facilitating their escape from the primary tumor site and starting the metastatic process. One study, for example, has shown that excess hyaluronic acid synthesis and processing directly promotes metastasis of prostate cancer (Bharadwaj et al. 2009). Hence it is reasonable to speculate that this may be related to the mechanical forces generated by the increased content of hyaluronic acid around the cancer cells, leading to the disconnection between the host cells and other cells, as well as between the cells and their basement membrane (discussed in Sect. 10.1).

10.4.2 Prevention from Programmed Cell Death

Analyses of gene expression data of CTCs of breast cancers (see Table 10.3) show that their hyaluronic acid synthesis gene, *HAS2*, and exporter gene, *ABCC5*, are both up-regulated compared to the levels in their primary counterparts, indicating that hyaluronic acid is being synthesized, exported and possibly used in circulation. Hence, one can hypothesize that these circulating cells may use hyaluronic acid on their cell surfaces to prevent activation of the programmed cell death by anoikis as discussed earlier. This possibility clearly requires experimental validation.

10.4.3 Helping Adaptation to and Change of the New Microenvironments

Gene expression data of newly-arrived cancer cells in a secondary site show that the hyaluronic acid synthesis genes, *HAS1* and *HAS2*, and the exporter gene, *ABCC5*, are further up-regulated compared to CTCs, indicating that hyaluronic acid, in addition to its role in preventing programmed cell death, is also serving in a key role to assist the integration of the cells into the local environment. Previous studies have shown that hyaluronic acid and its cell-surface receptor *CD44* are important in changing the local environment to a more pro-metastatic environment by promoting the generation of various growth factors, e.g., *FGF* and *VEGF* for growth and tumor angiogenesis, respectively (Misra et al. 2011; Ween et al. 2011).

10.5 Concluding Remarks

Rapid progress has been made in the past decade in our overall understanding of the process of cancer metastasis, such as elucidation of the functional roles played by the EMT pathway and by the ECM compositional changes in metastasis. Building on this knowledge, numerous hypotheses have been developed regarding some important causal relationships in the overall process of cancer metastasis. Such knowledge, in conjunction with the increasing pool of *omic* data collected on metastases at different developmental stages, including primary cancers with different levels of local metastasis, CTCs, DTCs, micro-metastasis and full metastatic tumors, provides unprecedented opportunities for computational cancer biologists to develop and computationally validate causal models. Such models can, in turn, be directly validated experimentally, significantly accelerating research progress in gaining a full understanding of cancer metastasis.

By integrating the information provided in this chapter and the one in Chap. 5, one can possibly develop a full model in which hyaluronic acid and fragments serve as the information backbone for providing instructions for stress-responses and possibly guiding a cancer to evolve. This seems reasonable since the whole purpose of cancer cells is growth as Otto Warburg pointed out in the 1960s (Warburg 1966). For a growing machine, like cancer, living in a rapidly changing and highly stressful environment, all they need for survival is to interact with hyaluronic acid and its fragments, which already encode all the "instructions" related to tissue remodeling and repair under different conditions, and well-tuned through millions of years of evolution. It is this well-developed instruction set that guides the evolving cancer cells to find their means to survive and proliferate. This is the fundamental difference between our view and the current genome-centric views about cancer. That is: survival guided by a set of well-developed instructions *versus* survival by selecting genomic mutations that are offered to them by chance.

Appendix

Table 10.1 Data used for gene-expression analysis of circulating tumor cells

Data set	Tissue	Platform	#Samples
GSE31364	Breast	GPL14378	72
GSE18670	Pancreatic	GPL570	24

Table 10.2 Data used for gene-expression analysis for Fig. 10.5

Data set	Tissue	Platform	#Samples
GSE36895	Kidney	GPL570	76
GSE31048	Leukemia	GPL570	221
GSE41804	Liver	GPL570	40
GSE30219	Lung	GPL570	85

Table 10.3 Data used for gene-expression analysis for Fig. 10.6

Data set	Tissue	Platform	#Samples
GSE31364	Breast	GPL14378	72
GSE18670	Pancreatic	GPL570	24

The breast cancer set contains seven DTC samples, and the pancreatic cancer dataset has seven CTC samples

References

Aguirre-Ghiso JA (2007) Models, mechanisms and clinical evidence for cancer dormancy. Nature reviews Cancer 7: 834-846

Alberts B, Johnson A, Lewis J et al. (2002) The extracellular matrix of animals.

Alexander NR, Branch KM, Parekh A et al. (2008) Extracellular matrix rigidity promotes invadopodia activity. Curr Biol 18: 1295-1299

Alix-Panabieres C, Riethdorf S, Pantel K (2008) Circulating tumor cells and bone marrow micrometastasis. Clinical cancer research: an official journal of the American Association for Cancer Research 14: 5013-5021

Amirkhosravi A, Mousa SA, Amaya M et al. (2003) Inhibition of tumor cell-induced platelet aggregation and lung metastasis by the oral GpIIb/IIIa antagonist XV454. Thromb Haemost 90: 549-554

Artym VV, Zhang Y, Seillier-Moiseiwitsch F et al. (2006) Dynamic interactions of cortactin and membrane type 1 matrix metalloproteinase at invadopodia: defining the stages of invadopodia formation and function. Cancer research 66: 3034-3043

Barcellos-Hoff M, Dix TA (1996) Redox-mediated activation of latent transforming growth factor-beta 1. Molecular endocrinology (Baltimore, Md) 10: 1077-1083

Barkan D, Green JE, Chambers AF (2010) Extracellular matrix: a gatekeeper in the transition from dormancy to metastatic growth. European journal of cancer 46: 1181-1188

Barkan D, Kleinman H, Simmons JL et al. (2008) Inhibition of metastatic outgrowth from single dormant tumor cells by targeting the cytoskeleton. Cancer research 68: 6241-6250

Bendas G, Borsig L (2012) Cancer cell adhesion and metastasis: selectins, integrins, and the inhibitory potential of heparins. Int J Cell Biol 2012: 676731

Bharadwaj AG, Kovar JL, Loughman E et al. (2009) Spontaneous metastasis of prostate cancer is promoted by excess hyaluronan synthesis and processing. Am J Pathol 174: 1027-1036

Chen H, Paradies NE, Fedor-Chaiken M et al. (1997) E-cadherin mediates adhesion and suppresses cell motility via distinct mechanisms. Journal of cell science 110 (Pt 3): 345-356

Cho EH, Wendel M, Luttgen M et al. (2012a) Characterization of circulating tumor cell aggregates identified in patients with epithelial tumors. Phys Biol 9: 016001

Cho JA, Park H, Lim EH et al. (2012b) Exosomes from breast cancer cells can convert adipose tissue-derived mesenchymal stem cells into myofibroblast-like cells. Int J Oncol 40: 130-138

Chung C, Beecham M, Mauck RL et al. (2009) The influence of degradation characteristics of hyaluronic acid hydrogels on in vitro neocartilage formation by mesenchymal stem cells. Biomaterials 30: 4287-4296

Craig EA, Parker P, Camenisch TD (2009) Size-dependent regulation of Snail2 by hyaluronan: its role in cellular invasion. Glycobiology 19: 890-898

Dallas MR, Liu G, Chen WC et al. (2012) Divergent roles of CD44 and carcinoembryonic antigen in colon cancer metastasis. FASEB journal: official publication of the Federation of American Societies for Experimental Biology 26: 2648-2656

Derynck R, Zhang YE (2003) Smad-dependent and Smad-independent pathways in TGF-beta family signalling. Nature 425: 577-584

Dimitroff CJ, Descheny L, Trujillo N et al. (2005) Identification of leukocyte E-selectin ligands, P-selectin glycoprotein ligand-1 and E-selectin ligand-1, on human metastatic prostate tumor cells. Cancer research 65: 5750-5760

Douma S, Van Laar T, Zevenhoven J et al. (2004) Suppression of anoikis and induction of metastasis by the neurotrophic receptor TrkB. Nature 430: 1034-1039

Eccles SA, Welch DR (2007) Metastasis: recent discoveries and novel treatment strategies. Lancet 369: 1742-1757

Fidler IJ (2003) The pathogenesis of cancer metastasis: the 'seed and soil' hypothesis revisited. Nature reviews Cancer 3: 453-458

Fidler IJ, Poste G (2008) The "seed and soil" hypothesis revisited. Lancet Oncol 9: 808

Garofalo A, Chirivi RG, Foglieni C et al. (1995) Involvement of the very late antigen 4 integrin on melanoma in interleukin 1-augmented experimental metastases. Cancer research 55: 414-419

Gasic GJ (1984) Role of plasma, platelets, and endothelial cells in tumor metastasis. Cancer metastasis reviews 3: 99-114

Gasic GJ, Gasic TB, Stewart CC (1968) Antimetastatic effects associated with platelet reduction. Proc Natl Acad Sci U S A 61: 46-52

Gay LJ, Felding-Habermann B (2011) Contribution of platelets to tumour metastasis. Nature reviews Cancer 11: 123-134

Gui T, Sun Y, Shimokado A et al. (2012) The Roles of Mitogen-Activated Protein Kinase Pathways in TGF-beta-Induced Epithelial-Mesenchymal Transition. J Signal Transduct 2012: 289243

Gupta GP, Massague J (2006) Cancer metastasis: building a framework. Cell 127: 679-695

Gupta SC, Hevia D, Patchva S et al. (2012) Upsides and downsides of reactive oxygen species for cancer: the roles of reactive oxygen species in tumorigenesis, prevention, and therapy. Antioxid Redox Signal 16: 1295-1322

Hagedorn EJ, Sherwood DR (2011) Cell invasion through basement membrane: the anchor cell breaches the barrier. Current Opinion in Cell Biology 23: 589-596

Hagedorn EJ, Ziel JW, Morrissey MA et al. (2013) The netrin receptor DCC focuses invadopodia-driven basement membrane transmigration in vivo. The Journal of cell biology 201: 903-913

Hammes HP, Lin J, Wagner P et al. (2004) Angiopoietin-2 causes pericyte dropout in the normal retina: evidence for involvement in diabetic retinopathy. Diabetes 53: 1104-1110

Hansen LK, Albrecht JH (1999) Regulation of the hepatocyte cell cycle by type I collagen matrix: role of cyclin D1. Journal of cell science 112 (Pt 17): 2971-2981

Hay ED (1981) Extracellular matrix. The Journal of cell biology 91: 205s-223s

Hill A, McFarlane S, Mulligan K et al. (2006) Cortactin underpins CD44-promoted invasion and adhesion of breast cancer cells to bone marrow endothelial cells. Oncogene 25: 6079-6091

Hiraga T, Ito S, Nakamura H (2013) Cancer stem-like cell marker CD44 promotes bone metastases by enhancing tumorigenicity, cell motility, and hyaluronan production. Cancer research 73: 4112-4122

Hofbauer R, Frass M, Salfinger H et al. (1999) Propofol reduces the migration of human leukocytes through endothelial cell monolayers. Crit Care Med 27: 1843-1847

Horak CE, Lee JH, Marshall JC et al. (2008) The role of metastasis suppressor genes in metastatic dormancy. APMIS 116: 586-601

Hsu C, Morohashi Y, Yoshimura S et al. (2010) Regulation of exosome secretion by Rab35 and its GTPase-activating proteins TBC1D10A-C. The Journal of cell biology 189: 223-232

Ina K, Kitamura H, Tatsukawa S et al. (2011) Significance of alpha-SMA in myofibroblasts emerging in renal tubulointerstitial fibrosis. Histology and histopathology 26: 855-866

Jain M, Rivera S, Monclus EA et al. (2013) Mitochondrial Reactive Oxygen Species Regulate Transforming Growth Factor-β Signaling. Journal of Biological Chemistry 288: 770-777

Kahlert C, Kalluri R (2013) Exosomes in tumor microenvironment influence cancer progression and metastasis. J Mol Med (Berl) 91: 431-437

Kim YN, Koo KH, Sung JY et al. (2012) Anoikis resistance: an essential prerequisite for tumor metastasis. Int J Cell Biol 2012: 306879

Kothapalli CR, Ramamurthi A (2009) Biomimetic regeneration of elastin matrices using hyaluronan and copper ion cues. Tissue engineering Part A 15: 103-113

Kothapalli CR, Taylor PM, Smolenski RT et al. (2009) Transforming growth factor beta 1 and hyaluronan oligomers synergistically enhance elastin matrix regeneration by vascular smooth muscle cells. Tissue engineering Part A 15: 501-511

Langley RR, Fidler IJ (2011) The seed and soil hypothesis revisited–the role of tumor-stroma interactions in metastasis to different organs. Int J Cancer 128: 2527-2535

Lee JM, Dedhar S, Kalluri R et al. (2006) The epithelial–mesenchymal transition: new insights in signaling, development, and disease. The Journal of cell biology 172: 973-981

Li J, Zhao Z, Wang J et al. (2008) The role of extracellular matrix, integrins, and cytoskeleton in mechanotransduction of centrifugal loading. Mol Cell Biochem 309: 41-48

Lindqvist U (1997) Is serum hyaluronan a helpful tool in the management of patients with liver diseases? J Intern Med 242: 67-71

Liu RM, Gaston Pravia KA (2010) Oxidative stress and glutathione in TGF-beta-mediated fibrogenesis. Free Radic Biol Med 48: 1-15

Lu X, Mu E, Wei Y et al. (2011) VCAM-1 promotes osteolytic expansion of indolent bone micro-metastasis of breast cancer by engaging alpha4beta1-positive osteoclast progenitors. Cancer Cell 20: 701-714

Lyons RM, Keski-Oja J, Moses HL (1988) Proteolytic activation of latent transforming growth factor-beta from fibroblast-conditioned medium. The Journal of cell biology 106: 1659-1665

Martinez-Outschoorn UE, Pavlides S, Howell A et al. (2011) Stromal-epithelial metabolic coupling in cancer: integrating autophagy and metabolism in the tumor microenvironment. The international journal of biochemistry & cell biology 43: 1045-1051

Meadows GG (2005) Integration/Interaction of Oncologic Growth. Springer-Verlag Inc., New York

Meng S, Tripathy D, Frenkel EP et al. (2004) Circulating tumor cells in patients with breast cancer dormancy. Clinical cancer research: an official journal of the American Association for Cancer Research 10: 8152-8162

Migneco G, Whitaker-Menezes D, Chiavarina B et al. (2010) Glycolytic cancer associated fibroblasts promote breast cancer tumor growth, without a measurable increase in angiogenesis: evidence for stromal-epithelial metabolic coupling. Cell Cycle 9: 2412-2422

Misra S, Heldin P, Hascall VC et al. (2011) Hyaluronan-CD44 interactions as potential targets for cancer therapy. FEBS J 278: 1429-1443

Miyazawa K, Shinozaki M, Hara T et al. (2002) Two major Smad pathways in TGF-beta superfamily signalling. Genes Cells 7: 1191-1204

Molloy TJ, Roepman P, Naume B et al. (2012) A prognostic gene expression profile that predicts circulating tumor cell presence in breast cancer patients. PloS one 7: e32426

Montgomery N, Hill A, McFarlane S et al. (2012) CD44 enhances invasion of basal-like breast cancer cells by upregulating serine protease and collagen-degrading enzymatic expression and activity. Breast Cancer Res 14: R84

Montserrat N, Gallardo A, Escuin D et al. (2011) Repression of E-cadherin by SNAIL, ZEB1, and TWIST in invasive ductal carcinomas of the breast: a cooperative effort? Human pathology 42: 103-110

Nash KT, Welch DR, Nash K et al. (2006) The KISS1 metastasis suppressor: mechanistic insights and clinical utility. Frontiers in bioscience: a journal and virtual library 11: 647

Nataatmadja M, West J, West M (2006) Overexpression of transforming growth factor-beta is associated with increased hyaluronan content and impairment of repair in Marfan syndrome aortic aneurysm. Circulation 114: I371-377

Okahara H, Yagita H, Miyake K et al. (1994) Involvement of very late activation antigen 4 (VLA-4) and vascular cell adhesion molecule 1 (VCAM-1) in tumor necrosis factor alpha enhancement of experimental metastasis. Cancer research 54: 3233-3236

Ostrowski M, Carmo NB, Krumeich S et al. (2010) Rab27a and Rab27b control different steps of the exosome secretion pathway. Nat Cell Biol 12: 19-30; sup pp 11-13

Owen SC, Shoichet MS (2010) Design of three-dimensional biomimetic scaffolds. Journal of Biomedical Materials Research Part A 94A: 1321-1331

Paez D, Labonte MJ, Bohanes P et al. (2012) Cancer dormancy: a model of early dissemination and late cancer recurrence. Clinical cancer research: an official journal of the American Association for Cancer Research 18: 645-653

Palumbo JS, Talmage KE, Massari JV et al. (2005) Platelets and fibrin(ogen) increase metastatic potential by impeding natural killer cell-mediated elimination of tumor cells. Blood 105: 178-185

Pani G, Galeotti T, Chiarugi P (2010) Metastasis: cancer cell's escape from oxidative stress. Cancer metastasis reviews 29: 351-378

Parekh A, Ruppender NS, Branch KM et al. (2011) Sensing and modulation of invadopodia across a wide range of rigidities. Biophys J 100: 573-582

Pavlides S, Whitaker-Menezes D, Castello-Cros R et al. (2009) The reverse Warburg effect: aerobic glycolysis in cancer associated fibroblasts and the tumor stroma. Cell Cycle 8: 3984-4001

Peinado H, Aleckovic M, Lavotshkin S et al. (2012) Melanoma exosomes educate bone marrow progenitor cells toward a pro-metastatic phenotype through MET. Nat Med 18: 883-891

Peinado H, Ballestar E, Esteller M et al. (2004) Snail mediates E-cadherin repression by the recruitment of the Sin3A/histone deacetylase 1 (HDAC1)/HDAC2 complex. Molecular and cellular biology 24: 306-319

Prime SS, Davies M, Pring M et al. (2004) The role of TGF-beta in epithelial malignancy and its relevance to the pathogenesis of oral cancer (part II). Crit Rev Oral Biol Med 15: 337-347

Rifkin DB, Sheppard D (1999) The integrin v 6 binds and activates latent TGF 1: a mechanism for regulating pulmonary inflammation and fibrosis. Cell 96: 319-328

Schwartz MA (2010) Integrins and extracellular matrix in mechanotransduction. Cold Spring Harb Perspect Biol 2: a005066

Seoane J (2006) Escaping from the TGFbeta anti-proliferative control. Carcinogenesis 27: 2148-2156

Strell C, Entschladen F (2008) Extravasation of leukocytes in comparison to tumor cells. Cell Commun Signal 6: 10

Takeda K, Sakai N, Shiba H et al. (2011) Characteristics of high-molecular-weight hyaluronic acid as a brain-derived neurotrophic factor scaffold in periodontal tissue regeneration. Tissue engineering Part A 17: 955-967

ten Dijke P, Hill CS (2004) New insights into TGF-beta-Smad signalling. Trends Biochem Sci 29: 265-273

Thomas SN, Zhu F, Schnaar RL et al. (2008) Carcinoembryonic antigen and CD44 variant isoforms cooperate to mediate colon carcinoma cell adhesion to E- and L-selectin in shear flow. J Biol Chem 283: 15647-15655

Vong S, Kalluri R (2011) The role of stromal myofibroblast and extracellular matrix in tumor angiogenesis. Genes Cancer 2: 1139-1145

Wang G, Wang S, Li Y et al. (2013) Clinical study of disseminated tumor cells in bone marrow of patients with gastric cancer. Hepatogastroenterology 60: 273-276

Wang HS, Tung WH, Tang KT et al. (2005) TGF-beta induced hyaluronan synthesis in orbital fibroblasts involves protein kinase C betaII activation in vitro. Journal of cellular biochemistry 95: 256-267

Wang J, Rong W, Hu X et al. (2012) Hyaluronan tetrasaccharide in the cerebrospinal fluid is associated with self-repair of rats after chronic spinal cord compression. Neuroscience 210: 467-480

Warburg O (1966) The Prime Cause and Prevention of Cancer.

Waris G, Ahsan H (2006) Reactive oxygen species: role in the development of cancer and various chronic conditions. J Carcinog 5: 14

Ween MP, Oehler MK, Ricciardelli C (2011) Role of Versican, Hyaluronan and CD44 in Ovarian Cancer Metastasis. Int J Mol Sci 12: 1009-1029

Wikman H, Vessella R, Pantel K (2008) Cancer micrometastasis and tumour dormancy. APMIS 116: 754-770

Wipff PJ, Rifkin DB, Meister JJ et al. (2007) Myofibroblast contraction activates latent TGF-beta1 from the extracellular matrix. The Journal of cell biology 179: 1311-1323

Wong SY, Hynes RO (2006) Lymphatic or hematogenous dissemination: how does a metastatic tumor cell decide? Cell Cycle 5: 812-817

Xian X, Hakansson J, Stahlberg A et al. (2006) Pericytes limit tumor cell metastasis. The Journal of clinical investigation 116: 642-651

Xu J, Lamouille S, Derynck R (2009) TGF-beta-induced epithelial to mesenchymal transition. Cell Res 19: 156-172

Yu Q, Stamenkovic I (2000) Cell surface-localized matrix metalloproteinase-9 proteolytically activates TGF-β and promotes tumor invasion and angiogenesis. Genes & development 14: 163-176

Zhang YE (2009) Non-Smad pathways in TGF-beta signaling. Cell Res 19: 128-139

Chapter 11
Cancer After Metastasis:
The Second Transformation

It has long been recognized by oncologists and families of cancer patients that metastatic cancers are considerably different from their primary cancer counterparts. Their growth, for example, tends to be more explosive (Weiss et al. 1986; Blomqvist et al. 1993; Oda et al. 2001; Klein 2009), easier to spread and more difficult to stop; yet, very little is understood about the differences between the biology of metastatic cancer and that of primary cancer. Current research efforts have mostly focused on understanding the mechanisms of metastatic processes by addressing questions like the following: *What triggers cancer cells to spread*? *How do they circumvent the body's defenses*? and *How does one prevent metastasis*? In comparison, relatively little can be found in the literature about the unique biology of metastatic cancer. For example, the molecular and genetic mechanisms responsible for driving the more explosive growth of metastatic cancers are virtually unknown. This may have reflected a belief widely held by cancer researchers that metastatic cancer is a terminal illness and hence not much can be done to stop its progression once a cancer has metastasized, which may have influenced the priorities in studying the underlying biology of metastasis.

Like primary cancers, metastatic cancers in different organs and of different origins tend to have different growth patterns and respond differently to the same chemotherapy treatments; and yet they exhibit certain clinical commonalities, which are distinct from those of primary cancers, suggesting that metastatic cancers may share some common biology. As discussed in Chap. 10, the common biology of different metastatic cancers may be driven by common challenges that the cancer cells must overcome when adapting to their new microenvironments that are considerably different from their old habitats. In this chapter, we present some common differences between the two different microenvironments; and discuss how metastatic cancers

© Springer Science+Business Media New York 2014
Y. Xu et al., *Cancer Bioinformatics*, DOI 10.1007/978-1-4939-1381-7_11

may have responded to the new challenges induced by their new microenvironments, which has led to common or similar clinical behaviors among all metastatic cancers, e.g., accelerated cell proliferation compared to their primary counterparts.

11.1   Characteristics Shared Among Metastatic Cancers But Distinct from Primary Cancers

The first study on elucidation of molecular signatures common to different types of metastatic cancers, which are distinct from primary cancers in general, was done by Ramaswamy et al. (2003). By analyzing transcriptomic data of 12 metastatic adeno-carcinomas of different origins (breast, colorectal, lung, ovary, prostate, and uterus) and 64 unmatched primary adenocarcinomas with similar organ distribution, the authors identified 128 genes whose expression patterns distinguished quite well the metastatic cancers from the primary cancers, where 64 of these genes are up-regulated and 64 are down-regulated in the metastatic cancers. Further analysis led to the identification of a subset consisting of 17 genes exhibiting strong discerning power between the two classes of cancers. Of the 17 genes, 8 are up-regulated in metastatic cancers (*SNRPF, EIF4EL3, HNRPAB, DHPS, PTTG1, COL1A1, COL1A2, LMNB1*), and 9 are down-regulated (*ACTG2, MYLK, MYH11, CNN1, HLA-DPB1, R4A1, MT3, RBM5, RUNX1*). Of the eight up-regulated genes, four (*SNRPF, EIF4EL3, HNRPAB, DHPS*) are related to mRNA (pre)processing and translation initiation; three (*COL1A1, COL1A2, LMNB1*) are related to extracellular matrix compositions, and one (*PTTG1*) is probably relevant to blocking of the *P53* activity. Among the nine down-regulated genes, four (*ACTG2, MYLK, MYH11, CNN1*) are relevant to actin binding; two (*HLA-DPB1, R4A1*) are involved in immune responses; two (*MT3, RBM5*) are involved in induction of apoptosis; and one (*RUNX1*) is a core transcription binding protein. While these gene-expression patterns are clearly consistent with our general understanding about the phenotypic differences between metastatic and primary cancers, the functions of these genes do not provide any obvious information for explaining why metastatic cancers grow substantially faster than their primary counterparts.

A number of additional studies have appeared on the differences in gene-expression patterns between metastatic and primary cancers, which have been focused more on specific cancer types rather than the general differences between the two classes of cancers. For example, a panel of 70 genes was published in 2002 for predicting breast cancer with strong metastatic potential (van 't Veer et al. 2002), and another panel of six genes (*DSC2, TFCP2L1, UGT8, ITGB8, ANP32E, FERMT1*) was published around the same time for distinguishing between breast cancer with strong potential for lung metastasis and those without (Landemaine et al. 2008). Similar studies focused on other cancer types can be found in the litera-ture, including melanoma (Bittner et al. 2000; Onken et al. 2004; Winnepenninckx et al. 2006), breast cancer (Minn et al. 2005; Wang et al. 2005), prostate cancer (Dhanasekaran et al. 2001), gastric cancer (Oue et al. 2004), pancreatic cancer

(Stratford et al. 2010; Van den Broeck et al. 2012), colon cancer (Bertucci et al. 2004), and squamous cell carcinoma of the head and neck (HNSCC) (Ginos et al. 2004; O'Donnell et al. 2005). While these signature-gene sets are clearly useful as predictive tools, none of them have provided new insights explaining why metastatic cancers so different from primary cancers in general.

11.1.1   Understanding the Difference Between Microenvironments of Metastatic Versus Primary Cancers

Here we examine the microenvironments of metastatic *versus* primary cancers, with the aim of understanding the common differences between these two types of tumor microenvironments, extending the brief analysis started in Chap. 10. The rationale is that microenvironments are known to have essential roles in the initiation, progression and metastasis of primary cancers. Specifically, the interplay between cancer cells and their microenvironment probably serves as the most important factor in determining the evolutionary trajectories of a cancer. The microenvironment of a cancer refers to the physical and chemical pericellular environment of a cancer, which includes (1) the composition and the physical properties of the ECM, (2) the stromal cell populations and relative sizes, (3) various chemical properties such as the levels of ROS, hypoxia and pH, and (4) the collection of a variety of signaling molecules such as chemokines and cytokines. While it is rather difficult to experimentally measure the micro-environmental factors of a cancer, one can possibly estimate them through analyses of gene-expression data of the collected cancer tissue samples. This is possible because cancer-affecting environmental factors should induce changes in expression patterns of some genes. By detecting and characterizing the relationships between the environmental changes and the expression changes in specific genes, one should be able to infer changes in a microenvironment. This is the basis for inference of micro-environmental changes through gene-expression analysis of cancer tissues.

We have carried out comparative analyses of transcriptomic data on a number of metastatic *versus* primary cancer types in the same organs, hence referred to as *corresponding* cancers, in order to determine, as best as possible, the common differences between the microenvironments of metastatic cancers *versus* those of the corresponding primary cancers. The reason that we compare microenvironments in the same organs (not from the same patients) is that the background microenvironmental factors should be similar for the same type of organ, hence making direct comparisons between the microenvironments of metastatic and primary cancers more meaningful.

In the analysis, genes that are related to the following environmental factors are examined: (1) hypoxia, (2) ROS, (3) ECM composition, and (4) immune responses. A few datasets covering primary and metastatic cancers in four organs: brain, liver, lung and ovary, were retrieved from the GEO database (See Table 11.2) and used in our analysis, with the detailed information on these datasets given in Table 11.2. Throughout all the analyses in this chapter, the expression data are all normalized in a sample-centric manner so the total expression level over all genes in each sample is the same. With this normalization for correcting possible batch effects (Lazar et al. 2013), expression-levels across different samples, including both meta-static and of primary cancers, are directly comparable.

11.1.2   Reactive Oxygen Species

Two genes, *NRF2* (gene name: *NFE2L2*) and *INRF2* (*KEAP1*), are examined, which are known marker genes of the ROS level. Specifically *NRF2* is the main regulator of the antioxidant response pathway (Nguyen et al. 2009), and *INRF2* inhibits the activation and promotes the degradation of *NRF2* when it binds to *NRF2* (Kaspar et al. 2009). Both genes show lower expression levels in metastatic cancers *versus* the corresponding primary cancers across all four organs (see Fig. 11.1), strongly indicating that metastatic cancers have lower intracellular ROS levels than the corresponding primary cancers.

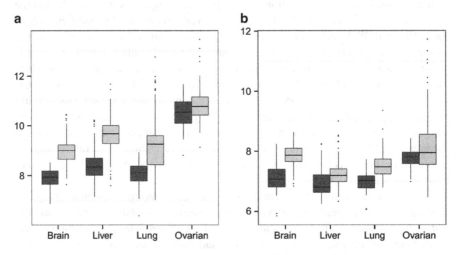

Fig. 11.1 Boxplots for comparisons between the (log-2 transformed) expression levels of two ROS marker genes, *NFE2L2* (**a**) and *KEAP1* (**b**), across multiple samples in metastatic *versus* corresponding primary cancers. Each dot for each gene in each cancer type is the normalized expression level of the gene in a specific sample. In both (**a**) and (**b**), four groups of metastatic and corresponding primary cancer samples are shown, from *left* to *right*: (metastatic brain (*left*), primary brain (*right*)), (metastatic liver, primary liver), (metastatic lung, primary lung) and (metastatic ovarian, primary ovarian). The *bottom* and the *top lines* of each *box* are the first and the third quartiles, respectively, and the *line inside the box* is the mean

In addition, we have also examined the expression data of 23 antioxidant enzyme genes, consisting of *HMOX1, HMOX2, GSR, GCLC, GCLM, NQO1*, and a few additional genes selected in three families: *SOD* (superoxide dismutase), *GPX* (glutathione peroxidase) and *TXN* (thioredoxin). It is noted that 12 out of the 23 genes show substantially lower expressions in metastatic *versus* the corresponding primary cancers across all four organs, namely *HMOX1, GCLC, GCLM, SOD1, SOD2, GPX1, GPX3, GPX4, PRDX3, PRDX4, PRDX6* and *TXNL1*, while only three genes, *NQO1, GPX2* and *PRDX2*, show over-expression in metastases *versus* the primary cancers. These observations provide additional supporting evidence that metastatic cancers have lower ROS levels than those in primary cancers across all four organs. Since these four were randomly picked, it is reasonable to expect that all solid tumors have this property, which is consistent with a recent study suggesting that metastasis is cancer's way to escape from the high ROS levels in their primary sites (Pani et al. 2010).

11.1.3 Hypoxia

Two hypoxia-induced factor genes, *HIF1α* and *HIF2α*, were examined in metastatic *versus* corresponding primary cancers. Both genes are established marker genes for the intracellular hypoxic level (Koukourakis et al. 2002), i.e., the higher the *HIF* gene expression, the higher the hypoxic level. Figure 11.2 shows that both *HIF1α* and *HIF2α* are consistently under-expressed in metastatic *versus* primary cancers, indicating that metastatic cancer in general has higher oxygen levels than primary cancer across all four organ types. This is expected since metastatic cancers tend to develop in blood-rich environments. In addition, genes involved in the TCA cycle (also referred to as the Krebs cycle) and the oxidative phosphorylation pathway are found to be consistently overexpressed in metastatic cancer compared to the corresponding primary cancer. These results provide independent supporting evidence for the increased oxygen levels since these two processes require oxygen.

One interesting observation here is that the majority of the glycolysis genes are known to be positively regulated by *HIF1α* in primary cancers in general (Denko 2008), such as the glucose transporters *GLUT1* and *GLUT3* and glucose enzymes *HK2* and *LDHA*. However, these genes are consistently up-regulated in metastatic *versus* corresponding primary cancers, even though *HIF1α* is down-regulated in metastatic cancers as shown in Fig. 11.2. This suggests that metastatic cancers may employ a regulatory mechanism different than *HIF1α* to maintain the glycolytic activity level high, which is different from that used in primary cancers. We anticipate that well-designed statistical analyses may lead to the identification of transcriptional factors showing co-expression patterns with the above glycolysis genes and possibly even to the discovery of an unknown regulatory system used by metastatic cancers, which does not involve *HIF1α*.

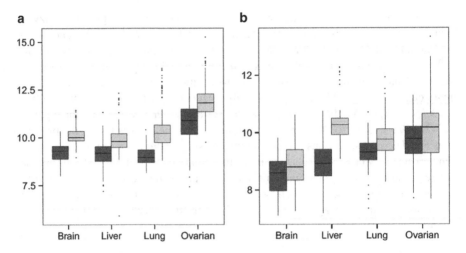

Fig. 11.2 Boxplots for comparisons between the average expression levels of two marker genes, *HIF1α* (**a**) and *HIF2α* (**b**), for hypoxic levels across multiple samples in metastatic *versus* corresponding primary cancers in four organs: brain, liver, lung and ovary. All the definitions in the figure are the same as in Fig. 11.1

11.1.4   Acidic pH Level

Both extracellular and intracellular acidity levels are known to be associated with the development of cancer. The expression levels of acid-sensing ion channel gene *ASIC3*, previously used as the marker genes for the extracellular pH (Waldmann et al. 1997; Delaunay et al. 2012), have been found to be up-regulated in metastatic *versus* primary cancers. Figure 11.3 shows one *ASIC* gene, *ASIC3*, with increased expression levels in ovarian, liver and lung metastases *versus* their corresponding primary cancers, respectively, suggesting that these metastatic cancers have a lower pH than the corresponding primary cancers. This is consistent with the above observation that glycolysis is up-regulated in metastatic cancers. Brain metastases behave differently for this particular gene, and further analyses are clearly needed.

11.1.5   Immune Response

Immune responses are indispensable components throughout cancer development, including initiation, progression and metastasis as discussed in Chap. 8. Of interest however, immune responses play dual roles relevant to cancer evolution. Specifically, immune surveillance and immunoediting are essential for detecting and inhibiting cancer development while some immune responses, such as chronic inflammation and the partnership between cancer and macrophages (Condeelis and Pollard 2006; Mantovani et al. 2006a), are probably required for cancer development (see Chap. 8).

A number of marker genes for the activity levels of TAMs and T-cells are examined here. It is noteworthy that gene-expression data collected on a cancer tissue

Fig. 11.3 Boxplots for comparisons between expression levels of one gene, *ASIC3*, reflecting the pH level in metastatic and primary cancers in brain, liver, lung and ovary. The definitions of the plot are the same as in Fig. 11.1

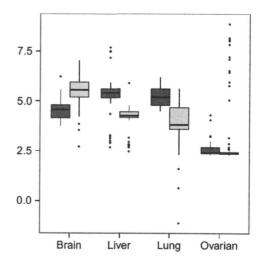

sample is likely to have contributions from all cell types contained in the specimen, including immune cells, since complete removal of non-cancerous cells from a specimen requires a considerable amount of effort and is rarely done. On a positive note, however, the non-cancerous cells remaining in a tissue sample may prove beneficial to cancer research since they provide additional and important information about the microenvironment of a cancer. As discussed in Chap. 2, derivation of the detailed contribution from each cell type to each gene's expression level generally requires substantial effort, including characterization of each cell type in terms of its signature genes and gene-expression covariance matrix, as well as a de-convolution algorithm (Ahn et al. 2013). Here we do not attempt to derive accurate contributions to each gene's expression by individual cell types. Instead, it will suffice to make a rough estimate about the immune-specific genes that are up- or down-regulated in metastases and those genes that may have altered expression levels as a response. With these caveats, only expression data of immune-specific genes are examined in metastatic *versus* the corresponding primary cancers, in an attempt to derive new insights into the differences in gene activity levels in immune cells associated with metastatic cancer *versus* those with primary cancer.

Genes selected for analysis include the following. *IL4* is an immune-cell gene and known to promote differentiation of helper T-cells to type-2 helper T-cells (Th2) and activate macrophage M1 cells to become M2 cells (Sokol et al. 2008; Ho and Sly 2009; Martinez et al. 2013). Moreover, the combined signals of *IL4* from T-cells and *CSF1* (hematopoietic growth factor) from cancer cells can stimulate the biosynthesis and release of the potent growth factor *EGF* by TAMs. *ST2* (suppression of tumorigenicity 2) and *OX40* (also known as *CD134*) are marker genes of Th2 cells. *CD4* (cluster of differentiation 4) is an immune cell-derived glycoprotein, and has been found on the surface of T helper cells and macrophages. *CD8* and *GZMA* are both marker genes of cytotoxic T-cells, and *KIR* is a marker gene of natural killer cells.

Comparative expression data analyses of these genes revealed the following. *IL4, IL4R, ST2, OX40, CD4* and *KIR* are consistently up-regulated in brain, liver and

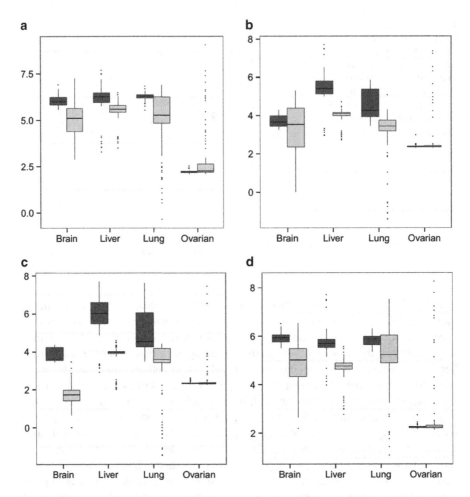

Fig. 11.4 Boxplots for comparisons between expression levels of genes, *CSF1* (**a**), *IL4* (**b**), *CD4* (**c**) and *CTLA4* (**d**), relevant to immunity in metastatic and primary cancers in brain, liver, lung and ovary. The definitions of *boxes* in the plot are the same as in Fig. 11.1

lung metastases compared to their corresponding primary cancers. Hence, one can infer that metastatic cancer tissues tend to have increased concentrations of helper T-cells, tumor-associated macrophages and natural killer cells.

CD8 and *GZMA* are consistently down-regulated in metastatic cancers *versus* the corresponding primary cancers across all the datasets examined. This is consistent with a previous report that cytotoxic T-cells can be inhibited by regulatory T-cells through *CTLA-4* (cytotoxic T-lymphocyte antigen 4), and by tumor associated macrophages through the production of *TGFβ* and *IL10* (Vesely et al. 2011). This mechanism allows metastatic cancer cells to escape immune controls.

Figure 11.4 shows the expression patterns of a few genes selected from the above immune-cell gene list. It is worth emphasizing that ovarian cancer, both metastatic and primary, seems to have rather distinct expression patterns compared to the other metastases examined here as can be seen from Fig. 11.4. The reason could be one of the following: (1) ovarian cancer may use a parallel set of genes to accomplish the same functions as these genes do; and (2) ovarian cancer may have rather distinct immune responses from the other cancer types, which seems unlikely.

11.1.6 Changes in Extracellular Matrix

The physical properties of ECM have key roles in cancer initiation, development and metastasis as discussed in Chaps. 4, 6 and 10. For example, stiffer ECMs tend to increase the effectiveness of growth factors in driving the proliferation of cancer cells upwards of 100-fold. In addition, higher ECM stiffness is the cue for the differentiation of myofibroblasts (Hinz 2009). Here we examine if the ECMs of metastatic cancers have distinct properties *versus* those of primary cancers. Since the physical and chemical properties of ECMs have not been directly measured, the approach taken here was to examine the expression data of a few genes whose proteins contribute to various components of the ECM.

Recall from Chaps. 1 and 6 that the ECM is composed of collagen and hyaluronic acid fibrils, proteoglycans, fibronectins, laminins, elastins and other linker proteins (see Fig. 1.6). Out of these components, the levels of collagen, elastins and laminins are known to be directly related to the stiffness of an ECM (Bruel and Oxlund 1996; An et al. 2009; Ng and Brugge 2009), while relatively little is known about the contribution of the other components to the physical properties. A simple data analysis revealed that there is a fairly large gene pool that contributes to the assembly of ECM: (1) at least 43 genes in the human genome encode collagen proteins, which together have at least 322 known splicing isoforms according to the ACEVIEW database (Thierry-Mieg and Thierry-Mieg 2006); and (2) genes encoding ECM linker proteins such as fibronectin, laminin and elastin have numerous splicing variants, not to mention the diversity of proteoglycans found in ECM. Hence, one can infer that the number of different combinations of different components comprising an ECM is exceedingly large, thus resulting in a rather large number of different functional states, each of which may represent a distinct signal in cell-ECM interactions. Specifically, any subtle change in the composition of an ECM may lead to changes in the physical properties of the matrix, hence sending different signals to the corresponding cells through the cell-matrix adhesion, which directly connects with the cell nucleus (Ingber 2006).

From an examination of the expression levels of a few genes that are involved in the synthesis of hyaluronic acid and collagen fibrils, and the biosynthesis of fibronectins, laminins, elastins and cell focal adhesion, we were able to garner some understanding of how the expression levels of various contributing genes differ between primary and metastatic cancers. This represents but a first step, and in order to obtain a more accurate picture of the diversities of ECM compositions and properties, a much larger and more sophisticated analyses are needed.

Figure 11.5 shows the expression pattern changes of five genes, *CD44*, *ITGB7*, *LAMC2*, *COL6A3* and *CTTN*, which were selected somewhat arbitrarily but,

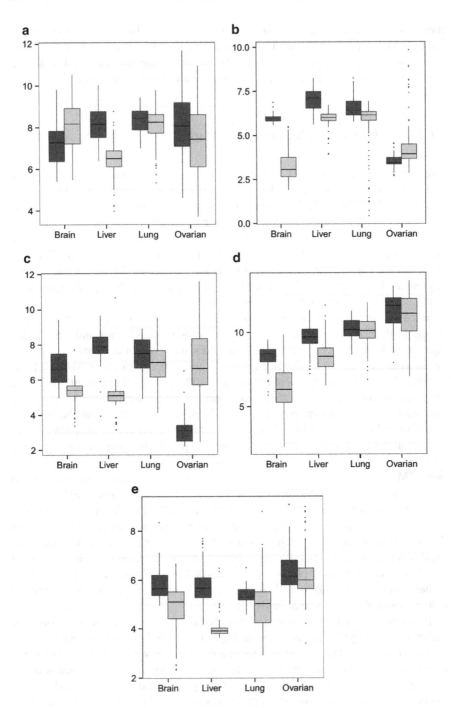

Fig. 11.5 Boxplots for comparisons between expression levels of genes relevant to the composition of ECM components in metastatic *versus* primary cancers in brain, liver, lung and ovary, where the marker genes are (**a**) *CD44*, (**b**) *ITGB7* (integrin beta-7), (**c**) *LAMC2* (laminin subunit gamma-2), (**d**) *COL6A3* and (**e**) *CTTN* (cortactin). The definitions of *boxes* in the plot are the same as in Fig. 11.1

importantly, cover five different areas of ECM: hyaluronic acid export, the laminin level, the collagen composition and level, and actin rearrangement, respectively. *CD44* is found to be up-regulated across all metastatic cancers *versus* their corresponding primary cancers, which is consistent with the increased proliferation rates of metastatic cancers. The same can be said about *CTTN* (cortactin), which has increased expression levels in all metastatic cancers compared to the primary ones, and is also consistent with the increased proliferation rates observed in metastatic cancers. While the expression levels of these two genes reflect the overall proliferation rates, the other three genes reflect just the concentration levels of the individual proteins. *ITGB7* is up-regulated in three metastatic cancer types, brain, liver and lung *versus* the corresponding primary cancers, while it is down-regulated in ovarian metastatic cancer. This is consistent with the ovarian genes shown in Figs. 11.3 and 11.4, where the gene expression patterns in ovarian cancer are different from those in cancers of brain, liver and lung. A specific type of collagen, *COL6A3*, is up-regulated in cancers of these three organs, while the gene shows an opposite expression pattern in ovarian cancer.

From the above analyses, one can infer that the microenvironments of metastatic cancers tend to: (1) be less hypoxic; (2) be of lower ROS level; (3) be more acidic; (4) have increased involvement by macrophages and T-cells; and (5) exhibit increased laminin content in ECM in three of the four metastatic cancer types, suggesting the possibility of increased stiffness of the ECM. A natural question to ask here is: *What responses do these environmental changes cause in the cancer cells when they migrate into new environments?* The answer to this question could potentially lead to novel and possibly deep insights into the unique biology of metastatic cancer. To address this issue, further functional analyses have been conducted to infer the biological pathways where expression patterns are statistically correlated with the expression-changes of the micro-environmental factor genes.

11.2 Cellular Responses to Altered Microenvironments

In order to derive causal relationships between the observed changes in the microenvironment of a metastatic cancer and in other cellular activities, one must first identify statistical associations between the environmental changes and changes in other processes. One can then apply general knowledge about the relationships among different cellular processes to infer possible causality. Specifically, the following procedure can be used to infer statistical associations between gene-expression changes in two gene-sets or pathways. For each gene g in a gene-expression dataset D, Spearman correlation is calculated between the expression levels of g and every other gene in D. If multiple gene-expression datasets are considered, one can use the Fisher transformation (Fisher 1921) to combine the calculated correlations across different datasets. The GSEA algorithm (Subramanian et al. 2005a) can then be used for identification of pathways enriched with non-microenvironmental genes having statistical correlations with micro-environmental genes in terms of their gene-expression patterns.

All the major micro-environmental factors were examined in a preliminary analysis, and it was found that changes in the following environmental factors have consistent and strong correlations with the various cellular processes.

11.2.1 Pathways Showing Strong Correlations with Reduced Hypoxia Level

The following pathways were found to be highly correlated with the reduced hypoxic level as reflected by the reduced *HIF1α* expression in metastatic cancer: *glucose metabolism* and *cellular responses to stress and extracellular stimuli* are negatively correlated with the expression level of *HIF1α*; in addition, *aerobic respiration, fatty acid oxidation, TCA cycle, protein synthesis, cholesterol synthesis and uptake, metabolism* and *primary bile acid synthesis* pathways show negative correlations with *HIF1α*. As mentioned earlier, glycolysis is probably regulated by factors other than *HIF1α*, unlike that in primary cancers. Further data analysis could potentially lead to the identification of the distinct regulatory mechanism(s) of glycolysis in metastatic cancers.

In addition, a large number of immune response-related genes were observed to be positively correlated with *HIF1α*, including the *immune response, leukocyte, lymphocyte* and *myeloid cell differentiation*. In comparison, a number of ECM-related pathways were found to be negatively correlated with *HIF1α*, including the *biosynthesis pathways of proteoglycans, hyaluronic acid* and *collagens* and the *cell-shape regulation pathway*. These data indicate that metastatic cancers tend to have reduced immune responses and increased activities of ECM component production and cell morphogenesis.

11.2.2 Pathways Showing Strong Correlations with the Activation of Helper T-Cells and TAMs

A few pathways are positively correlated with the activation signals of the two cell types: *cell cycle, TCA cycle, cellular pH regulation, detection of external stimuli, cholesterol synthesis, primary bile acid synthesis, biosynthesis of ECM components* and *cell adhesion*.

11.2.3 Pathways Showing Strong Correlations with Altered ECM Compositions

The following pathways are up-regulated and strongly correlated with changes in the synthesis of the key ECM components: genes involved in *cell cycle G_2/M and S phase controls, actin filament-based movement, glucose metabolism, IL10 pathway, platelet adhesion to exposed collagen*, and *protein complex disassembly*, suggesting an increased cell proliferation rate.

Table 11.1 lists all the pathways that are either up-regulated or down-regulated consistently across all the metastatic cancers in comparison with the corresponding primary cancers. Some of these pathways are found to have strong correlations with the changes of the environmental factors identified in Sect. 11.1, while a few do not

Table 11.1 Differentially expressed pathways in metastatic cancers

(a) Up-regulated pathways in metastatic cancers identified through enrichment analyses against pathways in the KEGG, BIOCARTA, REACTOME and Msigdb (Subramanian et al. 2005b)

Inflammation	INTRINSIC_PATHWAY
	CLASSIC_PATHWAY
Immune response	DC_PATHWAY
	IL10_PATHWAY
	IL4_PATHWAY
	POSITIVE_REGULATION_OF_CYTOKINE_SECRETION
	POSITIVE_REGULATION_OF_DEFENSE_RESPONSE
	POSITIVE_REGULATION_OF_LYMPHOCYTE_ACTIVATION
Cell cycle	RB_PATHWAY
	CELL_CYCLE_G2_M_HASE
Cell growth signal	EPIDERMAL_GROWTH_FACTOR_RECEPTOR_SIGNALING_PATHWAY
	MTA3_PATHWAY
Detection of external stimuli	DETECTION_OF_ABIOTIC_STIMULUS
	DETECTION_OF_CHEMICAL_STIMULUS
	DETECTION_OF_EXTERNAL_STIMULUS
	DETECTION_OF_STIMULUS_INVOLVED_IN_SENSORY_PERCEPTION
Metabolism	CELLULAR_PROTEIN_COMPLEX_DISASSEMBLY
	COENZYME_BIOSYNTHETIC_PROCESS
	COFACTOR_BIOSYNTHETIC_PROCESS
	ALPHA_LINOLENIC_ACID_METABOLISM
	ETHER_LIPID_METABOLISM
	PANTOTHENATE_AND_COA_BIOSYNTHESIS
	STEROID_BIOSYNTHESIS
	THYROID_CANCER
	VALINE_LEUCINE_AND_ISOLEUCINE_BIOSYNTHESIS
	MACROMOLECULAR_COMPLEX_DISASSEMBLY
	PEPTIDE_METABOLIC_PROCESS
	POSITIVE_REGULATION_OF_PROTEIN_SECRETION
	PROTEIN_COMPLEX_DISASSEMBLY
	PROTEIN_EXPORT_FROM_NUCLEUS
	PROTEOGLYCAN_BIOSYNTHETIC_PROCESS
	PROTEOGLYCAN_METABOLIC_PROCESS
	PYRIMIDINE_CATABOLISM
	RNA_ELONGATION
	TRICARBOXYLIC_ACID_CYCLE_INTERMEDIATE_METABOLIC_PROCESS

(continued)

Table 11.1 (continued)

Cell junction and cell membrane	ACTIN_FILAMENT_BUNDLE_FORMATION
	ACTIN_FILAMENT_ORGANIZATION
	FOCAL_ADHESION_FORMATION
	NEGATIVE_REGULATION_OF_CELL_MIGRATION
	CELL_EXTRACELLULAR_MATRIX_INTERACTIONS
	PLATELET_ADHESION_TO_EXPOSED_COLLAGEN
	REGULATION_OF_GTPASE_ACTIVITY
	REGULATION_OF_RAS_GTPASE_ACTIVITY
	REGULATION_OF_RAS_PROTEIN_SIGNAL_TRANSDUCTION
	REGULATION_OF_RHO_GTPASE_ACTIVITY
	REGULATION_OF_RHO_PROTEIN_SIGNAL_TRANSDUCTION
Microenvironment	REGULATION_OF_PH
(b) Down-regulated pathways in metastatic cancers identified through enrichment analyses	
Inflammation	ACUTE_INFLAMMATORY_RESPONSE
Immune response	ST_IL_13_PATHWAY
	REACTOME_IL_6_SIGNALING
	REGULATION_OF_IMMUNE_EFFECTOR_PROCESS
Cell response to external stimuli	CELLULAR_RESPONSE_TO_EXTRACELLULAR_STIMULUS
	CELLULAR_RESPONSE_TO_NUTRIENT_LEVELS
	CELLULAR_RESPONSE_TO_STRESS
	DEFENSE_RESPONSE_TO_VIRUS
	ST_TYPE_I_INTERFERON_PATHWAY
	REGULATION_OF_INTERFERON_GAMMA_BIOSYNTHETIC_PROCESS
Metabolism	FATTY_ACID_BIOSYNTHETIC_PROCESS
Metastasis	EPITHELIAL_TO_MESENCHYMAL_TRANSITION
Microenvironment	SUPEROXIDE_METABOLIC_PROCESS

show obvious statistical associations with any of the five factors we checked. This observation suggests two possibilities: (1) there are other micro-environmental or intracellular factors such as genomic mutations that could drive the changes in expression of these pathways as observed in metastatic cancer, or (2) they result from combinations of some micro-environmental factors whose relationship with the observed responses are non-linear and consequently much more difficult to identify. In either case, this list provides a good starting point for more in-depth analyses of the transcriptomic and other types of data to infer the causes of the observed changes.

Based on the above analysis, we infer that metastatic cancers have: (1) increased oxygen availability and consumption; (2) increased activities of regulatory T-cells and TAMs; and (3) increased ECM component synthesis. These may represent the basic causes for the changes of the other cellular processes listed in Table 11.1 since most of them show strong correlations with at least one of these three changes. Moreover, the three major changes are highly correlated with each other. While these associations do not necessarily prove causal relationships, they provide a good starting point for inference of possible causality when used in conjunction with additional biological information.

11.3 Understanding the Accelerated Growth of Metastatic Cancer: A Data Mining Approach

Results from the above analyses show that, when relocating to a new site, metastatic cancer cells indeed face new challenges for survival, which are distinct from those of the primary cancer. These new challenges force various cellular responses, such as those delineated in Table 11.1, to protect the cancer cells and maintain their viability. Unfortunately, some of these responses add (new) fuel to the fire, and drive the metastatic cancers to grow faster than their primary cancers. In the following, one model is presented regarding how the migrated cells respond to the increased oxygen level by accelerating their growth, based on a more in-depth analysis of the data from Sects. 11.1 and 11.2.

We have examined 16 sets of genome-scale transcriptomic data from the GEO database (Barrett et al. 2005), covering the following 11 types of primary → metastatic cancers: breast to liver, colon to liver, pancreas to liver and prostate to liver metastases; bone to lung, breast to lung, colon to lung, kidney to lung and pancreas to lung metastases; breast to brain metastases; and prostate to bone metastases. Detailed information on these datasets is given in the Appendix. The main question addressed here is: "*Which genes are consistently up-regulated in metastatic cancer versus the corresponding primary cancer?*" Among the identified genes, the following groups are particularly interesting, i.e., genes that encode: (1) the uptake and metabolism of cholesterol towards the production of steroidal metabolites; (2) nuclear receptors; (3) growth factor receptors; and (4) cell proliferation markers. Specifically, the following pertinent observations were made:

(1) In the majority of the metastatic cancers examined, at least one of the receptor genes for HDL (high density lipoprotein), LDL (low density lipoprotein) and VLDL (very low density lipoprotein) is up-regulated while the remaining metastatic cancers, most of them being brain metastases, have their cholesterol biosynthesis pathway up-regulated. Specifically, *SRB1* (scavenger receptor class B1), which can transport HDL and oxidized LDL particles (with their cargo cholesterol) into cells, is up-regulated in 44 % of the metastatic cancers examined. *LDLR* and *VLDLR*, receptors for LDL/chylomicrons and VLDL, respectively, are up-regulated in 50 % and 19 % of the metastatic cancers, respectively (Cao et al. 2014). After entering the metastatic cancer cells, cholesterol and cholesteryl esters will be shed from the lipoprotein carriers in lysosomes (Fielding and Fielding 1997; Ioannou 2001).

Brain metastases tend to have up-regulated *de novo* cholesterol biosynthesis, probably because circulating cholesterol cannot enter brain due to the blood-brain barrier (Orth and Bellosta 2012). Interestingly, a substantial fraction of metastatic cancer samples examined here use more than one mechanism to increase the influx of cholesterol. All these observations strongly suggest that metastatic cancers have an increased (seemingly urgent) need for cholesterol. One natural question emerges: *Is the increased need solely due to the increased proliferation rates in metastatic cancer, knowing that cholesterol is a key component of cell membranes?*

To address this question, we have calculated the statistical correlation between expression levels of the marker genes of cell proliferation, namely cyclins, *CDKs* (cyclin-dependent kinases) and *MCM* (DNA replication licensing factor) (Alison et al. 2002; Wheeler et al. 2008; Peurala et al. 2013), and the cholesterol receptor and synthesis genes. The rationale is that if cholesterol is needed solely for synthesizing membranes in support of cell proliferation, there should be a strong correlation between these two sets of genes. It was found that this is exactly the case for primary cancers in general, but the correlation is substantially weaker for metastatic cancers. This observation suggests that there are additional reasons for cholesterol uptake other than their important role in cell membrane synthesis in support of cell proliferation in metastases. In addition, another key piece of supporting data is that different modes of increasing cholesterol influx, including different lipoprotein receptors and/or *de novo* biosynthesis, are used in different tissue of the same primary to metastatic cancer type, strongly suggesting that the increased cholesterol influx is not regulated by a well-designed program such as accelerated proliferation, but instead it is the result of responses to cellular stress.

(2) In each metastatic cancer examined, at least one *CYP* (cytochrome P450) gene is up-regulated, whose main function is to oxidize cholesterol (or sterol) in the steroid pathway or to synthesize steroid hormones. *CYP27A1*, *CYP3A4*, *CYP17A1* and *CYP19A1* account for the majority of the up-regulated *CYP* genes across all the metastatic cancers considered. *CYP27*, sterol 27-hydroxylase, is a key enzyme involved in the conversion of cholesterol to bile acid, and *CYP17*, steroid 17-α-monooxygenase, is a key enzyme in the steroidogenic pathway that produces progestins, mineralocorticoids, glucocorticoids, androgens and estrogens. Bile acids are among the metabolites of *CYP*-encoded enzymes, along with other metabolites such as 27-hydroxycholesterol and 4β-hydroxycholesterol. Of interest, 27-hydroxycholesterol has recently been shown to promote tumor growth and metastasis in mammary tumors of mice, attributed to its role as a partial agonist for the estrogen receptor (Nelson et al. 2013). In addition, mass spectrometry-based metabolomic analyses have shown that metastatic cancers also tend to have substantially increased auto-oxidized cholesterol products such as α-EPOX, β-EPOX and 7-ketocholesterol.

A number of enzymes that can further metabolize the oxidized cholesterols towards steroidal products are also up-regulated in various metastatic cancers in an organ-specific manner. For example, *HSD11* (11β-hydroxysteroid dehydrogenase type 1) and *SRD5A1-2* are up-regulated in multiple metastatic cancer types such as colon metastases in liver and lung, and pancreatic metastases in liver. Steroid hormone synthesis as a whole is significantly up-regulated across virtually all metastatic cancer types. Figure 11.6 shows the relationships between cholesterol and a number of its key metabolic products. Particular attention should be paid to the relationships between oxidized cholesterols and various steroids hormones.

(3) In each metastatic cancer sample examined, some nuclear receptors (*NRs*) were over-expressed. *NRs* are transcription factors that can be activated by oxysterols, specific hormones or vitamins, and then regulate genes involved in <u>development and homeostasis</u> of various types. Among the most commonly up-regulated *NRs* across all the metastatic cancer samples examined are *FXR* (farnesoid X receptor) and *HNF4A* (hepatocyte nuclear factor 4α). The natural ligand for *FXR* is bile acid, whose synthesis pathway is up-regulated across all metastases. *LXR* (liver X receptor) and *RXR* (retinoid X receptor) are also up-regulated in most metastases, which can be activated by their agonists such as 4β-hydroxycholesterol and 27-hydroxycholesterol, and retinoic acid, whose metabolism is up-regulated across all metastases examined. Other up-regulated *NRs* include estrogen receptors, *ESR1*, and androgen receptor *AR*.

(4) In each metastatic cancer sample examined, some growth-factor receptors (*GFRs*) are up-regulated. The most commonly up-regulated *GFRs* are *EGFR* and *FGFR4*, which are consistent with the previous observations that *ESR1* and *FXR* are common regulators for activating *EGFR* (Levin 2003; Razandi et al. 2003; Sukocheva et al. 2006) and *FGFR4* (Chiang 2009), respectively. Interestingly, for some up-regulated *GFRs*, their natural ligand growth-factors are not up-regulated, suggesting two possibilities. One is that these *GFRs* are activated by non-native ligands; and another is that their cognate growth factors may be mostly produced by the neighboring stromal cells such as macrophages, instead of by the cancer cells, a phenomenon that has been widely observed and reported in the cancer literature (Qian and Pollard 2010; Hanahan and Weinberg 2011).

(5) Cell proliferation marker genes such as cyclins, *CDKs* and *MCM* genes are consistently up-regulated in metastases, indicating accelerated cell proliferation in the secondary tumors in general.

Some of the key observations made above have been validated experimentally using cancer cell lines, which include: (1) cell proliferation of metastatic cancer cells when exposed to cholesterol-containing HDL or oxidized LDL in the culture media, while no growth is observed without such treatment, thus indicating that metastatic cancer cells can grow on cholesterol from these two types of lipoproteins; (2) increased abundance of the protein product of the *SRB1* gene; (3) dimerization of nuclear receptors of multiple types, strongly suggesting that they have moved into the nucleus and serve as transcription factors; and (4) increased protein abundance of a number of growth factors such as *EGFR*. One particular exciting piece of data is the observation that metastatic cancers tend to have substantially higher oxysterol level than primary cancers as shown in Fig. 11.7.

These analyses and validation results provide strong evidence that oxidized cholesterol has an important role in accelerating the growth of metastatic cancers, while further studies are clearly needed to derive more detailed mechanistic information.

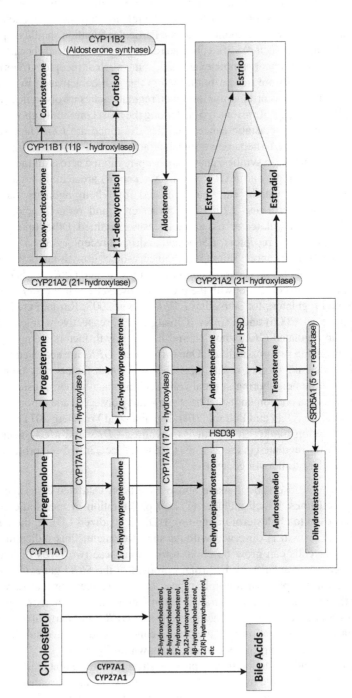

Fig. 11.6 The steroidogenic pathways showing the precursor role of cholesterol

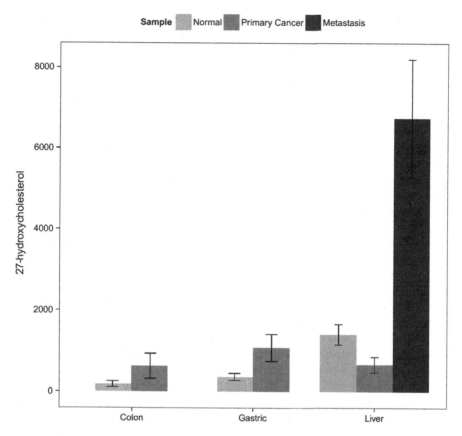

Fig. 11.7 The abundance of 27-hydroxycholesterol in five samples each of: primary colon cancer and matching normal tissue, primary gastric cancer and matching normal tissue, and primary liver cancer and matching normal tissue, along with liver metastases from 13 primary colon and breast cancers. Data on samples of normal and primary cancer tissues are shown in *light* and *dark gray*, respectively, and the results on liver metastases from colon and breast cancer tissues are given in *black*. Adapted from (Cao et al. 2014)

It is important to realize that too much oxysterol can be lethal to the host cancer cells as a recent study has observed (de Weille et al. 2013). Hence, we speculate that metastatic cancers may have employed some mechanism(s) to export the excess oxysterols. Interestingly, each of the 16 datasets is found to have at least one up-regulated efflux transporter for cholesterol or oxidized cholesterol, including *ABCA1*, *ABCG1*, *ABCG5* and *ABCB11*, strongly suggesting that unused cholesterols and oxidized cholesterols are released to the extracellular space.

Now the question is: *Why does metastatic cancer in general have increased needs for cholesterol compared to primary cancers?* The answer to this question could potentially lead us to the real root of this disease, which is generally considered as terminal. The following analysis, although not yet providing an answer, may potentially lead to new insights about answering this question.

Previous studies strongly suggested that cholesterol (or sterols in general) co-emerged with O2 during early evolution around 2.7–2.2 billion years ago as a way to protect obligate anaerobic (prokaryotic) cells against the toxic O2 (Galea and Brown 2009b). Further investigations have found that cholesterol may serve as a possible regulator of O_2 entry into cells and a primitive defense against ROS (Subczynski et al. 1989; López-Revuelta et al. 2006; Murphy and Johnson 2008; Galea and Brown 2009a). When linking this information to the following findings: (1) the human plasma-membrane cholesterol levels are found to be negatively correlated with the amount of changes in cellular O_2 levels of red blood cells when the blood O_2 level changes (Buchwald et al. 2000); and (2) higher membrane cholesterol-phospholipid ratios leads to lower O_2 permeability of cellular membranes, we are inspired to ask: *Is it possible that when primary cancer cells leave their hypoxic environment and migrate to a blood-vessel rich environment, and hence oxygen-rich, a yet-to-be-understood mechanism is triggered by the higher O_2 level to increase the cholesterol concentration in their cell membrane and hence the need for increased uptake and/or synthesis of cholesterol?*

This hypothesis is supported by some recent studies that identified the known regulator of cholesterol uptake and synthesis, *SREBP* in fission yeast, can be triggered by the O_2 level (Hughes et al. 2005)! This strongly suggests that while cholesterol has evolved to have many functional and signaling roles in human cells, their oldest function as a defense against O_2 may have been retained during evolution and used opportunistically by metastatic cancers. Yet, another key piece of data to solve this puzzle could lie in the knowledge that membrane cholesterols (and phospholipids) can be oxidized through (continuous) lipid peroxidation (Halliwell and Chirico 1993) in an environment high in oxidative stress, which is typical for a metastatic cancer environment (Cao et al. 2014), hence causing membrane damages and possibly a continuous need for cholesterol to replace that damaged in the membrane. Again two pieces of data provide supporting evidence for damaged plasma membranes on a continuous basis: (1) regulators in response to membrane damages are constantly up-regulated; and (2) the catabolism of the oxidized products of phospholipids such as arachidonic acid, linoleic acid and linolenic acid, a key component of cell membrane, is up-regulated.

11.3.1   A Model for Cholesterol Uptake/Synthesis, Metabolism and Accelerated Proliferation

Based on the above analyses and discussion, a model has been developed to depict how cancer cells utilize the cholesterol metabolites for their accelerated growth (Cao et al. 2014). The model consists of all the key observations discussed above. In addition, our experimental data suggest that metastatic cancer cells may use growth factors from one or both of the following two sources: (a) growth factors that are released by the cancer cells and then act on the growth factor receptors on the surface of cancer cells, i.e. an autocrine mechanism; and (b) growth factors that are released by TAMs, a paracrine mechanism, as has been reported in the literature (Mantovani et al. 2006b; Hao et al. 2012) (Fig. 11.8).

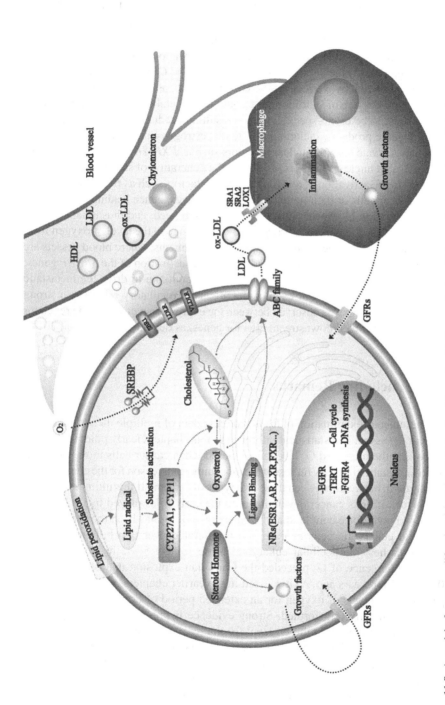

Fig. 11.8 A model of cancer proliferation driven by cholesterol uptake and oxidation products, which may be the result of cellular responses to increased oxygen levels in the metastatic sites. The increased O_2 level may trigger two separate results, increased cholesterol uptake (or synthesis in brain metastases) and increased lipid peroxidation of cellular membrane

In brief, the model consists of the following key steps: (1) cholesterol is taken up by metastatic cells through various lipoprotein receptors such as *SRB1*, *LDLR* or *VLDLR* (or synthesized *de novo*) and exists in the free cholesterol form; (2) multiple *CYP* genes are up-regulated possibly induced, for example, by oxidative stress via nuclear receptors; (3) the cholesterol is oxidized by the *CYP*-encoded enzymes and further metabolized by various other enzymes such as *HSD11* and *SRD5A1-2* towards the production of steroids and steroidogenic metabolites, where the detailed enzymes seem to be organ-specific; (4) the resulting products include a wide range of steroidogenic products, including oxysterols, estrogens and androgens; (5) these metabolites activate various nuclear receptors such as *FXR*, *HNF4A*, *LXR* and *RXR*; (6) the activated nuclear receptors, as well as the generation of steroidogenic metabolites, lead to the activation of a variety of growth factors such as *EGFR* and *FGFR4* either in cancer cells or their neighboring macrophages, hence leading to the accelerated proliferation by cancer cells; (7) the increased need, and thus uptake or synthesis of cholesterol seems to be triggered by the change in the level of oxygen from highly hypoxic environments to oxygen-rich environments, where blood vessels are plentiful, which is probably coordinated through the activation of the *SREBP* genes; and (8) there may be multiple vicious cycles that continue to drive the metastatic cancer cells to proliferate. For each of the predicted regulatory relations, a strong statistical relationship was observed between the expression patterns of the regulator genes and the predicted downstream effector genes, as detailed in (Cao et al. 2014).

11.4   Concluding Remarks

The simultaneous examination of transcriptomic data of multiple metastatic cancer types and the associated statistical inference in this chapter clearly point to the supposition that the increased O_2 level encountered when cancer cells move from primary to metastatic sites may represent the most important reason for the considerably altered growth behavior between these two types of cancers. If this ultimately proves to be correct, it links cancer development to the early roles played by O_2 in evolution. It is known, for example, that the emergence of O_2 has been the fundamental reason that eukaryotic cells evolved and proliferated over two billion years ago while the earliest multicellular eukaryotes arose some 1.5–1.2 billion years ago. Thus, the emergence of O_2 preceded the Cambrian explosion that occurred about 540–525 million years ago. As discussed in the earlier chapters, when human cells experience low levels of oxygen for an extended period of time, they are prone to become malignantly transformed. Strong evidence is presented herein suggesting that the sudden increase in the O_2 level for the migrated cancer cells in their new environment(s), after having already adapted to the hypoxic environment, may have forced these cells to go through a second transformation for their survival. This fundamentally novel perspective of viewing the development of metastatic cancer may lead to unorthodox and novel approaches for more effective ways to terminate or at least diminish the development and progression of the disease.

Appendix

Table 11.2 A list of transcriptomic data used in the analyses in Sect. 11.1

Data set	Tissue	Platform	Primary cancer	Metastasis cancer	#Samples
GSE20565	Ovary	GPL570	*	*	129
GSE14407	Ovary	GPL570	*		24
GSE42952	Liver	GPL96		*	33
GSE41258	Liver	GPL96		*	47
GSE14020	Liver	GPL96		*	5
GSE29721	Liver	GPL570	*		10
GSE14323	Liver	GPL570	*		53
GSE14323	Liver	GPL96	*		11
GSE14020	Lung	GPL570		*	4
GSE14020	Lung	GPL96		*	16
GSE41258	Lung	GPL96		*	20
GSE33356	Lung	GPL570	*		60
GSE27262	Lung	GPL570	*		25
GSE31547	Lung	GPL96	*		25
GSE14020	Brain	GPL570		*	7
GSE14020	Brain	GPL96		*	15
GSE14108	Brain	GPL96		*	28
GSE8692	Brain	GPL96	*		12
GSE4271	Brain	GPL96	*		100

The first column shows the name of each dataset used in our analysis, and each * denotes that the corresponding dataset contains the relevant cancer type, primary or metastatic

Table 11.3 A list of transcriptomic data sets used in Sect. 11.3

Dataset	Metastatic cancer type	# Primary cancer samples	# Metastatic cancer samples
GSE14297	Colon → liver	18	18
GSE6988	Colon → liver	53	29
GSE41258	Colon → liver	186	47
GSE26338/GPL5325	Breast → liver	19	5
GSE34153	Pancreas → liver	14	20
GSE42952	Pancreas → liver	12	7
GSE6752	Prostate → liver	10	5
GSE8511	Prostate → liver	12	6
GSE41258	Colon → lung	186	20
GSE26338/GPL1390	Breast → lung	201	6
GSE34153	Pancreas → lung	14	8
GSE14359	Bone → lung	10	8
GSE22541	Kidney → lung	24	24
GSE26338/GPL5325	Breast → brain	19	9
GSE26338/GPL1390	Breast → brain	201	8
GSE32269	Prostate → bone	22	29

References

Ahn J, Yuan Y, Parmigiani G et al. (2013) DeMix: deconvolution for mixed cancer transcriptomes using raw measured data. Bioinformatics 29: 1865-1871

Alison MR, Hunt T, Forbes SJ (2002) Minichromosome maintenance (MCM) proteins may be pre-cancer markers. Gut 50: 290-291

An SS, Kim J, Ahn K et al. (2009) Cell stiffness, contractile stress and the role of extracellular matrix. Biochem Biophys Res Commun 382: 697-703

Barrett T, Suzek TO, Troup DB et al. (2005) NCBI GEO: mining millions of expression profiles—database and tools. Nucleic acids research 33: D562-D566

Bertucci F, Salas S, Eysteries S et al. (2004) Gene expression profiling of colon cancer by DNA microarrays and correlation with histoclinical parameters. Oncogene 23: 1377-1391

Bittner M, Meltzer P, Chen Y et al. (2000) Molecular classification of cutaneous malignant melanoma by gene expression profiling. Nature 406: 536-540

Blomqvist C, Wiklund T, Tarkkanen M et al. (1993) Measurement of growth rate of lung metastases in 21 patients with bone or soft-tissue sarcoma. British journal of cancer 68: 414-417

Bruel A, Oxlund H (1996) Changes in biomechanical properties, composition of collagen and elastin, and advanced glycation endproducts of the rat aorta in relation to age. Atherosclerosis 127: 155-165

Buchwald H, O'Dea TJ, Menchaca HJ et al. (2000) Effect of plasma cholesterol on red blood cell oxygen transport. Clinical and Experimental Pharmacology and Physiology 27: 951-955

Cao S, Zhang C, Liu C et al. (2014) Oxidized Cholesterol Plays a Key Role in Driving the Accelerated Growth of Metastatic Cancer. Unpublished results.

Chiang JY (2009) Bile acids: regulation of synthesis. J Lipid Res 50: 1955-1966

Condeelis J, Pollard JW (2006) Macrophages: obligate partners for tumor cell migration, invasion, and metastasis. Cell 124: 263-266

de Weille J, Fabre C, Bakalara N (2013) Oxysterols in cancer cell proliferation and death. Biochemical pharmacology 86: 154-160

Delaunay A, Gasull X, Salinas M et al. (2012) Human ASIC3 channel dynamically adapts its activity to sense the extracellular pH in both acidic and alkaline directions. Proceedings of the National Academy of Sciences of the United States of America 109: 13124-13129

Denko NC (2008) Hypoxia, HIF1 and glucose metabolism in the solid tumour. Nature Reviews Cancer 8: 705-713

Dhanasekaran SM, Barrette TR, Ghosh D et al. (2001) Delineation of prognostic biomarkers in prostate cancer. Nature 412: 822-826

Fielding CJ, Fielding PE (1997) Intracellular cholesterol transport. Journal of lipid research 38: 1503-1521

Fisher RA (1921) On the probable error of a coefficient of correlation deduced from a small sample. Metron 1: 3-32

Galea AM, Brown AJ (2009a) Special relationship between sterols and oxygen: were sterols an adaptation to aerobic life? Free radical biology & medicine 47: 880-889

Galea AM, Brown AJ (2009b) Special relationship between sterols and oxygen: Were sterols an adaptation to aerobic life? Free Radical Bio Med 47: 880-889

Ginos MA, Page GP, Michalowicz BS et al. (2004) Identification of a Gene Expression Signature Associated with Recurrent Disease in Squamous Cell Carcinoma of the Head and Neck. Cancer Research 64: 55-63

Halliwell B, Chirico S (1993) Lipid peroxidation: its mechanism, measurement, and significance. The American journal of clinical nutrition 57: 715S-724S

Hanahan D, Weinberg RA (2011) Hallmarks of cancer: the next generation. Cell 144: 646-674

Hao NB, Lu MH, Fan YH et al. (2012) Macrophages in tumor microenvironments and the progression of tumors. Clin Dev Immunol 2012: 948098

Hinz B (2009) Tissue stiffness, latent TGF-β1 activation, and mechanical signal transduction: implications for the pathogenesis and treatment of fibrosis. Current rheumatology reports 11: 120-126

Ho VH, Sly L (2009) Derivation and Characterization of Murine Alternatively Activated (M2) Macrophages. In: Reiner NE (ed) Macrophages and Dendritic Cells, vol 531. Methods in Molecular Biology™. Humana Press, New York. pp 173-185

Hughes AL, Todd BL, Espenshade PJ (2005) SREBP pathway responds to sterols and functions as an oxygen sensor in fission yeast. Cell 120: 831-842

Ingber DE (2006) Cellular mechanotransduction: putting all the pieces together again. FASEB journal: official publication of the Federation of American Societies for Experimental Biology 20: 811-827

Ioannou YA (2001) Multidrug permeases and subcellular cholesterol transport. Nature reviews Molecular cell biology 2: 657-668

Kaspar JW, Niture SK, Jaiswal AK (2009) Nrf2:INrf2 (Keap1) signaling in oxidative stress. Free radical biology & medicine 47: 1304-1309

Klein CA (2009) Parallel progression of primary tumours and metastases. Nature reviews Cancer 9: 302-312

Koukourakis MI, Giatromanolaki A, Sivridis E et al. (2002) Hypoxia-inducible factor (HIF1A and HIF2A), angiogenesis, and chemoradiotherapy outcome of squamous cell head-and-neck cancer. International Journal of Radiation Oncology* Biology* Physics 53: 1192-1202

Landemaine T, Jackson A, Bellahcene A et al. (2008) A six-gene signature predicting breast cancer lung metastasis. Cancer Res 68: 6092-6099

Lazar C, Meganck S, Taminau J et al. (2013) Batch effect removal methods for microarray gene expression data integration: a survey. Briefings in bioinformatics 14: 469-490

Levin ER (2003) Bidirectional signaling between the estrogen receptor and the epidermal growth factor receptor. Molecular endocrinology 17: 309-317

López-Revuelta A, Sánchez-Gallego JI, Hernández-Hernández A et al. (2006) Membrane cholesterol contents influence the protective effects of quercetin and rutin in erythrocytes damaged by oxidative stress. Chemico-biological interactions 161: 79-91

Mantovani A, Schioppa T, Porta C et al. (2006a) Role of tumor-associated macrophages in tumor progression and invasion. Cancer and Metastasis Reviews 25: 315-322

Mantovani A, Schioppa T, Porta C et al. (2006b) Role of tumor-associated macrophages in tumor progression and invasion. Cancer metastasis reviews 25: 315-322

Martinez FO, Helming L, Milde R et al. (2013) Genetic programs expressed in resting and IL-4 alternatively activated mouse and human macrophages: similarities and differences. Blood 121: e57-69

Minn AJ, Gupta GP, Siegel PM et al. (2005) Genes that mediate breast cancer metastasis to lung. Nature 436: 518-524

Murphy RC, Johnson KM (2008) Cholesterol, reactive oxygen species, and the formation of biologically active mediators. The Journal of biological chemistry 283: 15521-15525

Nelson ER, Wardell SE, Jasper JS et al. (2013) 27-Hydroxycholesterol links hypercholesterolemia and breast cancer pathophysiology. Science 342: 1094-1098

Ng MR, Brugge JS (2009) A Stiff Blow from the Stroma: Collagen Crosslinking Drives Tumor Progression. Cancer Cell 16: 455-457

Nguyen T, Nioi P, Pickett CB (2009) The Nrf2-antioxidant response element signaling pathway and its activation by oxidative stress. The Journal of biological chemistry 284: 13291-13295

O'Donnell RK, Kupferman M, Wei SJ et al. (2005) Gene expression signature predicts lymphatic metastasis in squamous cell carcinoma of the oral cavity. Oncogene 24: 1244-1251

Oda T, Miyao N, Takahashi A et al. (2001) Growth rates of primary and metastatic lesions of renal cell carcinoma. International journal of urology: official journal of the Japanese Urological Association 8: 473-477

Onken MD, Worley LA, Ehlers JP et al. (2004) Gene expression profiling in uveal melanoma reveals two molecular classes and predicts metastatic death. Cancer research 64: 7205-7209

Orth M, Bellosta S (2012) Cholesterol: its regulation and role in central nervous system disorders. Cholesterol 2012: 292598

Oue N, Hamai Y, Mitani Y et al. (2004) Gene Expression Profile of Gastric Carcinoma Identification of Genes and Tags Potentially Involved in Invasion, Metastasis, and Carcinogenesis by Serial Analysis of Gene Expression. Cancer research 64: 2397-2405

Pani G, Galeotti T, Chiarugi P (2010) Metastasis: cancer cell's escape from oxidative stress. Cancer metastasis reviews 29: 351-378

Peurala E, Koivunen P, Haapasaari KM et al. (2013) The prognostic significance and value of cyclin D1, CDK4 and p16 in human breast cancer. Breast Cancer Res 15: R5

Qian BZ, Pollard JW (2010) Macrophage diversity enhances tumor progression and metastasis. Cell 141: 39-51

Ramaswamy S, Ross KN, Lander ES et al. (2003) A molecular signature of metastasis in primary solid tumors. Nat Genet 33: 49-54

Razandi M, Pedram A, Park ST et al. (2003) Proximal events in signaling by plasma membrane estrogen receptors. The Journal of biological chemistry 278: 2701-2712

Sokol CL, Barton GM, Farr AG et al. (2008) A mechanism for the initiation of allergen-induced T helper type 2 responses. Nat Immunol 9: 310-318

Stratford JK, Bentrem DJ, Anderson JM et al. (2010) A six-gene signature predicts survival of patients with localized pancreatic ductal adenocarcinoma. PLoS Med 7: e1000307

Subczynski WK, Hyde JS, Kusumi A (1989) Oxygen permeability of phosphatidylcholine–cholesterol membranes. Proceedings of the National Academy of Sciences 86: 4474-4478

Subramanian A, Tamayo P, Mootha VK et al. (2005a) Gene set enrichment analysis: A knowledge-based approach for interpreting genome-wide expression profiles. Proceedings of the National Academy of Sciences of the United States of America 102: 15545-15550

Subramanian A, Tamayo P, Mootha VK et al. (2005b) Gene set enrichment analysis: a knowledge-based approach for interpreting genome-wide expression profiles. Proceedings of the National Academy of Sciences of the United States of America 102: 15545-15550

Sukocheva O, Wadham C, Holmes A et al. (2006) Estrogen transactivates EGFR via the sphingosine 1-phosphate receptor Edg-3: the role of sphingosine kinase-1. The Journal of cell biology 173: 301-310

Thierry-Mieg D, Thierry-Mieg J (2006) AceView: a comprehensive cDNA-supported gene and transcripts annotation. Genome biology 7: S12

van 't Veer LJ, Dai H, van de Vijver MJ et al. (2002) Gene expression profiling predicts clinical outcome of breast cancer. Nature 415: 530-536

Van den Broeck A, Vankelecom H, Van Eijsden R et al. (2012) Molecular markers associated with outcome and metastasis in human pancreatic cancer. J Exp Clin Cancer Res 31: 68

Vesely MD, Kershaw MH, Schreiber RD et al. (2011) Natural innate and adaptive immunity to cancer. Annu Rev Immunol 29: 235-271

Waldmann R, Champigny G, Bassilana F et al. (1997) A proton-gated cation channel involved in acid-sensing. Nature 386(6621):173-177

Wang Y, Klijn JG, Zhang Y et al. (2005) Gene-expression profiles to predict distant metastasis of lymph-node-negative primary breast cancer. The Lancet 365: 671-679

Weiss L, Grundmann E, Torhorst J et al. (1986) Haematogenous metastatic patterns in colonic carcinoma: An analysis of 1541 necropsies. The Journal of Pathology 150: 195-203

Wheeler LW, Lents NH, Baldassare JJ (2008) Cyclin A-CDK activity during G1 phase impairs MCM chromatin loading and inhibits DNA synthesis in mammalian cells. Cell Cycle 7: 2179-2188

Winnepenninckx V, Lazar V, Michiels S et al. (2006) Gene expression profiling of primary cutaneous melanoma and clinical outcome. Journal of the National Cancer Institute 98: 472-482

Chapter 12
Searching for Cancer Biomarkers in Human Body Fluids

One of the important lessons learned about cancer survival from the past few decades' experience in cancer treatment is: <u>early detection is the key</u>. It has now become common knowledge that as a cancer progresses from early to more advanced stages, it gradually changes from a local and relatively simple problem to a very complex health issue involving the body at large. Once a cancer has metastasized, tumors in the new locations tend to grow substantially faster and metastasize further and much more rapidly than the primary counterpart, hence making the disease considerably more difficult to control and treat. The available statistics show that the survival rate of a cancer patient drops substantially when an encapsulated tumor spreads to the neighboring tissue and then to distant locations. For example, the 5-year survival rate drops from 99 to 66 % and then down to 9.4 % when a colorectal cancer is localized, has spread to only the local tissue and then to distant organs, respectively. Similar survival statistics hold for virtually all cancers. It is particularly worth emphasizing that the 5-year survival rates tend to drop to single digits or low tens of percentages for most of the cancers when they have spread to distal organs.

From these alarming and unsettling statistics, an urgent question begs to be answered: *Can one reliably detect cancer in its early stage?* This issue clearly has significant implications in saving the lives of cancer patients. Based on our current understanding of cancer, one is tempted to say: Yes, but the technical challenges to achieve this goal are substantial. It is noted that a basis for early detection lies in the observation that cancers at different stages tend to have distinct molecular signatures, as discussed throughout the earlier chapters. For example, it was shown in Chap. 3 that the expression patterns of some genes tend to strongly correlate with the stage of a cancer. From a computational perspective, the following technical problems must be solved before reliable early detection of cancer can become a reality. The essential questions to be addressed are: (1) *The abundances of which biomolecules accurately reflect the early stage of a specific cancer type?* (2) *Which of these putative markers or their products can be secreted into circulation and possibly enter into other body fluids?* (3) *Of the biomolecules identified in (2), which*

© Springer Science+Business Media New York 2014
Y. Xu et al., *Cancer Bioinformatics*, DOI 10.1007/978-1-4939-1381-7_12

may serve as reliable biomarkers for a specific cancer based on their half-life and detectability in circulation or other body fluids?

Disease detection through blood or urine tests have long been used for various noncancerous illnesses such as virus infection, diabetes, hepatitis, kidney disease and now even Alzheimers (Leidinger et al. 2013), but reliable blood or urine biomarkers are not yet available for accurate cancer diagnosis, particularly for early detection. In this chapter, some general ideas are presented about how computational approaches can aid to make this a reality.

12.1　A Historical Perspective of Biomarker Identification for Disease Diagnosis

The earliest diagnostic technique for human diseases can probably be traced back to a few thousand years ago when Chinese physicians determined the nature of an illness by checking a patient's pulse, inspecting the tongue coating, examining the urine color and odor, and smelling the stool. Clearly, our ancestors had learned long ago that there were signals in our body wastes that were informative for disease diagnoses.

Urine was probably the earliest body fluid that was used for medical diagnoses in a systematic manner, particularly in ancient Greece. By examining the urine color, ancient Greek physicians believed that they could tell the nature of the illness of a patient. Hippocrates (460–370 BC), who is considered as the founding father of Western medicine, even suggested that "no other organ system or organ of the human body provides so much information by its excretion as does the urinary system" (Scholkopf et al. 2001). Interestingly, the first cancer biomarker was also found in urine rather than in blood. It was observed in 1848 that 75 % of myeloma patients had elevated levels of immunoglobulin in their urine and hence could be used as a biomarker for the diagnosis of this illness (Jones 1848).

The use of blood biomarkers for disease detection on a large scale began in the 1950s for two main reasons: (1) a better understanding of numerous human illnesses at the molecular level had been gained; and (2) isotope-based analytical chemistry techniques had matured and become widely available, mainly in the area of radio-immunoassays. A large number of blood biomarkers have subsequently been discovered and widely used in clinics since then, including from a long list the use of the: (1) blood glucose level for diagnosis and control maintenance of diabetes, (2) blood transaminase levels for detection of liver damage; (3) creatinine level for detection of kidney illness, and (4) blood levels of 12 microRNAs for the detection of Alzheimers, as recently reported (Leidinger et al. 2013).

A number of blood biomarkers for cancer have also been proposed and reported in the literature in the past decade, some of which have been used clinically. These include: (1) *PSA* (prostate specific antigen) for prostate cancer; (2) *CEA* (carcinoembryonic antigen) for colon cancer, which is also found to be informative for detection of gastric, pancreatic, lung and breast cancers; (3) *AFP* (alpha-fetoprotein) for liver cancer; (4) *S100* for melanoma; (5) *CA 125* (cancer antigen, also known as *MUC 16*) for ovarian cancer; (6) *CA 19-9* (carbohydrate antigen 19-9) for pancreatic cancer; (7) *BCR-ABL* for chronic myeloid leukemia; plus a number of multi-protein diagnostic panels such as (8) a 21-protein panel for early non-small cell lung cancers (NSCLC), which includes multiple interleukins, *TFGα* and interferon γ, (9) a 7-protein panel (*P53, MYC, HER2, CTAG1, BRCA1, BRCA2, MUC1*) for breast cancer; metabolite-based biomarkers such as (10) a panel of six glycolytic metabolites (lactate, alanine, succinate, glutamate, citrate, aspartate) for lung cancer; and microRNA-based biomarkers such as (11) miR-25 and miR-223 for NSCLC. In addition to these and other diagnostic markers, a number of prognostic biomarkers have also been proposed and tested on limited samples such as a panel of five proteins (*DUSP6, MMD, STAT1, ERBB3, LCK*) for prognostic prediction of NSCLC patients (Sanchez-Cespedes 2008) and a 5-protein panel (*AANG2, CRP, ICAM1, IGFBP1, TSP2*) for prognostic prediction of advanced pancreatic cancer, as recently reported (Nixon et al. 2013).

As of now, numerous cancer-related biomarkers have been developed and are available as clinical tests, but the reality is that the majority of these biomarkers cannot be considered as reliable indicators for cancer detection in terms of both their detection sensitivity and specificity, even though some of them have been widely used in clinics. Among these biomarkers, only one has been FDA approved, which is *PSA* used for both diagnosis and prognosis of prostate cancer. Even for this widely-used blood biomarker, its predictive value is far below those used for diagnoses of noncancerous diseases such as the blood glucose level for diabetes or the transaminase level for hepatitis. A previous assessment of the PSA as a prostate-cancer biomarker involving 5,112 patients found that the cancer-identification specificity and sensitivity rates were 93 % and 24 %, respectively, when using a PSA threshold ≥ 4 ng/ml (Thompson et al. 2006). In a different large-scale assessment using a lower PSA cutoff, >3.0–3.99 ng/ml (Lilja et al. 2008), the detection sensitivity improved to 33 % but at the cost of substantially lowering the detection specificity. Experience gained over the years has shown that it is the amount of change over time of a biomarker score, rather than the score itself, that is more informative. Hence the value for PSA and other cancer biomarkers lies in monitoring the progress of a cancer or assessing the effectiveness of a specific drug or treatment regimen (Bhatt et al. 2010) rather than for pure diagnostic purposes.

There are multiple reasons for the subpar performances of the proposed cancer diagnostic biomarkers. First and the foremost is that the field is still in the early stage in searching for effective strategies for tackling the very challenging problem of finding reliable biomarkers for cancer. Second, by examining how the majority of the current cancer biomarkers have been discovered, one will note that they have been typically identified through comparative proteomic (or other *omic*) analyses of blood (or urine) samples between cancer patients of a specific cancer type and healthy controls. This has been typically done by searching for molecular species that show consistent abundance differences in blood samples of cancer patients *versus* healthy controls.

While in principle this approach should lead to the discovery of molecules with discerning power between the two sample pools, it turns out that this is an exceedingly challenging technical problem in reality. Using protein biomarkers as an example, one needs to be aware that the dynamic range in protein concentrations in human blood spans more than 10 orders of magnitude (Anderson and Anderson 2002), orders of magnitude larger than those of the current analytical techniques. Furthermore, the target proteins, in general, tend to be among the proteins with the lowest concentrations in blood, basically at the noise level since they arise from tumors of relatively small sizes in small quantities compared to the concentration levels of the native and constitutively produced blood proteins such as albumin and other proteins secreted from major organs, particularly liver. In addition, a very large number of different peptides are present as a result of protein degradation, much of the fragmentation occurring in cells, on cell surfaces and in circulation.

"Searching for a needle in a haystack" is probably a significant understatement of the challenging nature of this search problem. More realistically it could be modified to "searching for a needle in a field of haystacks" or from a similar saying in Chinese, "searching for a needle in an ocean" may better reflect the level of difficulty in solving the problem. The currently identified biomarkers tend to be for tumors having reached certain sizes so that the concentration of the "biomarkers" released are already relatively significant. In addition, some of the current biomarker molecules may not necessarily be secreted from the tumor cells through secretion pathways, instead they have leaked from the damaged cancer cells, which tends to be associated with cancers already in an advanced stage. A more fundamental issue is that the identified biomarkers may work well for a sample set that was used to identify the markers originally and often do not generalize well to larger sample sets.

Some of the recent biomarkers have been identified through searches in a more informed manner, specifically guided using *omic* data collected on the relevant cancer tissues. For example, a urine biomarker for gastric cancer, epithelial lipase (EL), has been identified using transcriptomic data of the gastric cancer tissues as guidance (Hong et al. 2011). The idea is that *omic* data analyses of cancer *versus* control tissues can reveal which biomolecules, such as proteins or microRNAs, are consistently and differentially expressed, particularly over-expressed, in cancer tissues *versus* controls. This approach provides a candidate list for biomarker searches rather than searches being conducted in a blind fashion.

Clearly, such an approach is in the right direction, but it tends to generate a rather sizable candidate list that taxes experimentalists, even those with high-throughput assays. In addition, the putative biomarkers so identified may not be the optimal ones when taken in a larger context. For example, consideration must be given to achieve the best discerning power, not only between the cancer samples and the appropriate controls, but also between the target cancer and other diseases in general. Overall, this will require a significant effort to deal with the issue in a more informed manner than that of just comparing two pools of blood samples. More comprehensive information regarding the specifics of a particular cancer type needs to be included. In addition, information about the ability of cells to secrete biomolecules extracellularly, the half-life of the biomolecule or its fragments in circulation and their detectability using the existing analytical techniques also need to be considered. Then and only then can one ensure accurate prediction of bio-markers, which can help to minimize the experimental effort in successfully finding highly informative biomarkers. Fortunately, there are a variety of data resources on the Internet, the mining of which could lead to the information needed for effective biomarker searches.

As a starting point, it is necessary to carefully study the metabolism of a specific type of cancer to identify those aspects of the cancer cells that distinguish them from the metabolism of normal cells, as well as from the metabolism of other cancer types and possibly even noncancerous diseases in the same organ types. A pathway-level analysis can guide the search to focus on biomolecules associated with the unique metabolic activities of a target cancer type. Then one will need to address the three questions asked in the beginning of the chapter. Such a study could provide a list of biomarker candidates, specific to a cancer type, which can be checked to determine if they are indeed present in circulation and have differential abundances compared to the control blood-sample pool. This could be done for example by using targeted detection methods such as antibody-based approaches. Such a method should largely bypass the demanding issue of dealing with the very com-plex composition of the blood proteome or other *omes*, such as metabolomics, as well as the abundance issue discussed earlier. Further analyses regarding the stabil-ity and detectability of specific biomolecules could help to rank the candidate list in an informative manner to further diminish the search to the most promising candi-dates for a specific cancer.

In the following sections, various technical issues are discussed that must be overcome to make this strategy a reality. In essence, this means solving the cancer biomarker search problem in a systematic manner, guided by our understanding of cancer metabolism unique to specific cancer types, our current knowledge of bio-molecules in circulation and the technical strengths and limitations of the current analytical techniques.

12.2   Search for Biomarkers Using a Top-Down Approach

The basic questions to be addressed here are: *Can one identify biomolecules whose combination and abundances are unique to: (a) cancer tissues in general; (b) tissues of specific cancer types; and (c) tissues of specific cancer types in early stages?* If answers to these questions are generally yes, half of the cancer-biomarker search problem will be solved. The second half of the problem is: *Can one predict if these biomolecules are released from the cells through normal channels and enter into blood circulation, and then, if such molecules are detectable using the existing analytical techniques.* These important questions will be discussed in the next sec-tion. In the following, proteins are used as an example to explain the basic ideas, while a discussion of how to search for metabolites and microRNAs as potential biomarkers will be given in Sect. 12.5. In addition, we will use gene-expression levels to approximate protein abundances since there is no whole-cell protein expression data publicly available for cancer tissues. Furthermore, it is known that while protein and gene expressions do not always have the same correlations across different classes of proteins, it is true that increased gene expression implies increased protein expression, at least qualitatively, in steady state (Vogel and Marcotte 2012).

Here we demonstrate how to predict the initial candidates for three scenarios: (1) general cancer biomarkers; (2) biomarkers for a specific cancer type; and (3) biomarkers for a specific cancer type in the early stage.

12.2.1 Biomarker Prediction for Cancers in General

While no general biomarkers for cancer, i.e., for detecting if someone has cancer or not, have been reported in the literature, we believe that identifying such biomarkers is plausible. The rationale is that cancers have numerous characteristics that are very distinct from normal and noncancerous diseased tissues, and it takes some unique combinations of certain biomolecules to realize these distinctive phenotypes. These molecules can potentially be used as biomarkers if they are secreted into blood circulation. The following gives a partial list of such unique features of cancer in general:

1. Unique characteristics in cell proliferation: The key characteristic of cancer cells is that they continuously proliferate, and their proliferation is fundamentally different from that associated with normal tissue development and remodeling. When a normal tissue is signaled to develop (or remodel), the development involves coordinated activities between cell division and their ECM. As discussed in Chaps. 5 and 10, tissue development involves: (1) changes in the shape and the physical properties of the underlying ECM; (2) multiple signals to the relevant cells in terms of biomass growth, cell division and cell survival, along with angiogenesis; and (3) continuous interplay between the cells and their ECM. As repeatedly stressed elsewhere in the book, normal cells have direct communication with their ECM via ECM-binding integrins, actin filaments and connections to their chromatin (Xu et al. 2009), thus facilitating rapid cellular responses to ECM changes. In contrast, cancer tissue development seems to be triggered by cellular pressures related to proliferation signals but without top-down signaling that coordinates different aspects of tissue development, as discussed in Chap. 5. While cell division is probably facilitated by signals from the hyaluronic acid fragments, there seems to be either no or inadequate corresponding signaling to the ECM for their changes. This hypothesis is strongly supported by the large number of genomic mutations in ECM-building proteins in precancerous tissues as shown in Chap. 4. The implication is that cancer uses a very distinct set of proteins to initiate tissue development (or the repair process),

facilitated by abnormal production of hyaluronic acid (and fragments) and by genetic mutations in multiple ECM-building proteins. Hence, one can expect that careful expression-pattern analyses of genes involved in cell division and EMC-building in cancer *versus* normal tissues will lead to the identification of very distinct gene-expression patterns between cancers in general and normal tissues, as well as possibly all noncancerous tissues, which should be generally shared across the majority of the cancer types.

In addition, one can expect that various other genes may show similar expression patterns among cancer tissues of different types, but distinct from normal tissues. This list may include certain oncogenes such as MYC, which is generally up-regulated in many cancer types.

2. Unique characteristics in metabolism: As discussed in the previous chapters, cancer cells have distinct characteristics in a few metabolic systems such as energy metabolism and ECM-associated metabolism, as well as angiogenesis. The unique aspects of cancer energy metabolism include the tendency to: (1) have an up-regulated glycolytic fermentation pathway, either instead of or in conjunction with aerobic respiration for ATP production; (2) use of glutaminolysis as a mode of energy production; and (3) have substantially higher metabolic rates compared to normal cells. To determine which specific genes or gene groups may show similar expression patterns across multiple cancer types, all being different from normal tissues, one needs to examine a large number of gene-expression data of various cancer tissues *versus* control non-cancerous tissues.

In addition, healthy tissues should not have highly activated angiogenesis, a biological process that is active for all solid cancers. Large scale gene-expression analyses between cancer and noncancerous tissues should reveal which angiogenesis genes may make strong biomarkers based on the gene expression data. Furthermore, cancer tissues have highly active genes involved in changing the morphology and the physical properties of their ECMs as discussed above. These considerations suggest that analyses focused on ECM-building genes in cancer *versus* normal growing tissues will lead to interesting discoveries for biomarker candidates.

3. Responses to the unique microenvironments of tumors: Knowing that cancers have a very distinct microenvironment such as hypoxia, elevated ROS levels, increased acidity and altered ECM properties, one can expect that a variety of genes will respond to these changes by an adjustment in their expression levels. Earlier we discussed (cf. Chaps. 5, 9, 10 and 11) those genes that may respond to changes in hypoxia, ROS, pH and the increased rigidity of an ECM, respectively, but in order to obtain a comprehensive list of such genes across different cancer types, systematic comparative analyses of gene-expression data of cancer *versus* noncancerous tissues are needed. Intuitively genes may need to be grouped together with similar or complementary functions as different genes may respond to the same environmental change in different cancers. Hence, it is expected that future biomarkers for cancer diagnosis (and prognosis) will be in the form of a collection of gene groups instead of individual genes; indeed, some such gene

groups are currently being used clinically. This approach permits one to better capture the reality that the responses will be at the pathway level instead of at the individual gene level. Also, different genes in the same pathway or functional group may respond to the same environmental changes in different cancers, again emphasizing the need for gene group-based biomarkers.

4. Unusual regulatory behaviors: Genes involved in cell survival, e.g., prevention of the activation of apoptosis and necrosis, is another area deserving attention, particularly since all cancers utilize some mechanism(s) to avoid the activation of apoptosis while normal tissue generally does not. Genes involved in DNA repair is another area recommended for study since cancer genomes tend to have large numbers of mutations as discussed in Chap. 4, while noncancerous tissues generally do not. Hence, one can expect that the DNA repair genes will be up-regulated in cancers generally.

In addition, cancers tend to have increased epigenomic activities as discussed in Chap. 9, making this another fruitful area in which to search for candidate marker genes. It has also been well established that cancer cells tend to have altered circadian rhythms. Consequently, some genes involved in this area may have very distinct expression patterns compared to noncancerous tissues.

Basically, multiple pathways or gene groups are expected to have similar expression patterns among tissues across multiple cancer types, which are not shared by noncancerous tissues in general. By accurately identifying these genes or gene groups with common or similar expression patterns across different cancer types, but distinct from noncancerous tissues, one can potentially identify gene groups as likely biomarkers for cancer in general, based on cancer tissue information. It is worth noting that the pathways and genes discussed here are not intended to be a complete list. Instead, they serve as a starting point for our readers to seek and identify their own candidate genes inspired by the discussion here.

12.2.2 Biomarker Prediction for a Specific Cancer Type

Different cancers may have very different phenotypes. Some cancers in general grow much faster than other cancers, and some cancers tend to respond very well to a specific treatment scheme, while other cancers do not. Some cancers may have a long dormancy time before they begin growing very rapidly, while other cancers may grow fast from the onset. The question addressed here is: *How does one find the distinguishing characteristics of a specific cancer type from all the other cancers?*

It should be noted that reliable identification on biomarkers for a specific cancer type represents a different type of problem from, and a potentially more challenging problem than, the identification of biomarkers for cancers in general. The goal of the latter is to find a group of genes whose expression pattern(s) can distinguish cancers from all the other tissues, while that of the former requires the identification of genes whose expression pattern(s) can distinguish a specific cancer from all the other cancers. For the latter, we know that there are such genes, as assured by the hallmark events of cancers and other unique activities of cancers discussed in the above subsection, and the problem is to find a small subset of them that serve as reliable markers. In comparison, the former requires one to identify subtle differences between one specific cancer type and all the other cancers.

One brute-force way would be to find a set of genes whose combined expression patterns are shared by the available samples of one particular cancer type, but different from all the other cancer types (based on publicly available data). While conceptually simple, such an approach would not be productive for at least two reasons. First, to do this effectively, one would need to examine all K-gene combinations for large K's in order to cover genes involved in multiple functional groups or pathways, say $K = 30$, out of a few hundred-to-thousand differentially expressed genes in cancer *versus* normal tissues, a problem much too large for the current computers. Second, considering expression levels of individual genes may be too simplistic to capture the commonalities among the cancer tissues of the same type, which are distinct from all other cancer types. Some information about the higher-order relationships among the expression levels of some relevant genes, such as the covariance relationship among the expression levels of all the differentially expressed genes, may be needed.

In essence, a more careful design is needed to tackle this very challenging problem. Recall from Chap. 3 that each cancer type always has its distinct and defining phenotypes, shared only by samples of this type, which should be reflected through the distinct expression patterns of some genes. The key is how to find these type-defining gene-expression patterns for each cancer type. Figure 2.2 provides an encouraging example, which shows that each of the nine cancer types in the figure has somewhat distinct expression patterns in terms of their pathway genes in glycolysis. One can expect that the same observation should also apply to some other pathways. Basically, for each cancer type, one needs to systematically examine all cancer-related (or early cancer-related) pathways to: (1) identify those for which the cancer type has an outstandingly different expression pattern from all the other cancers, and then (2) find the combination of such pathways to maximize the difference between a specific cancer type and the remainder. In a sense, it is somewhat analogous to examining Fig. 2.2, only a much larger one with all the cancer-relevant pathways listed along the y-axis and all the major cancer types listed along the x-axis. The goal for each cancer type is to find all the pathways for which the target cancer type has gene-expression patterns distinct from all or the majority of the other cancers. In addition, one may also choose to consider some tissue-specific genes, i.e., genes that are expressed only in the underlying tissue for the target cancer. The idea is that some of the tissue-specific genes may continue to be expressed in the cancer cells. The inclusion of such genes in the analyses may help to better distinguish one cancer type from the others.

This approach clearly represents a new area for cancer biomarker identification through systematic analyses of gene-expression data. Knowing the challenging nature of the problem, advanced statistical analysis techniques may be needed here to capture all the information discussed and possibly more.

12.2.3   Biomarker Prediction for a Specific Cancer Type in an Early Stage

To search for biomarkers for early-stage cancers (including precancerous tissues), one initially needs to determine which pathways tend to be active in the early stage of a specific cancer type. Recall from Chap. 5 that one of the two events may be the very early driver for cancer initiation, namely persistent hypoxia or elevated ROS levels. These early events lead to the accumulation of glycolytic metabolites, possibly along with some other metabolites that build up along with the glucose metabolic pathway, including part of the TCA cycle and links with fatty acid metabolism and amino acid metabolism. These congestions may lead to the synthesis and export of hyaluronic acid as discussed in Chap. 6, which will lead to signaling for a variety of tissue repair pathways such as inflammation, cell survival, cell proliferation and possibly other pathways. In addition, ECM remodeling is activated based on gene-expression data of early stage cancers. Other than this general information, very little is known about the differences among early stage cancers of different types based on the published data. One possible starting point to look for potential biomarkers is to conduct an analysis similar to that in the above subsection, but focused on the few pathways known to be activated in early stage cancers. The hope is that among the hundreds of genes involved in the above mentioned pathways, some distinctions can be identified for each cancer type.

Clearly, the earlier the stage a neoplastic tissue is, the more challenging it is to find a distinct biomarker for a specific cancer type, mainly because the initial responses to hypoxia and ROS may be limited to just a few pathways across different organs while the divergence in cancer evolutionary trajectories occurs later on. This suggests that focusing on organ-specific genes may prove to be a productive way to pursue this problem.

12.3   Prediction of Secretome and Circulating Proteins: A Data-Mining Approach

The main issue addressed in this section is: Given a set of proteins such as those predicted to be biomarkers for a specific cancer type (as in Sect. 12.2), *is it possible to predict which of them may be good biomarkers in (blood) circulation?* This problem will be tackled by solving the following two technical issues: (a) predict if a protein can be secreted extracellularly into circulation; and (b) assess if a protein in circulation may be relatively stable with a reasonable half-life (Fig. 12.1).

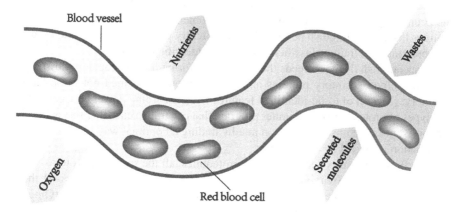

Fig. 12.1 A schematic of blood circulation and transportation of different molecular species to and from cells

12.3.1   Prediction of Blood-Secretory Proteins

A number of computational methods have been developed for predicting if a protein can be secreted (out of a cell), based on the identification of specific signal peptides in the protein sequence (Bendtsen et al. 2004a, b; Yu et al. 2010). However, such programs do not solve our problem here since not all secreted proteins enter into blood circulation. For example, some proteins may act on the cell surface or become part of the ECM after being secreted, while others may indeed enter into circulation. The current understanding is very limited about which proteins in the extracellular space may gain entry into circulation. We have previously developed a computational method for predicting if a secreted protein is likely to enter into circulation using a data classification technique (see Chap. 2 for definition) (Cui et al. 2008). This type of classification technique has been widely used for solving a variety of biological data analysis problems such as predicting if a protein is a membrane or soluble protein (Lo et al. 2008; Fuchs et al. 2009; Mishra et al. 2010) or if a protein is an enzyme or not (Fernandez et al. 2010).

The basic idea of a (2-class) data classification problem is to search for features with discerning power between two given, non-overlapping sets of proteins (or any objects) and then *train* a classifier to best separate the two sets of proteins using an optimal combination of the identified features. Once such a classifier is trained on *training* data, it can be used to predict if a new protein belongs to the first or the second set based on its feature values. One simple example of a data classification problem is the following. For a group of dogs and a group of cats, can one identify a few distinguishing features between the two groups and train a classification function, e.g., a weighted combination of the features, to best distinguish the dogs from the cats. This can then be used to predict if a new furry animal is a cat or dog?

Potentially such features could include the shape of the head and the frequency of the sound(s) produced by the animals. Our problem here, however, is to identify features that can distinguish between blood-secretory proteins and non-blood secretory proteins, and train a classifier based on the identified features.

To accomplish this, one needs the following information and capability: (a) a set of known blood secretory proteins and a set of proteins deemed not to be blood secretory; (b) a set of to-be-identified features with discerning power between these two types of proteins; (c) a computer program that can train a classifier that achieves the lowest possible misclassification rate by finding an optimal way to combine the information provided by each feature.

For (a), using results generated from proteomics, all the secretory proteins were collected from the Swissprot and SPD databases (Chen et al. 2005) and then compared with the proteins in the Plasma Proteome Project (PPP) database (Omenn et al. 2005), which contains over 16,000 proteins that have been identified in human plasma. From this analysis 305 proteins were found to belong to both sets, and hence considered as blood secretory proteins; these are denoted as the *positive training data*. The *negative training set* was generated by including one representative from each Pfam family (Bateman et al. 2002), the most popular protein family classification database, that does not overlap with the PPP proteins.

For (b), since no information is known about what features may possibly have discerning power between the two sets of proteins, some 50 protein features were considered, which fall into four categories: (1) sequence features such as amino acid and di-peptide compositions; (2) physicochemical properties such as solubility, disordered regions and charges, (3) structural properties such as secondary structural content and solvent accessibility, and (4) specific functional domains and motifs such as signal peptides, transmembrane regions and the twin-arginine signal peptide motif (TAT).

For (c), a support vector machine program was used to train a classifier based on the 50+ protein features to distinguish the positive from the negative dataset (Platt 1999; Keerthi et al. 2001). To determine which features may not have any discerning power for the classification, a feature-selection procedure, called *recursive feature elimination*, was employed to remove features that do not contribute to the classification accuracy. At the end of the training process, 12 of the most informative features were retained, which included transmembrane regions, charges, the TatP motif, solubility, signal peptides and the O-linked glycosylation motif. Based on these 12 features, the trained classifier achieved high classification accuracies on both the training data and on large independent evaluation sets (Cui et al. 2008).

The first application of this classifier for prediction of blood secretory proteins was on a set of gene-expression data collected on 80 pairs of gastric cancer tissues and neighboring noncancerous tissues (Cui et al. 2010). Out of the ~20,000 human genes, 715 genes were found to be differentially expressed consistently between cancer and the matching control tissues. Of these genes 136 were predicted to encode blood-secretory proteins by this classifier. A combined proteomic analysis using both mass spectrometry and Biotin label-based antibody arrays detected 81 of the 136 predicted proteins in serum samples, indicating the high-quality of the trained classifier. Based on specific selection criteria, 18 proteins were selected for detailed analyses. Of the 18 proteins, 5 (*COL10A1, GKN2, LIPG, MUC13, TOP2A*) were found to have differential concentrations in blood samples of cancer patients *versus* healthy controls, suggesting the potential of these proteins as blood biomarkers for gastric cancer detection (Cui et al. 2010).

This represents only a first attempt to develop one such predictor. To make it fully useful for reliable identification of blood biomarkers for gastric cancer, a number of improvements need to be made, including: (a) improved training data containing substantially larger datasets; (b) improved feature identification based on improved understanding of what other features may be relevant in distinguishing between proteins that are secreted into blood circulation and proteins that are not; and (c) most importantly, a new capability for predicting in what form a protein will remain in circulation in steady state.

12.3.2   Towards Prediction of the Half-Life of a Protein and Fragments in Circulation

Not all the predicted blood secretory proteins from cancer tissue may serve as good biomarkers since they may not necessarily maintain their intact forms in circulation very long because of the following reasons: (1) some proteins may be partially digested by proteases in circulation or on cell surfaces, which may be particularly abundant in the blood in the vicinity of the tumor(s) of advanced stage cancer patients (Chan et al. 2002; Woo et al. 2012); (2) some proteins may become fragmented due to the mechanical and sheer forces of the blood flow (Di Stasio and De Cristofaro 2010); and (3) many of these proteins will be degraded quickly by the liver. Hence, one needs to be able to predict which fragments of a protein may remain in circulation in steady state in order to develop a highly reliable predictor for biomarkers in circulation. This is clearly a very challenging and also very important problem. As of now, no solutions to this vexing problem have been reported.

One possible way to approach this problem is to study the peptide patterns of proteins detected in blood using information provided by the Plasma Proteome Project database (Omenn et al. 2005). This repository is the most comprehensive database for proteins found in human plasma, with most of the data obtained from mass spectrometry (MS). The data are all in the form of peptides, resulting from protein degradation, e.g., by circulating or cell-bound proteases, while in circulation or the enzymatic fragmentation (mainly by trypsin, a protease that cleaves peptide bonds C-terminal to lysine and arginine residues) required for mass spectrometric identification. From the PPP data, one can see that some proteins have significantly higher peptide coverage than others, hence providing some information about which proteins may possibly serve as better biomarkers since they have more fragments detectable in circulation. By carefully examining the peptide patterns, one could possibly derive which fragmentations result from *in vivo* cleavage and which are due to MS-associated digestions. This information could then be used to construct a model for predicting the circulating peptide patterns for each candidate protein in its steady state in circulation. Such information can then be used to guide further experimental validation and suggest antibodies that are specific for the peptides instead of the intact proteins from which they were derived.

12.4 Searching for Biomarkers in Other Human Body Fluids

Human blood is clearly the most information-rich source for detecting human diseases in different organs since this is the central transportation system to and from cells. However, detection of specific signals is often challenging due to the enormous complexity in terms of the composition and the dynamic range of blood proteins and metabolites as discussed earlier. Other human body fluids may also provide highly informative signals. For example, recent large-scale urinary proteomic analyses revealed that urine is also an information-rich source for disease detection. In addition, urine tests are clearly less invasive compared to blood tests, as is saliva.

12.4.1 Searching for Biomarkers in Urine

Recent proteomic analyses of human urine has identified thousands of different protein species (Pang et al. 2002; Weissinger et al. 2007; Zimmerli et al. 2008), which is somewhat surprising as it was previously thought that very little protein is in the urine of a healthy person (Adachi et al. 2006). This is true but only in terms of protein abundances, not as much in terms of protein diversities.

Since urine is formed via filtration of blood through the kidneys, some blood-borne proteins can be filtered and fail to be reabsorbed, thus becoming excreted into the urine. Compared to serum, the challenge in searching for urinary biomarkers is that urinary proteins/peptides tend to be of substantially lower abundances, while the advantage is that the dynamic range of the protein/peptide abundances is substantially smaller than that in blood.

We have previously developed a computational method for predicting proteins that can be excreted into urine through secretion from diseased human tissues (Hong et al. 2011), using essentially the same idea and the same three key steps, (a), (b) and (c), as used in predicting blood secretory proteins and outlined in Sect. 12.3. For (a), a positive training dataset was formed by including 1,313 proteins found in urine samples of healthy people by a recent large proteomic study (Adachi et al. 2006), and the negative dataset consisted of 2,627 proteins selected in a similar fashion to that in the blood secretory protein prediction.

For (b), a large number of potentially relevant features were examined. For this data-classification problem, more information was used to guide the feature search. For example, electronic charge and the rigidity of a protein structure were included in the feature selection, knowing that the glomerular walls in kidneys are negatively charged and with pores of relatively small size. In the end, 18 features were found to be informative for prediction of urine-excretory proteins. For (c), a similar support vector machine-based classifier was trained as before, which achieved a 78 % prediction sensitivity and a 92 % prediction specificity on an independent test set consisting of 460 known urine excretory proteins and 2,148 non-urine excretory proteins (Hong et al. 2011).

This classifier was then applied to the 715 differentially expressed genes in gastric cancer *versus* matching control tissues for prediction of urine excretion (Hong et al. 2011). This analysis yielded six proteins that were predicted to be candidates as urinary biomarkers for gastric cancer, (*AZGP1, COL10A1, EL, LIPF, MMP3, MUC13*), for experimental validation on urine samples of 21 gastric cancer patients (mostly advanced stages) and 21 age/gender-matched healthy individuals. Of the six proteins, five (*AZGP1, COL10A1, EL, LIPF, MUC13*) were detected by Western blots of urine samples, indicating the high-quality of the prediction classifier for urine-excretory proteins. *MMP3* was not found in any urine sample, possibly due to its low abundance or a false prediction by the classifier. The Western blots for *EL* (epithelial lipase) showed a substantial reduction in its abundance in urine samples of the 21 gastric cancer patients compared to the 21 control samples. Specifically, the majority of the control samples were found to have *EL*, whereas most of the gastric cancer samples had either relatively low amounts or no detectable *EL* (see Fig. 12.2). The same pattern was observed on a substantially larger sample set, achieving a classification AUC value = ~96 % (Hong et al. 2011), which also includes 30 urinary samples from patients of other cancer types that show similar abundance patterns to those of healthy controls. This result suggests that *EL* could prove to be a highly promising urinary biomarker specifically for gastric cancers.

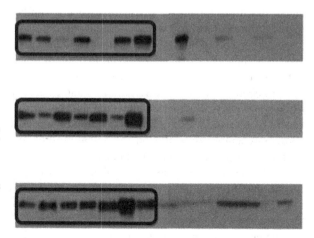

Fig. 12.2 Western blots showing three independent gels of EL on urine samples of 21 gastric cancer patients and of 21 controls. Results from the 21 controls are given in the 21 lanes *outlined* in *black* (seven lanes per gel) on the *left portion* of each gel. Data from the patients are shown in the *right portion* of each gel (seven lanes per gel). Adapted from (Hong et al. 2011)

12.4.2   Searching for Biomarkers in Saliva

We have recently extended this line of study to salivary biomarkers. Compared to human blood and urine, saliva has not been used for detection of human disease until recently. The impetus for this approach was the report that salivary proteomic analyses suggested that human saliva is rich in proteins (Denny et al. 2008), some of which are derived from the blood and hence can potentially serve as an information pool for disease biomarker identification for distal organs. Some salivary proteins have been used for disease detection such as salivary kallikrein for breast cancer and gastrointestinal cancer (Jenzano et al. 1986), *PSA* for prostate cancer (Turan et al. 2000) and *HER2* and *P53* for breast cancer (Streckfus et al. 2000). However, no general methods for systematic searches for biomarkers in saliva have been reported.

Three mechanisms have been identified for biomolecules to travel from circulation into saliva (Wong 2006; Pfaffe et al. 2011): (1) active transport for some proteins such as secretory IgA and immunoglobulin E; (2) passive transport for drugs and steroids; and (3) ultrafiltration for small polar molecules such as creatinine. In addition, salivary glands may secrete their own proteins into saliva in response to specific proteins in blood circulation. For such cases, however, bioinformatics cannot be fruitfully applied yet since there are no stimuli-response data available for salivary glands, the basis on which a predictor can be developed.

Since proteins that circulate in blood can be detected in saliva, we have developed a predictor for salivary biomarkers, in a very similar manner to the urine marker prediction study (Wang et al. 2013). The main difference here is that the amount of training data is sparse. Consequently, an extensive literature search was carried out, which led to the identification of 62 experimentally validated human salivary proteins coming from circulation. These proteins were used as the positive training dataset. The negative dataset included 6,816 proteins that were deemed to be absent in saliva, based on procedures similar to those used in the development of blood and urine biomarker predictors. Then a similar procedure was applied to search for protein features with discerning power between the positive and the negative datasets. A classifier was then trained as in the previous sections. Following this approach, the outcome was identification of the following features that were found to be most informative for the classification: the radius of gyration of the protein, hydrophobicity, Geary autocorrelation, amino acid composition, dipeptide composition, secondary structure composition and polarity. These selected features are generally consistent with our understanding of secretory proteins and salivary proteins. For example, the diffusion coefficient for proteins is inversely proportional to the protein structural feature, the radius of gyration (Brandtzaeg 1971).

The trained classifier was then applied to a set of gene-expression data of breast cancer, and 31 proteins were predicted as biomarker candidates in the patients' saliva, including *TIMP2*, *CFD*, *CCL14* and *FBLN* (Wang et al. 2013). These proteins are known to be involved in wound response, acute inflammatory response, complement and coagulation cascades, cell adhesion, biological adhesion and immune response, all of which are related to cancer development.

12.5   Searching for Biomarkers Among Other Molecular Species

To this point, all the discussions have been focused on identification of proteins as potential biomarkers for cancer, but there is no reason to limit the search to proteins only. Other molecular species such as microRNAs and metabolites could potentially serve the same purpose and may even be more effective. For example, one advantage of using microRNAs as blood or other body fluid biomarkers is that they tend to remain in circulation much longer than proteins since they are not degraded by the hepatic enzymes (Etheridge et al. 2011). As of now, a number of microRNAs have been used as potential cancer biomarkers. For example, a combination of miR-145 and miR-451 was recently proposed as a blood biomarker for breast cancer (Wimberly et al. 2013), and a panel of five microRNAs (miR-200a, miR-100, miR-141, miR-200b, miR-200c) has been recommended as a biomarker for ovarian cancer. In addition, numerous other microRNA combinations have been suggested for a variety of cancers in the past few years (Cho 2007; Bartels and Tsongalis 2009; Kosaka et al. 2010). A number of metabolite-based cancer biomarkers has also been proposed in the past few years, including metabolites involved in glycolysis, glutamine utilization, fatty acid synthesis and mitochondrial function (Chiaradonna et al. 2012).

12.5.1   MicroRNAs as Cancer Biomarkers

MicroRNAs are small, single-strand non-coding RNAs (18–22 nucleotides) that are estimated to comprise 1–3 % of the human genome (Zhao and Srivastava 2007). This group of RNAs has been found to play crucial roles in maintaining the normal levels of ~30 % of human mRNAs. The recent discovery of the key roles played by microRNAs in epigenomic regulation and execution (Choudhry and Catto 2011; Kunej et al. 2011) should make them attractive candidates for cancer biomarkers. As discussed in Chap. 9, increased epigenomic activities may represent a key transition from proliferating cells to malignant cells. Hence if the associated microRNAs are detected in circulation, it may give a strong indication of the stage of a cancer.

While the mechanism of microRNA release from cells into circulation is largely unknown, it is well accepted that some microRNAs are released into circulation (Chen et al. 2008). Zhang et al. suggested that cells may selectively package microRNAs into micro-vesicles and then secrete them (Zhang et al. 2010). The resistance of microRNAs to *RNase-A* digestion (Chen et al. 2008) suggests that serum microRNAs might be modified from cellular microRNAs or they may be in complex with a vesicle or proteins.

In order to predict blood secretory or urine-excretory microRNAs, a classifier need to be trained as in the previous sections, but the key is to identify a set of different features that have been derived from microRNAs rather than proteins.

A number of features have been found to be useful. For example, recent studies revealed that the major forms of circulating microRNAs involve a complex with the *AGO* (argonaute) proteins; in cells this is part of the RNAi silencing complex. The stability of the microRNAs in circulation may be attributable to the formation of the *AGO2* complex. Complexation with protein seems to protect the microRNAs from *RNase* degradation (Arroyo et al. 2011; Turchinovich et al. 2011), although the mechanism of the miRNA-*AGO2* complex secretion remains to be understood. In order for non-renal microRNAs to be excreted into urine, they must first enter into circulation and then be excreted through the renal glomerulus. Therefore, the discriminating features for excretory microRNAs may be linked to the microRNA uptake and release mechanism by transport vesicles or the association with *AGO* proteins. Thus, features related to binding of a microRNA need to be examined, including the secondary structures and sequence-level features. For example, a recent study showed that strand-bias selection exists for microRNAs in incorporation into the *RISC* complex, and highly expressed strands tend to have nucleotide G-bias and U-bias at the 5′ end (Hu et al. 2009). This suggests that microRNAs enriched with G and U nucleotides at the 5′ end may be more likely to bind to the *AGO2* protein and form a *RISC* complex. Among those features, a few were identified that can collectively make known urine-excretory microRNAs conspicuous among all known human microRNAs, strongly suggesting the possibility of identifying microRNAs as serum or urine biomarkers for cancer.

A classifier was recently developed based on a set of features showing discerning power between 325 microRNAs that have been found in serum, but not in urine, and a set of 100 microRNAs that have been identified in the urine of healthy individuals (unpublished data). When the trained classifier was applied to 138 microRNAs that have been reported to be present in urine, the prediction accuracy was ~70 %. This result is clearly encouraging for developing a reliable predictor for microRNA excretion. We expect that, as increasingly more microRNAs are detected in serum and urine and posted in the miRBase database (Griffiths-Jones et al. 2006; Kozomara and Griffiths-Jones 2011), a highly reliable predictor will surface soon.

12.5.2 Metabolites as Cancer Biomarkers

Most of the current metabolite-based cancer biomarkers tend to be those derived from glycolysis, glutamine utilization, fatty acid synthesis and mitochondrial function (Chiaradonna et al. 2012). This should not be surprising since the various cancers are known to have distinct activities in these metabolic pathways. In order to expand the number of potential cancer biomarkers, one should consider the metabolites associated with ECM development and remodeling. It is known that the composition of the ECM, and hence its physical properties, change continually during cancer tissue development; yet, the metabolites derived from these processes have received little attention (Lu et al. 2012). One group of such metabolites was discussed in Chaps. 6 and 10, namely those fragments formed from hyaluronic acid.

These fragments clearly deserve systematic analyses for their potential to serve as biomarkers for different types of cancer, with emphasis on grading, staging and possibly drug-resistance properties. In addition to hyaluronic acid fragments, there are other types of metabolites involved in ECM such as: (a) heparan sulfates that are active in tissue development, angiogenesis and cancer metastasis (Vlodavsky and Friedmann 2001); (b) chondroitin sulfates that affect the tensile strength of a matrix; and (c) keratan sulfates, known to be involved in development and scar formation after injury (Zhang et al. 2006). Most of these metabolites are relatively small and water soluble, and it is expected that most, if not all, can readily enter into blood circulation.

12.6 Concluding Remarks

With the availability of multiple types of *omic* data collected on various body fluids of healthy individuals, it is a most propitious time to develop powerful computation-guided search paradigms for highly effective biomarkers in different body fluids for cancer diagnosis and prognosis. Differing from the traditional approaches that essentially try the proverbial "search for a needle in a haystack(s)" with little guidance, hence often leading to predicted markers with subpar prediction capabilities, the combination of: (1) the current knowledge of cancer biology at its different developmental stages; (2) available *omic* data collected on cancer *versus* control tissues and *omic* data collected from different body fluids; and (3) availability of statistics-based data-mining approaches that allow realistic and reliable model building for predicting biomarker candidates, followed by target-based experimental validation. It is expected that the full execution of such ideas could lead to the efficient identification of highly reliable biomarkers for cancers of different types, at different developmental stages and different levels of malignancy. Success in these endeavors would lead to a fundamental improvement in our current capabilities in cancer detection, especially in the early stages. In addition, we believe that the information revealed through such body-fluid based biomarker searches will not only be useful for cancer diagnosis, but also helpful in selecting the most effective therapeutic strategies for individualized cancer treatment using the biomarkers to monitor effectiveness.

References

Adachi J, Kumar C, Zhang Y et al. (2006) The human urinary proteome contains more than 1500 proteins, including a large proportion of membrane proteins. Genome Biol 7: R80

Anderson NL, Anderson NG (2002) The human plasma proteome: history, character, and diagnostic prospects. Mol Cell Proteomics 1: 845–867

Arroyo JD, Chevillet JR, Kroh EM et al. (2011) Argonaute2 complexes carry a population of circulating microRNAs independent of vesicles in human plasma. Proc Natl Acad Sci U S A 108: 5003–5008

Bartels CL, Tsongalis GJ (2009) MicroRNAs: novel biomarkers for human cancer. Clin Chem 55: 623–631

Bateman A, Birney E, Cerruti L et al. (2002) The Pfam protein families database. Nucleic acids research 30: 276–280

Bendtsen JD, Jensen LJ, Blom N et al. (2004a) Feature-based prediction of non-classical and leaderless protein secretion. Protein Eng Des Sel 17: 349–356

Bendtsen JD, Nielsen H, von Heijne G et al. (2004b) Improved prediction of signal peptides: SignalP 3.0. J Mol Biol 340: 783–795

Bhatt AN, Mathur R, Farooque A et al. (2010) Cancer biomarkers - current perspectives. Indian J Med Res 132: 129–149

Brandtzaeg P (1971) Human secretory immunoglobulins. II. Salivary secretions from individuals with selectively excessive or defective synthesis of serum immunoglobulins. Clin Exp Immunol 8: 69–85

Chan JM, Stampfer MJ, Ma J et al. (2002) Insulin-like growth factor-I (IGF-I) and IGF binding protein-3 as predictors of advanced-stage prostate cancer. J Natl Cancer Inst 94: 1099-1106

Chen X, Ba Y, Ma L et al. (2008) Characterization of microRNAs in serum: a novel class of biomarkers for diagnosis of cancer and other diseases. Cell Res 18: 997–1006

Chen Y, Zhang Y, Yin Y et al. (2005) SPD–a web-based secreted protein database. Nucleic Acids Res 33: D169–173

Chiaradonna F, Moresco RM, Airoldi C et al. (2012) From cancer metabolism to new biomarkers and drug targets. Biotechnol Adv 30: 30–51

Cho WC (2007) OncomiRs: the discovery and progress of microRNAs in cancers. Mol Cancer 6: 60

Choudhry H, Catto JW (2011) Epigenetic regulation of microRNA expression in cancer. Methods Mol Biol 676: 165–184

Cui J, Chen Y, Chou WC et al. (2010) An integrated transcriptomic and computational analysis for biomarker identification in gastric cancer. Nucleic Acids Res 39(4):1197–207

Cui J, Liu Q, Puett D et al. (2008) Computational Prediction of Human Proteins That Can Be Secreted into the Bloodstream. Bioinformatics 24(20): 2370–2375

Denny P, Hagen FK, Hardt M et al. (2008) The proteomes of human parotid and submandibular/sublingual gland salivas collected as the ductal secretions. J Proteome Res 7: 1994–2006

Di Stasio E, De Cristofaro R (2010) The effect of shear stress on protein conformation: Physical forces operating on biochemical systems: The case of von Willebrand factor. Biophysical chemistry 153: 1–8

Etheridge A, Lee I, Hood L et al. (2011) Extracellular microRNA: a new source of biomarkers. Mutat Res 717: 85–90

Fernandez M, Ahmad S, Sarai A (2010) Proteochemometric recognition of stable kinase inhibition complexes using topological autocorrelation and support vector machines. J Chem Inf Model 50: 1179–1188

Fuchs A, Kirschner A, Frishman D (2009) Prediction of helix-helix contacts and interacting helices in polytopic membrane proteins using neural networks. Proteins 74: 857–871

Griffiths-Jones S, Grocock RJ, van Dongen S et al. (2006) miRBase: microRNA sequences, targets and gene nomenclature. Nucleic Acids Res 34: D140–144

Hong CS, Cui J, Ni Z et al. (2011) A computational method for prediction of excretory proteins and application to identification of gastric cancer markers in urine. PLoS One 6: e16875

Hu HY, Yan Z, Xu Y et al. (2009) Sequence features associated with microRNA strand selection in humans and flies. BMC Genomics 10: 413

Jenzano JW, Courts NF, Timko DA et al. (1986) Levels of glandular kallikrein in whole saliva obtained from patients with solid tumors remote from the oral cavity. J Dent Res 65: 67–70

Jones HB (1848) On a new substance occurring in the urine of a patient with mollifies ossium. Philosophical Transactions of the Royal Society 138: 55–62

Kosaka N, Iguchi H, Ochiya T (2010) Circulating microRNA in body fluid: a new potential biomarker for cancer diagnosis and prognosis. Cancer Sci 101: 2087–2092

Kozomara A, Griffiths-Jones S (2011) miRBase: integrating microRNA annotation and deep-sequencing data. Nucleic Acids Res 39: D152–157

Kunej T, Godnic I, Ferdin J et al. (2011) Epigenetic regulation of microRNAs in cancer: an integrated review of literature. Mutat Res 717: 77–84

Leidinger P, Backes C, Deutscher S et al. (2013) A blood based 12-miRNA signature of Alzheimer disease patients. Genome Biology 14: R78

Lilja H, Ulmert D, Vickers AJ (2008) Prostate-specific antigen and prostate cancer: prediction, detection and monitoring. Nat Rev Cancer 8: 268–278

Lo A, Chiu HS, Sung TY et al. (2008) Enhanced membrane protein topology prediction using a hierarchical classification method and a new scoring function. J Proteome Res 7: 487–496

Lu P, Weaver VM, Werb Z (2012) The extracellular matrix: a dynamic niche in cancer progression. The Journal of cell biology 196: 395–406

Mishra NK, Agarwal S, Raghava GP (2010) Prediction of cytochrome P450 isoform responsible for metabolizing a drug molecule. BMC Pharmacol 10: 8

Nixon AB, Pang H, Starr MD et al. (2013) Prognostic and predictive blood-based biomarkers in patients with advanced pancreatic cancer: results from CALGB80303 (Alliance). Clin Cancer Res 19: 6957–6966

Omenn GS, States DJ, Adamski M et al. (2005) Overview of the HUPO Plasma Proteome Project: results from the pilot phase with 35 collaborating laboratories and multiple analytical groups, generating a core dataset of 3020 proteins and a publicly-available database. Proteomics 5: 3226–3245

Pang JX, Ginanni N, Dongre AR et al. (2002) Biomarker discovery in urine by proteomics. J Proteome Res 1: 161–169

Pfaffe T, Cooper-White J, Beyerlein P et al. (2011) Diagnostic potential of saliva: current state and future applications. Clin Chem 57: 675–687

Platt JC (1999) Fast Training of Support Vector Machines using Sequential Minimal Optimization. In: Advances in kernel methods: support vector learning. MIT Press Cambridge, MA, USA, pp 185 – 208

S. S. Keerthi, S. K. Shevade, C. Bhattacharyya,K. R. K. Murthy (2001) Improvements to Platt's SMO Algorithm for SVM Classifier Design Neural Computation 13: 637–649

Sanchez-Cespedes M (2008) The impact of gene expression microarrays in the evaluation of lung carcinoma subtypes and DNA copy number. Arch Pathol Lab Med 132: 1562–1565

Scholkopf B, Platt JC, Shawe-Taylor J et al. (2001) Estimating the support of a high-dimensional distribution. Neural Computation 13: 1443–1471

Streckfus C, Bigler L, Tucci M et al. (2000) A preliminary study of CA15-3, c-erbB-2, epidermal growth factor receptor, cathepsin-D, and p53 in saliva among women with breast carcinoma. Cancer Investigation 18: 101–109

Thompson IM, Chi C, Ankerst DP et al. (2006) Effect of finasteride on the sensitivity of PSA for detecting prostate cancer. J Natl Cancer Inst 98: 1128–1133

Turan T, Demir S, Aybek H et al. (2000) Free and total prostate-specific antigen levels in saliva and the comparison with serum levels in men. Eur Urol 38: 550–554

Turchinovich A, Weiz L, Langheinz A et al. (2011) Characterization of extracellular circulating microRNA. Nucleic Acids Res 39: 7223–7233

Vlodavsky I, Friedmann Y (2001) Molecular properties and involvement of heparanase in cancer metastasis and angiogenesis. J Clin Invest 108: 341–347

Vogel C, Marcotte EM (2012) Insights into the regulation of protein abundance from proteomic and transcriptomic analyses. Nat Rev Genet 13: 227–232

Wang J, Liang Y, Wang Y et al. (2013) Computational Prediction of Human Salivary Proteins from Blood Circulation and Application to Diagnostic Biomarker Identification. PLoS ONE 8: e80211

Weissinger EM, Schiffer E, Hertenstein B et al. (2007) Proteomic patterns predict acute graft-versus-host disease after allogeneic hematopoietic stem cell transplantation. Blood 109: 5511–5519

Wimberly H, Shee C, Thornton PC et al. (2013) R-loops and nicks initiate DNA breakage and genome instability in non-growing Escherichia coli. Nature communications 4: 2115

Wong DT (2006) Salivary diagnostics powered by nanotechnologies, proteomics and genomics. J Am Dent Assoc 137: 313–321

Woo Y, Hyung WJ, Obama K et al. (2012) Elevated high-sensitivity C-reactive protein, a marker of advanced stage gastric cancer and postgastrectomy disease recurrence. J Surg Oncol 105: 405–409

Xu R, Boudreau A, Bissell MJ (2009) Tissue architecture and function: dynamic reciprocity via extra- and intra-cellular matrices. Cancer Metastasis Rev 28: 167–176

Yu L, Guo Y, Zhang Z et al. (2010) SecretP: a new method for predicting mammalian secreted proteins. Peptides 31: 574–578

Zhang H, Uchimura K, Kadomatsu K (2006) Brain keratan sulfate and glial scar formation. Ann N Y Acad Sci 1086: 81–90

Zhang Y, Liu D, Chen X et al. (2010) Secreted monocytic miR-150 enhances targeted endothelial cell migration. Mol Cell 39: 133–144

Zhao Y, Srivastava D (2007) A developmental view of microRNA function. Trends Biochem Sci 32: 189–197

Zimmerli LU, Schiffer E, Zurbig P et al. (2008) Urinary proteomic biomarkers in coronary artery disease. Mol Cell Proteomics 7: 290–298

Chapter 13
In Silico Investigation of Cancer Using Publicly Available Data

Cancer is a very complex disease, far more multifaceted than the traditional views rooted in the thinking that cancer is a genomic disease, at least for solid tumors as discussed in the previous chapters. The disease is a rapidly evolving biological system drifting away from normal cellular metabolism and homeostasis to adapt to the also evolving, increasingly more challenging and unfamiliar microenvironment. It may start from some, in and of itself, seemingly harmless metabolic changes in response to a stressful local condition such as persistent hypoxia and/or elevated ROS, which leads to gradual and continuing changes in the microenvironment, hence producing pressure for the underlying cells to evolve. The observed cell proliferation may represent a feasible and efficient route for the affected cells to escape from these pressures. The similar growth patterns and other common characteristics across different cancer types, referred to as hallmark activities, strongly suggest that the survival pathway of the affected cells is a well-coordinated process, possibly guided by signaling instructions manifested by hyaluronic acid and fragments as discussed in Chaps. 6 and 9. The continuous coadaptation and coevolution between the changing microenvironment and the altered cellular metabolism may drive the evolving cells to utilize whatever cellular capabilities encoded in their genomes via the increasingly more relaxed epigenomic regulations or random mutations confer for their survival.

As the evolution and natural selection of the affected cells continue, the local disease gradually becomes a holistic illness, not only because the cancer cells migrate to and colonize distant locations, but also because they have evolved to make a generalized impact on the body of the individual so affected. For example, they have gained capabilities to evade destruction by the immune system, learned to utilize increasingly more resources that a normal physiological system can offer, and consumed a substantial amount of energy compared to the rest of the body. Hence, one may posit that it is the micro-environmental pressures that drive the cancer to grow. From this perspective, one can argue that the other changes, including those at the functional execution, epigenomic and genomic levels, are facilitators for their survival through proliferation. Consequently, to treat cancer effectively, one needs to accurately identify the specific type of cellular pressure(s) that drives the current proliferation, as well as the root source of the pressures.

Since the discoveries of oncogenes and tumor suppressor genes in the 1970s, a substantial amount of knowledge has been gained about molecular and cellular level mechanisms of cancer, typically accomplished using model systems. This rich background of knowledge serves as the foundation for computational and systems biologists to study this disease as an evolving system in its full complexity. It is the availability of the variety of *omic* data collected on both model systems and human cancer tissues that makes such studies possible. This chapter reviews some of the popular and publicly available data resources, as well as computational tools related to cancer research, and illustrates the types of questions that can be addressed *in silico* through computational analyses of such data.

13.1 Questions Potentially Addressable Through Mining Cancer *Omic* Data

A few illustrative examples are outlined in this section to demonstrate the types of questions that can be addressed through computational data mining and statistical inference, with the aim of providing new information and approaches for cancer studies.

13.1.1 *Characterization of Tumor Microenvironments*

The importance of the microenvironment to the development of cancer has been well established in the literature. Throughout this book, emphasis has been placed on the role of the microenvironment in driving and facilitating the disease to evolve to overcome pressures, different types at different developmental stages, cast on the neoplastic cells. This environment can be defined in terms of a number of measures, such as the oxygen level, the oxidative stress level, the acidity level, the composition and mechanical properties of the local ECM, and the various signaling molecules released into the extracellular space by the local stromal cells. While it is vitally important to collect such data for cancer studies, it has proved to be a very challenging problem to use experimental techniques for *in situ* studies, clearly unrealistic for large-scale studies. Fortunately, computational analyses of gene-expression data can generate useful information in probing micro-environmental conditions.

The basic idea is that when the micro-environmental factors change, some genes will respond by altering their expression levels. A number of marker genes have been identified which respond to specific environmental changes, such as *HIF1 versus* cellular oxygen level, *SOD1 versus* ROS level, and the genes responding to acidity-level changes as discussed in Chap. 8. Through statistical association analyses, one can possibly identify genes that may respond to or define a specific environmental factor, and even genes that may respond to combinations of environmental stressors. When applied to gene-expression data collected on cell lines under specific conditions, such as exposure to different levels of oxygen, one can possibly derive quantitative relationships between gene-expression levels and the cellular oxygen level. Such analyses can lead to the development of predictive models, defining the detailed micro-environmental conditions in specific cancer samples and hence enabling cancer *omic* data analysis to be undertaken in a more informed manner. When linking such information with cancer clinical data, one may be able to drive new insights regarding the cellular mechanisms by which various environmental conditions affect the clinical outcome of a cancer, such as the growth rate, mortality rate, responses to various treatments and the potential for metastasis.

13.1.2 Identification of Key Transition Events and Possible Causes Throughout Cancer Development

A number of hallmark events of cancer have been identified (Hanahan and Weinberg 2011) such as reprogrammed energy metabolism, autonomous signaling for growth, and angiogenesis across all the (solid tumor) cancers, as introduced in Chap. 1. *Do these hallmark events take place independent of each other or do some need to precede the others?* It is foreseeable that analyses of transcriptomic data of cancer tissues, ordered according to their developmental stages, could address this issue. For example, each hallmark event can be characterized in terms of the expression patterns of a specific set of genes. By examining how the expression patterns of these genes change as a cancer evolves, one should be able to detect the transition points such as the starting point or monotonic change *versus* fluctuations of the activity level of each hallmark event, as well as the relative order among different hallmark events. One challenging issue in carrying out such analyses is that of determining the relative developmental stages among collected cancer tissues at a resolution finer than the four discrete stages typically used in the clinical setting, hence allowing one to use "time" on a finer and more useful scale and then study hallmarks or other events along the more accurate "time" line.

So the question is: *How much finer can one possibly make the clock based on the available omic data*, which will enable age comparisons among cancer tissues from different patients? Actually there are a few obvious measures that can potentially be used for such a clock. For example, the number of cancer-related mutations per genome could be a candidate. However, normalizations of such a measure across different tissues may represent a non-trivial issue since it has been reported that different cancer types may have vastly different median numbers of mutations per genome (see Chap. 4), suggesting that some micro-environmental factors, such as hypoxia and ROS levels, need to be taken into consideration when doing the normalization. Another possible clock could be related to the number of differentially expressed genes in cancer *versus* the matching control tissue as used in Fig. 2.1 (also see (Xu et al. 2010)). Then again subtle issues could arise that may need to be considered, e.g., gene-expression levels of individual tissues may not be as informative as the average gene-expression levels across multiple tissues as discussed in Chap. 3.

If one can identify or develop one such "universal" clock for measuring cancer progression, many important questions could be addressed, enabled by such a clock. For example, one should be able to address: (1) the possible dependence among different hallmark events, (2) possible triggering conditions in the microenvironment of each hallmark event; and (3) potential causal relationships among various events, including hallmark events, observed during cancer development.

13.1.3   Understanding the Effects of Redox States on Cancer Development

Oxidative stress has been known for some time to be associated with cancer development (Pani et al. 2010; Reuter et al. 2010), which is clearly a key driver for cancer initiation as discussed in Chap. 5 and possibly serves a driving role in cancer metastasis as discussed in Chaps. 10 and 11. While a number of reviews have been published on ROS and cancer, there are no published studies on redox (**reduction-oxi**dation) fluxes at the cellular level as a function of cancer progression. Such information could potentially reveal the root sources of oxidative stress surges observed during the development of a cancer. In comparison, a few metabolic flux analyses, e.g., on carbon and nitrogen fluxes at the pathway or cellular level, have been published on cancer (Hirayama et al. 2009; Achreja et al. 2013). Some interesting observations have been forthcoming, such as the accumulation of glucose metabolites throughout the entire glycolytic pathway in cancer (Hirayama et al. 2009). Analyses of the redox flux, i.e., flux of electrons, are likely to reveal even more exciting and surprising information since electron fluxes are not as intuitive as metabolic fluxes. One can expect that such studies could reveal how redox states, specifically oxidative stress, contribute to specific activities and hallmark events throughout the entire process of tumorigenesis. While it is not easy to directly collect electron potential data on a large scale, one can possibly model electron flow by examining enzymatic reactions that involve electron-charge changes in the metabolites on the two sides of each reaction and balance them with reducing/oxidizing agents, namely GSH/GSSG, $NAD^+/NADH$ and $NADP^+/NADPH$, whose quantities can be estimated based on the gene-expression levels of their relevant enzymes such as the NADH kinase and ferredoxin-$NADP^+$ reductase. This type of systems-level redox flux analysis could provide fundamentally novel information to the understanding of cancer initiation, progression and metastasis, and computational modelers have an opportunity to assume key roles in making this a reality.

13.1.4   Impact of Repeated Re-oxygenation on Selection of More Resilient Cancer Cells

As discussed in Chaps. 5 and 11, changes in cellular oxygen level have a fundamental impact on cancer initiation and metastasis. It is only natural to ask: *How does the repeated re-oxygenation induced by continuous tumor growth and new angiogenesis affect cancer evolution?*

It has been well established that as a tumor develops it becomes increasingly hypoxic, which will lead to the signaling for new angiogenesis, thus increasing the blood and oxygen supply. The increased blood supply allows the tumor to continue growth until it becomes hypoxic again due to the increased biomass. This cycle repeats and leads to multiple rounds of angiogenesis and the associated re-oxygenation. This alternation between hypoxia and normoxia may serve as a key measure for selecting robust cancer cells as the disease progresses, and possibly serve as a trigger to epigenomic changes that can be passed on to the offspring to better cope with the two extreme conditions.

Currently there are no published studies on how such a re-oxygenation cycle impacts cancer development at a cellular level. Using a bioinformatics approach, one can design studies based on the available transcriptomic data. One way to accomplish this is by grouping tissue samples into: (1) highly hypoxic ones before additional blood flows in via tumor angiogenesis; (2) ones with new blood flow; and (3) the remaining samples. This classification of samples can be achieved through the use of marker genes for hypoxia and angiogenesis. Then within each sample group, one can infer that genes or pathways whose expression levels exhibit strong correlations, either positively or negatively, with changes of the hypoxia level for samples. If successful, one should be able to identify cellular processes that are enhanced or diminished by repeated re-oxygenation, as well as by repeated hypoxia, thus gaining a detailed understanding of how re-oxygenation, in conjunction with repeated hypoxia, selects the most resilient cancer cells.

13.1.5   Triggering Events of Cancer Metastasis

A large number of studies have been published on cancer metastasis (see Chap. 10), and various events have been identified to facilitate metastasis, such as the activation of the EMT (epithelial-mesenchymal transition) pathway. However, the fundamental issue remains unanswered: *What triggers a cancer to metastasize* as the activation of the EMT is just a facilitator? It is possible that the driver for this critical event in cancer development arises from changes in the microenvironment as in the case of cancer initiation. Statistical association studies between metastasis-related events such as the activation of the EMT and expression changes of microenvironment-related genes can potentially provide useful suggestions about which micro-environmental changes serve as the main triggering events of metastasis, such as the intracellular ROS level going beyond some thresholds as suggested in Chap. 10.

13.1.6   Characterization of Cancers with Different Clinical Phenotypes Using Molecular Features

With the availability of large quantities of genomic, epigenomic and transcriptomic data, one can carry out association studies between cancers with similar phenotypes and their *omic* data. In time, a similar mission could be undertaken using proteomic, glycomic, metabolomic and other *omic* data. The goals are to identify common molecular-level characteristics among such cancers to obtain new and meaningful information connecting phenotypes with molecular changes. Examples of different phenotypes may include fast *versus* slow growing cancers, or readily metastasizing cancers *versus* cancers with low metastasis potential. Previous studies have been typically conducted on genomic data alone. For example, numerous GWAS (genome-wide association study) analyses have been employed to delineate type- or subtype-defining genes, but with only limited success in identifying common genetic traits as risk factors (Goldstein 2009). In their review article "Genetic heterogeneity in human disease" (McClellan and King 2010), the authors consider that "the vast majority of such [GWAS identified] variants have no established biological relevance to disease or clinical utility for prognosis or treatment". While the statement may be overly critical, it does point to one important issue that commonly observed genomic changes may not necessarily contain much disease-causing information. A generalized association study involving multiple *omic* data types needs to be done, which may require more general statistical frameworks than the GWAS type of analysis, hence creating challenges as well as opportunities for cancer statistical analysts.

13.2   Databases Useful for *In Silico* Investigation of Cancers

A few databases are deemed to be the "must have" resources for biological research, including cancer research. One is the NCBI database that provides access to a variety of databases for genomics, functional genomics and genetic information (NCBI 1988). For example, GenBank is the NCBI database of all publicly available DNA sequences with annotations (Benson et al. 2013), and the Gene database integrates a wide range of species information and provides more detailed functional annotation of each gene (Maglott et al. 2011). GeneCards (Rebhan et al. 1997), developed and hosted by the Weizmann Institute of Science, is another important database that covers more comprehensive information than just gene functions. It includes, for example, medical relevance, mutations, gene expression, protein interactions and pathway information for each gene in the database (Rebhan et al. 1998).

The most popular protein database is Swissprot, now a part of the Uniprot database (The-UniProt-Consortium 2014). As a manually curated database with high quality data, it provides comprehensive information about each experimentally validated protein, including its biological function, functional domains, subcellular localization and post-translational modifications, as well as links to many other databases on the Internet. A few other protein functional databases may also prove to be useful for cancer studies such as Interpro (Hunter et al. 2012), Pfam (Punta et al. 2012), the ENZYME database (Bairoch 2000) and the molecular interaction database DIP (Xenarios et al. 2002). HAPPI is an integrative database for protein-protein interactions collected from a few publicly available databases such as HPRD, BIND, MINT, STRING and OPHID. The database currently contains 142,956 non-redundant human protein-protein interactions among 10,592 human proteins (Chen et al. 2009). HINT is a collection of high-quality protein-protein interaction data for multiple organisms (Das and Yu 2012). For human data, the database covers 27,493 interactions and 7,629 protein complexes. A detailed review on protein-protein interaction databases can be found in (Lehne and Schlitt 2009).

A number of biological pathway databases may prove to be essential for functional studies of cancer such as pathway enrichment analyses. Among these databases, KEGG is probably the most frequently used, which consists of a collection of manually curated pathway models as well as molecular interactions for metabolism, regulation and signaling (Kanehisa and Goto 2000; Kanehisa et al. 2012). BIOCARTA is another popular pathway database for human and mouse. Consisting of 137 human pathways, PID (Pathway Interaction Database) covers 9,248 molecular species extracted from the NCI databases and an additional 322 pathways covering 7,575 interactions imported from the BIOCARTA and REACTOME databases. MsigDB is a set of gene groups, believed to be highly relevant to human cancer development, where each gene group potentially represents the components of a yet to be fully elucidated pathway (Subramanian et al. 2005). Table 13.1 summarizes a few popular pathway and molecular interaction databases, not limited to cancer studies.

In the remainder of this section, a number of more specialized databases covering different types of *omic* data are introduced, plus a number of cancer-specific databases. All these resources are freely available on the Internet and have been widely used by the research community.

Table 13.1 Pathway and molecular interaction databases

Database	Content	URL
KEGG	A collection of manually curated pathway models	www.genome.jp/kegg/pathway.html
BIOCARTA	A collection of pathways for more than 300 species	http://www.biocarta.com
PID	A collection of curated pathways for human signaling and regulatory processes	http://pid.nci.nih.gov/
Pathguide	A meta-database providing an overview of all web-accessible biological pathway and network databases	http://www.pathguide.org/
REACTOME	A resource of curated human pathways	http://www.reactome.org/
MSidDB	A collection of diseased-related gene sets	http://www.broadinstitute.org/gsea/msigdb/collections.jsp

13.2.1   *Human Cancer Genome and Gene Databases*

TCGA and ICGC are the two largest cancer genome sequencing projects. TCGA is a US-based project and its goal is to sequence and organize 10,000 cancer genomes, along with other matching *omic* data types, covering 25 cancer types (Cancer-Genome-Atlas-Research-Network et al. 2013). ICGC is a more international collaboration-oriented project, and its goal is to sequence and store 25,000 cancer genomes, along with other cancer-related *omic* data and information of 50 cancer types (International-Cancer-Genome-Consortium et al. 2010). Another large cancer genomic database is the COSMIC (Catalog of Somatic Mutations In human Cancer) database developed by the Wellcome Trust Sanger Institute (Forbes et al. 2001). Unlike the first two databases, COSMIC covers only genetic mutations and variants that are deemed to be critical to cancer development. The database currently contains 1,592,109 mutations identified in 25,606 human genes from 947,213 tumor samples.

In addition to these large general purpose databases, there are a few more specialized databases focused on specific cancer types or mutations in certain protein families. The Pediatric Cancer Genome Project Database is a comprehensive cancer genomic database for pediatric cancers (Downing et al. 2012). The database currently covers three classes of pediatric cancers, namely hematopoietic malignancies, brain tumor and solid tumors, consisting of 15 cancer types. The BIC (the Breast Cancer Information Core) database, hosted at the National Human Genome Research Institute, is a central repository for mutations and polymorphisms in breast cancer susceptibility genes (Szabo et al. 2000). CCDB (the Cervical Cancer Gene Database) is a manually curated database for genes involved in cervical carcinogenesis, currently covering 537 genes along with mutations in these genes and their protein products, including polymorphisms, methylations, genomic amplifications and gene expressions observed in cervical cancer samples (Agarwal et al. 2011).

The Cancer Gene Census database provides a catalog of mutations identified in over 400 cancer-related genes (Futreal et al. 2004). CanProVar is a database for single amino-acid alterations, including both germline and somatic variations in the human proteome (especially those related to human cancers), which is built based on information from the literature (Li et al. 2010). The IARC TP53 database consists of more than 10,000 variations in the P53 gene observed in human populations and cancer samples (Olivier et al. 2002). CDKN2A is a similar database but for the cyclin-dependent kinase inhibitor 2A gene (Murphy et al. 2004). The Androgen Receptor Gene Mutation database contains 374 mutations found in androgen receptor genes (Patterson et al. 1994).

There are a few general genomic mutation databases on the Internet, which are not specifically focused on human cancers but may prove to be useful to cancer research. For example, HGMD (Human Genome Mutation Database) contains 141,161 germline mutations that are associated with human inheritable diseases (Cooper et al. 1998). The dbSNP database (Single Nucleotide Polymorphism

Table 13.2 Human cancer genome and gene databases

Database	Content	URL
TCGA	A cancer *omic* data resource containing genomic, epigenomic and transcriptomic data sponsored by NIH	https://tcga-data.nci.nih.gov/tcga/
ICGC	A cancer *omic* data resource containing genomic, epigenomic and transcriptomic data sponsored by ICGC	http://icgc.org/
COSMIC	A catalog of somatic mutations in human cancers containing >50,000 mutations	http://www.sanger.ac.uk/perl/genetics/CGP/cosmic
Cancer gene census	A catalog of mutations in more than 400 cancer-related genes	www.sanger.ac.uk/genetics/CGP/Census/
SNP500 cancer database	A database for 3,400+ SNPs in cancer-related genes	http://snp500cancer.nci.nih.gov/
Breast cancer information core	A repository for all mutations and polymorphisms in breast cancer-related genes	http://research.nhgri.nih.gov/bic/
GAC	A database of gene mutations, loss of heterozygosity, and chromosome changes in human, mice or rat tumors	www.niehs.nih.gov/research/resources/databases/gac
HGMD	A database for germline mutations that are associated with heritable diseases	www.hgmd.org/
dbSNP	A catalog for genome variations	www.ncbi.nlm.nih.gov/projects/SNP/
MedRefSNP	A database with about 36,199 unique SNPs collected from the PubMed and OMIM databases	www.medclue.com/medrefsnp

Database) at the NCBI is another data archive for genetic variation data within and across different species, containing information on SNPs, short deletions and insertions, microsatellites, short tandem repeats and heterozygous sequences (Smigielski et al. 2000). The database has amassed more than 64 million distinct variants for 55 organisms since its formation in 1998. Table 13.2 provides some additional information about some of the above databases.

13.2.2 Human Epigenome Databases

Compared to genomic data, the human and cancer epigenomic data are probably two orders of magnitude smaller at this point due to a combination of (a) the current knowledge about human epigenomes is considerably less than that about human genomes, and (b) the technology development for massive production of epigenomic data is years behind that for genome sequencing. As mentioned above, both TCGA and ICGC have amassed quite a bit of cancer epigenome data. Specifically, TCGA aims to have a total of over 10,000 cancer epigenome datasets for 25 cancer types by 2015, and ICGC plans to generate 25,000 cancer epigenome datasets for 50 cancer types. The *ENCODE* (Encyclopedia of DNA Elements) project launched by the US National Human Genome Research Institute has a component on producing human epigenomic data (Encode-Project-Consortium et al. 2012). The project goals are to produce epigenomic profiles of 50 different human tissue types. In addition, a few focused epigenomics projects have produced a substantial amount of epigenomic data. For example, the NIH Roadmap Epigenomics Program was started in 2008 and aims to produce chromosome-binding histone modification data for over 30 types of modifications covering a variety of major human cell types (Chadwick 2012). The Human Epigenome Project at the Wellcome Trust Sanger Institute runs its own human epigenome database (Eckhardt et al. 2004). The database released 43 sets of human epigenome data for 12 different tissue types in 2006 and 32 sets of human epigenome data for 7 different tissue types in 2013.

There are also a few human (cancer) epigenome databases on a smaller scale deployed on the Internet. For example, MethyCancer is a database specifically designed for human cancer epigenome data (He et al. 2008), and the PubMeth database is also for human cancer epigenome data (Ongenaert et al. 2008). Table 13.3 lists a few epigenomic databases on the Internet.

Table 13.3 Human epigenome databases

Database	Content	URL
NIH roadmap epigenomics program	A database for human epigenomes now covering at least 23 cell types	http://www.roadmapepigenomics.org/data
Human epigenome project	A database for genome-wide DNA methylation patterns of all human genes in all major tissues	http://www.epigenome.org/
MethyCancer	A database for DNA methylation information in cancer-related genes, collected from public resource	http://methycancer.genomics.org.cn

13.2.3   *Transcriptomic Databases*

Transcriptomic data are by far the largest among all cancer *omic* data, at least one, possibly two orders of magnitude larger than the available genomic data. There are multiple large databases for cancer transcriptomic data on the Internet (Table 13.4). In addition to the data in the TCGA and the ICGC databases, large amounts of transcriptomic data are stored in the GEO (Gene Expression Omnibus) database at the NCBI and Arrayexpress at the EBI, the two most popular gene-expression databases. GEO currently has more than 32,000 sets of gene-expression data collected from 800,000 samples from more 1,600 organisms (Barrett et al. 2013). Arrayexpress consists of 1,245,005 sets of gene-expression data collected through 43,947 experiments, including both microarray and RNA-seq data. CGED (Cancer Gene Expression Database) currently consists of gene-expression data and associated clinical data collected on breast, colorectal, esophageal, gastric, glioma, hepatocellular, lung and thyroid cancers, predominantly using microarray techniques (Kato et al. 2005).

The ASTD database, although not specifically designed for transcriptomic data, may prove to be useful for cancer studies as it covers alternatively spliced isoforms encoded in human, mouse and rat genomes, derived based on gene-expression data (Koscielny et al. 2009).

Table 13.4   Gene expression databases

Database	Content	URL
NCBI GEO	A comprehensive collection of gene expression data	http://www.ncbi.nlm.nih.gov/gds
Arrayexpress	A database of functional genomics including gene expression data in both microarray and RNA-seq forms	http://www.ebi.ac.uk/arrayexpress/
SMD	Stanford microarray database for gene expression data covering multiple organisms	http://smd.stanford.edu/
CIBEX	A database for gene-expression data hosted at the National Institute of Genetics of Japan	http://cibex.nig.ac.jp/
Oncomine (research edition)	A commercial database for cancer transcriptomic and genomic data, with a free edition to academic and nonprofit organizations	https://www.oncomine.org/resource/login.html
ChipDB	A database for microarray gene-expression data hosted at the Whitehead Institute	http://chipdb.wi.mit.edu/chipdb/public/
ASTD	A database for human gene-expression data and derived alternatively spliced isoforms of human genes	http://drcat.sourceforge.net/astd.html

Table 13.5 MicroRNA databases

Database	Content	URL
miRecords	A database for animal miRNA-target interactions	http://mirecords.biolead.org
miRBase	A database for published miRNA sequences and annotations covering numerous species	http://www.mirbase.org
TargetScan	A database for microRNA targets	http://www.targetscan.org

13.2.4   MicroRNA and Target Databases

With the predicted intimate relationships between microRNAs and epigenomic activities (Chuang and Jones 2007), one can expect that microRNA-related cancer studies will increase significantly in the years to come, in addition to the functional roles that they have been found to play in cancer initiation, progression and metastasis (Calin et al. 2005; Yanaihara et al. 2006; Bloomston et al. 2007; Ambs et al. 2008; Garzon et al. 2008; Schetter et al. 2008; Croce 2009; Wyman et al. 2009). Two databases for experimentally validated human microRNAs, along with their validated and predicted target genes, are MiRecords and miRBase. Currently, miRecords consists of 2,705 interactions between 644 microRNAs and 1,901 target genes in nine animal species (Xiao et al. 2009). miRBase currently consists of 24,500+ mature microRNAs from 206 species (Griffiths-Jones et al. 2006). TargetScan (Lewis et al. 2005) and Miranda (Miranda et al. 2006) are two popular databases for microRNA target genes, including both experimentally validated and computationally predicted targets (Table 13.5).

13.2.5   Proteomic Databases

Compared with transcriptomic data, proteomic data, particularly quantitative proteomic data, are much more difficult to obtain on large scales because of: (1) the higher complexities of protein data due to posttranslational modifications on protein sequences and alternatively spliced isoforms, and (2) the substantially larger dynamic range of protein quantities than that of gene-expression levels. These difficulties have resulted in the current reality that only limited proteomic data are available for human cancers compared to transcriptomic data.

While there are no large-scale proteomic databases specifically for human cancers on the Internet, a few databases for general human proteomes are publicly available. For example, PeptideAtlas consists of a large collection of peptides experimentally validated from a number of organisms including human (Deutsch et al. 2008). PRIDE (PRoteomics IDEntificaitons) is a public repository of mass spectrometry (MS) based proteomic data (Martens et al. 2005). It currently consists of over 4 million identified proteins and 20 million peptides from 104 million

Table 13.6 Proteomic data resources

Database	Content	URL
PRIDE	A database for protein and peptide identification, post-translational modifications and supporting data	http://www.ebi.ac.uk/pride/
PeptideAtlas	A database of peptides from multiple organisms	http://www.peptideatlas.org/ http://www.peptideatlas.org/hupo/hppp/
Plasma Proteome Project database	A comprehensive resource for human plasma proteins, including protein splicing isoforms	www.plasmaproteomedatabase.org/
GeMDBJ	A collection of proteomic expression and identification data of surgically resected tissues and tissue-cultured cells of various malignancies	https://gemdbj.nibio.go.jp/dgdb/

spectral data, covering 60 species with human data accounting for the major portion of the data. PPP (Plasma Proteome Project) is another human proteomic database (Omenn et al. Proteomics 2005), focused on plasma proteins with quantitative information, which is currently hosted at the PeptideAtlas site. A database for raw proteomic data collected on surgically resected tissues and tissue-cultured cells of various malignancies, along with the associated biological and clinico-pathological data, is GeMDBJ Proteomics. The collected data were obtained using 2D PAGE and DIGE, with protein identification based on liquid chromatography and tandem mass spectrometry. Some detailed information about these databases is given in Table 13.6.

13.2.6 Metabolomic Databases

As of now, over 40,000 metabolites have been identified in human cells (Wishart et al. 2007), but our current ability in identifying metabolites in a (cancer) tissue is still rather limited, with a typical experiment using the most sensitive LC-MS/MS technology being able to identify only a few hundred up to 1,000 metabolites (Zhou et al. 2012). Compared to genomic and transcriptomic data, the available metabolomic data for human is substantially smaller (Table 13.7).

HMDB (the Human Metabolome Database) is a popular metabolite database, and currently consists of 41,514 metabolites identified in human (Wishart et al. 2009, 2013). A few databases contain the structural information of the known metabolites, such as the Fiehn GC-MS library (Kind et al. 2009), the GOLM metabolome library (Hummel et al. 2007) and the NIST library, derived from the Human Metabolome Library in HMDB. A few specialized databases for metabolites

Table 13.7 Metabolomic databases

Database	Content	URL
HMDB	A knowledgebase for the human metabolome	http://www.hmdb.ca
KEGG COMPOUND	A database of small molecules and other chemical substances relevant to enzymatic reactions	http://www.genome.jp/kegg/compound/
PubChem	A comprehensive database of chemical compounds	http://pubchem.ncbi.nlm.nih.gov/
UMDB	A comprehensive resource for confirmed human urine metabolites	http://www.urinemetabolome.ca
BRENDA	A database for enzymes and associated reaction kinetic parameters	http://www.brenda-enzymes.org/
SABIO-RK	A database for biochemical reactions and their kinetic parameters	http://sabio.h-its.org/
TECRDB	A database for enzyme thermodynamic data	http://xpdb.nist.gov/enzyme_thermodynamics/

detected in human body fluids can also be found on the Internet, such as Urine Metabolome Database (UMDB) that contains 2,651 urine metabolites.

In addition, there are a few databases with information useful for conducting metabolic flux analyses, including such information as biochemical reactions relevant to specific metabolites and the reaction kinetic parameters for the relevant enzymes. BRENDA is one such database with enzymes, turnover rates and reaction kinetic parameters such as Km, $Kcat$ and Ki values (Schomburg et al. 2004). Another database for biochemical reactions and the kinetic parameters is SABIO-RK (Rojas et al. 2007). TECRDB is a database for thermodynamic data of biochemical reactions (Goldberg et al. 2004), hence facilitating more realistic modeling studies.

13.2.7   Additional Cancer-Related Databases

There are various other cancer-related databases that may prove to be useful for cancer studies. For example, TANTIGEN is the most comprehensive database for human tumor T-cell antigens, which provides the information on validated T-cell epitopes and HLA ligands, antigen isoforms and sequence-level mutations (TANTIGEN 2009). The Cancer Immunity Peptide database contains 129 human tumor antigens all with defined T-cell epitopes (Novellino et al. 2005). The CT database provides information on cancer-testicular antigens.

Table 13.8 Specialized cancer-related databases

Database	Content	URL
TANTIGEN	A human tumor T-cell antigen database	http://cvc.dfci.harvard.edu/tadb/
Cancer mortality database	A database for cancer mortality statistics by country	http://www.who.int/healthinfo/statistics/mortality_rawdata/en/index.html
CGEM	A database for storing clinical information about tumor samples and microarray data	http://www.cangem.org

Databases for clinical reports on cancer are also available. For example, the cancer mortality database created by the International Arctic Research Center (IARC) contains cancer mortality statistics by country, and CGEM (Cancer GEnome Mine) contains cancer clinical information concerning tumor samples (Table 13.8).

13.3   Online Tools Useful for *In Silico* Cancer Studies

Numerous computational analysis and data mining tools have been published and deployed on the Internet, which can be used to analyze the databases presented in Sect. 13.2 or proprietary cancer data. A few are listed below as examples to illustrate the types of tools that one can find on the Internet. These tools are organized in a similar fashion to that in the above section, i.e., according to the data types on which they are used.

13.3.1   Human Genome Analysis Tools

As discussed in Chap. 4, cancer genomes tend to harbor various changes compared to the matching healthy genomes, including (1) single-point mutations; (2) copy number changes; and (3) structural variations such as reversals and translocations of genomic segments. Identification of such genomic variations can provide useful information about the evolutionary footprints of individual cancer samples, as well as infer common and distinct characteristics across cancers of different types. Such footprint information could possibly reveal bottlenecks that the underlying cancer needs to overcome at different developmental stages.

Among the many (freely available) cancer genome analysis tools/servers on the Internet, a few sites offer clusters of tools, such as the Cancer Genomics Hub at UCSC, the TCGA site, the ICGC site and the Cancer Genome Analysis suite at the Broad Institute. Specifically, the Cancer Genomics Hub (Cancer-Genomics-Hub 2013) is a good place to visualize cancer genomes mostly from the TCGA project and to retrieve simple analysis results such as genomic mutations.

Table 13.9 Tools for cancer genome analyses

Tool	Content	URL
Cancer genomics hub	A toolkit that provides analysis and visualization capabilities of cancer genomes	https://cghub.ucsc.edu/
Cancer genome analysis	A comprehensive suite of tools for identification of abnormalities in cancer genomes	http://www.broadinstitute.org/cancer/cga/
CREST	A downloadable toolkit for detecting genomic structural variations at base-pair resolution	http://www.stjuderesearch.org/site/lab/zhang

A number of tools provided at the Broad Institute's site (Cancer-Genome-Analysis 2013) may prove to be useful in doing the initial analysis of a cancer genome, with a few listed below. MuTect is a tool for identifying point mutations in a cancer genome in comparison with a matching control genome. Breakpointer is a tool that can pinpoint the breakpoints of genomic rearrangements in a cancer genome (Sun et al. 2012). To identify genomic rearrangements in a provided cancer genome *versus* a matching control genome, dRanger is recommended. Oncotator provides annotations of point mutations and indels in a cancer genome. Actually for each of these tools hosted at the Broad Institute's site, there are numerous other tools on the Internet, offering similar functions, some of which may fit better for specific analysis needs. For example, CREST is software tool, developed at St. Jude Children's Research Hospital, for mapping somatic structural variation in cancer genomes with high resolution (Wang et al. 2011). Table 13.9 lists a few tools and tool clusters useful for cancer genome analyses.

13.3.2 Human Epigenome Analysis Tools

Compared with the large number of genome analysis tools on the Internet, only a few tools for epigenomic analysis are there in the public domain, possibly reflecting the reality that the current understanding about human epigenome is far less than that about human genomes. This is not surprising, knowing that the current definition of "epigenetics" was not settled until a Cold Spring Harbor meeting in 2008 (Berger et al. 2009). A few tools have been published for identification of differentially methylated regions in a given genome in comparison with reference epigenomes. CHARM is an R package for making such identifications (Irizarry et al. 2008), and MethylKit is also an R package for identification and visualization of differential methylations across different epigenomes (Akalin et al. 2012). EpiExplorer is a similar R package for differential methylation analysis (Halachev et al. 2012); CpGassoc, also an R package, is for analysis of DNA methylation array data (Barfield et al. 2012) (Table 13.10).

Table 13.10   Tools for cancer epigenome analyses

Tool	Content	URL
CHARM	An early and widely used package for DNA methylation analysis	http://www.bioconductor.org/packages/release/bioc/html/charm.html
EpiExplorer	A web-based tool for identification of comparing epigenetic markers in a specific genome to reference human epigenomes	http://epiexplorer.mpi-inf.mpg.de/
methylKit	An R package for DNA methylation analysis based on bisulfite sequencing data	https://code.google.com/p/methylkit/

13.3.3   Human Transcriptome Analysis Tools

Our own experience has been that among the available cancer *omic* data, transcriptomic data are the most useful for gaining new insights about cancer biology. Such data have proven to be highly informative for addressing a wide range of cancer related problems, ranging from cancer typing, staging and grading to understanding the possible relationships among pathways with altered expression levels, further to elucidation of cancer initiation drivers and key facilitators, and then to inference of possible drivers for metastasis and understanding the main causes of accelerated growth of metastatic cancers. It should be emphasized that information derivation through transcriptomic data analysis can go substantially beyond the traditional association analyses among genes or pathways with correlated expression patterns. It can be used to infer: (a) possible flux distributions in a qualitative manner (see Chap. 6 for identification of hyaluronic acid synthesis as an exit for accumulated glucose metabolites); (b) possible causal relationships (see Chap. 11 for inference of reasons for increased cholesterol metabolisms in metastatic cancers *versus* matching primary cancers); (c) possible enzyme-encoding genes to fill gaps in metabolic pathways (see Chap. 6 for prediction of genes with specific enzymatic functions); and (d) inference of mechanistic models involving multiple genes (see for example Chap. 10). It is noteworthy that a recent study has demonstrated that one can infer the proteome from transcriptomic data (Evans et al. 2012).

Numerous analysis tools of transcriptomic data have been published and deployed on the Internet, some of which may prove to be useful for information derivation in cancer studies. These tools range from: (1) identification of differentially expressed genes in cancer *versus* matching control tissues such as edgeR (Robinson et al. 2010) and baySeq (Hardcastle and Kelly 2010); (2) identification of co-expressed genes or genes with correlated expression patterns such as WGCNA (Langfelder and Horvath 2008) and GeneCAT (Mutwil et al. 2008); (3) transcriptome-based protein identification (Evans et al. 2012); (4) inference of splicing variants from RNA-seq data such as CUFFLINK (Roberts et al. 2011); (5) inference of pathways enriched with up- or down-regulated genes such as DAVID; (6) elucidation of human signaling networks from gene-expression data (Brandenberger et al. 2004);

Table 13.11 Tools for transcriptomic data analyses

Tool	Content	URL
edgeR	An tool for detection of differentially expressed genes	http://www.genomine.org/edge/
CUFFLINK	A tool for transcript assembly and identification of splicing variants	http://cufflinks.cbcb.umd.edu/index.html
DAVID	A tool for pathways enriched with differentially expressed genes (or any specified set of genes)	http://david.abcc.ncifcrf.gov/

(7) de-convolution of gene expression data collected on tissue samples with multiple cell types to contributions from individual cell types (Ahn et al. 2013); and (8) development of predictive metabolic fluxes through integration of gene-expression data with flux balance analysis (Duarte et al. 2007), among numerous other tools. Table 13.11 lists a few of these tools.

13.3.4   Human Proteome Analysis Tools

Unlike cancer genomic, epigenomic, transcriptomic and metabolomic data, to the best of our knowledge there are currently no large-scale cancer proteomic data resources in the public domain. Hence the proteomic data analysis tools are also somewhat limited. There are a few websites that host clusters of proteomics-oriented analysis tools, such as those at the NCI Office of Cancer Clinical Proteomics Research (CPTAC 2013).

13.3.5   Human Metabolome Analysis Tools

Metabolic data are typically collected using mass spectrometry (MS) or nuclear magnetic resonance (NMR) techniques as discussed in Chap. 2. They provide information highly complementary to that of transcriptomic data, and hence can be used as validation information for metabolic pathways derived based on transcriptomic data analyses. There are a few software suites focused on metabolic data analysis on the Internet. For example, the website of the Metabolomics Society hosts a number of analysis tools, including raw metabolic data processing and normalization, chemical structure identification based on raw NMR or MS data, and linking identified chemical structures to relevant biochemical processes (Metabolomics-Society 2014). The Human Metabolite Database (HMDB) hosts not only a large collection of metabolite data (see Table 13.7), but also a variety of tools in support of search and analysis of the underlying data (Wishart et al. 2007). In addition, an associated website at the Metabolomics Innovation Center provides a large collection of

Table 13.12 Tools for metabolome analyses

Tool	Content	URL
Metabolomics Society website	A suite of tools for metabolomics data processing, normalization, analysis and structural identification	http://www.metabolomicssociety.org/softwar
The metabolomics innovation center	A collection of tools more focused on metabolic data analysis	http://www.metabolomicscentre.ca/software
NuGo	A set of tools including both metabolic data analysis and mapping to metabolic networks	http://www.nugo.org/ metabolomics/34821/7/0/30

Table 13.13 Tools for biological pathway prediction and mapping

Tool	Content	URL
Pathway tools	The website provides a wide ranges of pathway-related tools, ranging from pathway construction, editing, prediction and flux analysis	http://bioinformatics.ai.sri.com/ptools/
PathoLogic pathway prediction	The toolkit supports automated prediction of metabolic pathways supported by BioCyc database	http://g6g-softwaredirectory.com/bio/ cross-omics/pathway-dbs-kbs/20235S RIPathoLogicPathwPredict.php
BioCyc and pathway tools	BioCyc database provides a list of reconstruction and analysis tools of metabolic pathways	http://biocyc.org/publications.shtml

metabolic data analysis tools (The-Metabolomics-Innovation-Centre 2014). The Nutrigenomics Organization website also hosts a collection of metabolic data analysis and metabolic network modeling tools (The-Nutrigenomics-Organization 2008). These tools may prove to be useful to cancer metabolomic data analysis and mapping them onto metabolic networks (Table 13.12).

13.3.6   Pathway Mapping and Reconstruction Tools

Various tools are currently available for pathway model construction, analysis and comparison on the Internet. Table 13.13 provides a few such tools.

13.3.7   Statistical Analysis Tools

In addition to the above data type-specific tools, there is a large collection of statistical analysis tools on the Internet, which have been widely used for analyses of different *omic* data types. The following sites provide a number of such tools. Bioconductor is a community-wide effort for developing and deploying open source bioinformatics software packages. All the deployed tools are written in the R statistical programming language. Currently the website has about 750 software tools, covering a wide range of analysis and inference capabilities (Gentleman et al. 2004). The Galaxy project is another website that hosts a large collection of genomic data analysis tools (Goecks et al. 2010). The popular Gene Ontology website also hosts a wide range of analysis tools (Gene-Ontology-Tools 2013).

13.3.8   Visualization Tools

Visualization tools may prove to be highly useful when analyzing complex biological data and inferring biological relationships among biomolecules or pathways. A number of visualization tools have been developed in support of such needs and made publicly available. Among these tools are CytoScape for visualizing molecular interaction networks (Shannon et al. 2003), PathView for biological data integration and visualization (Luo and Brouwer 2013) and iPATH for visualization, analysis and customization of pathway models (Yamada et al. 2011).

13.4   Concluding Remarks

This is an exciting time to study the extraordinarily complex problems in cancer biology, specifically those dealing with the key drivers and facilitators at different developmental stages such as cancer initiation, progression, metastasis and post-metastasis development in their full complexity. A tremendous amount of cancer *omic* data has been generated and continues to be produced; fortunately, these data are being made publicly available. Without question, a substantial amount of information can be derived by analyzing and mining these data, particularly after one has learned the types of questions that cancer *omic* data analyses can help to address. It is the authors' hope and trust that we have provided our readers with a general understanding of modern cancer biology and sufficient knowledge of the types of questions one can ask and expect to obtain useful information through *omic* data analyses. The examples given in Sect. 13.1 represent only a small set of example questions that one can address and possibly solve. Using the knowledge gained herein as a starting point, one can examine many questions about cancer, which of course have to be guided by state-of-the-art reviews and original papers concerning individual topics that the readers will need to review before delving deeply into a specific problem. The data resources and tools listed in this chapter represent but a small set of all the available tools and databases of which we are aware. A recommended reference to find a more comprehensive list of the relevant databases and tools is the special database and server issues published annually by Nucleic Acids Research. There are undoubtedly many good problems for our readers to think about. A recommended start would be the list of the problems used by the authors as examples in Sect. 13.1. If the efforts expended in writing this book inspires computational scientists to tackle some of the challenging cancer problems and make progress that contributes not only to basic cancer biology, but importantly to early detection, more successful treatment and better prevention of cancer, the authors will feel gratified that their time has been well spent.

References

Achreja A, Yang L, Zhao H et al. (2013) Integrated energetics and flux analysis reveals differential metabolic reprogramming in highly and less invasive cancer cells. In: Proceedings of the 104th Annual Meeting of the American Association for Cancer Research 73:

Agarwal SM, Raghav D, Singh H et al. (2011) CCDB: a curated database of genes involved in cervix cancer. Nucleic Acids Res 39: D975-979

Ahn J, Yuan Y, Parmigiani G et al. (2013) DeMix: deconvolution for mixed cancer transcriptomes using raw measured data. Bioinformatics 29: 1865–1871

Akalin A, Kormaksson M, Li S et al. (2012) methylKit: a comprehensive R package for the analysis of genome-wide DNA methylation profiles. Genome biology 13: R87

Ambs S, Prueitt RL, Yi M et al. (2008) Genomic profiling of microRNA and messenger RNA reveals deregulated microRNA expression in prostate cancer. Cancer Res 68: 6162–6170

Bairoch A (2000) The ENZYME database in 2000. Nucleic Acids Res 28: 304–305

Barfield RT, Kilaru V, Smith AK et al. (2012) CpGassoc: an R function for analysis of DNA methylation microarray data. Bioinformatics 28: 1280–1281

Barrett T, Wilhite SE, Ledoux P et al. (2013) NCBI GEO: archive for functional genomics data sets–update. Nucleic Acids Res 41: D991–995

Benson DA, Cavanaugh M, Clark K et al. (2013) GenBank. Nucleic Acids Res 41: D36–42

Berger SL, Kouzarides T, Shiekhattar R et al. (2009) An operational definition of epigenetics. Genes & development 23: 781–783

Bloomston M, Frankel WL, Petrocca F et al. (2007) MicroRNA expression patterns to differentiate pancreatic adenocarcinoma from normal pancreas and chronic pancreatitis. JAMA 297: 1901–1908

Brandenberger R, Wei H, Zhang S et al. (2004) Transcriptome characterization elucidates signaling networks that control human ES cell growth and differentiation. Nature biotechnology 22: 707–716

Calin GA, Ferracin M, Cimmino A et al. (2005) A MicroRNA signature associated with prognosis and progression in chronic lymphocytic leukemia. N Engl J Med 353: 1793–1801

Cancer-Genome-Analysis (2013) ABSOLUTE.

Cancer-Genome-Atlas-Research-Network, Weinstein JN, Collisson EA et al. (2013) The Cancer Genome Atlas Pan-Cancer analysis project. Nature genetics 45: 1113–1120

Cancer-Genomics-Hub (2013) Cancer Genomics Hub.

Chadwick LH (2012) The NIH Roadmap Epigenomics Program data resource. Epigenomics 4: 317–324

Chen JY, Mamidipalli S, Huan T (2009) HAPPI: an online database of comprehensive human annotated and predicted protein interactions. BMC Genomics 10 Suppl 1: S16

Chuang JC, Jones PA (2007) Epigenetics and microRNAs. Pediatric research 61: 24R–29R

Cooper DN, Ball EV, Krawczak M (1998) The human gene mutation database. Nucleic Acids Res 26: 285–287

CPTAC (2013) Clinical Proteomic Technologies for Cancer initiative.

Croce CM (2009) Causes and consequences of microRNA dysregulation in cancer. Nat Rev Genet 10: 704–714

Das J, Yu H (2012) HINT: High-quality protein interactomes and their applications in understanding human disease. BMC systems biology 6: 92

Deutsch EW, Lam H, Aebersold R (2008) PeptideAtlas: a resource for target selection for emerging targeted proteomics workflows. EMBO reports 9: 429–434

Downing JR, Wilson RK, Zhang J et al. (2012) The Pediatric Cancer Genome Project. Nature genetics 44: 619–622

Duarte NC, Becker SA, Jamshidi N et al. (2007) Global reconstruction of the human metabolic network based on genomic and bibliomic data. Proceedings of the National Academy of Sciences of the United States of America 104: 1777–1782

Eckhardt F, Beck S, Gut IG et al. (2004) Future potential of the Human Epigenome Project. Expert review of molecular diagnostics 4: 609–618

Encode-Project-Consortium, Bernstein BE, Birney E et al. (2012) An integrated encyclopedia of DNA elements in the human genome. Nature 489: 57–74

Evans VC, Barker G, Heesom KJ et al. (2012) De novo derivation of proteomes from transcriptomes for transcript and protein identification. Nature methods 9: 1207–1211

Forbes SA, Bhamra G, Bamford S et al. (2001) The Catalogue of Somatic Mutations in Cancer (COSMIC). In: Current Protocols in Human Genetics. John Wiley & Sons, Inc., Hoboken, NJ

Futreal PA, Coin L, Marshall M et al. (2004) A census of human cancer genes. Nature reviews Cancer 4: 177–183

Garzon R, Volinia S, Liu CG et al. (2008) MicroRNA signatures associated with cytogenetics and prognosis in acute myeloid leukemia. Blood 111: 3183–3189

Gene-Ontology-Tools (2013) Gene Ontology Tools.

Gentleman RC, Carey VJ, Bates DM et al. (2004) Bioconductor: open software development for computational biology and bioinformatics. Genome biology 5: R80

Goecks J, Nekrutenko A, Taylor J et al. (2010) Galaxy: a comprehensive approach for supporting accessible, reproducible, and transparent computational research in the life sciences. Genome biology 11: R86

Goldberg R, Tewari Y, Bhat T (2004) Thermodynamics of Enzyme-Catalyzed Reactions -a Database for Quantitative Biochemistry. Bioinformatics 20: 2874–2877

Goldstein DB (2009) Common genetic variation and human traits. The New England journal of medicine 360: 1696–1698

Griffiths-Jones S, Grocock RJ, van Dongen S et al. (2006) miRBase: microRNA sequences, targets and gene nomenclature. Nucleic Acids Res 34: D140–144

Halachev K, Bast H, Albrecht F et al. (2012) EpiExplorer: live exploration and global analysis of large epigenomic datasets. Genome biology 13: R96

Hanahan D, Weinberg Robert A (2011) Hallmarks of Cancer: The Next Generation. Cell 144: 646–674

Hardcastle TJ, Kelly KA (2010) baySeq: empirical Bayesian methods for identifying differential expression in sequence count data. BMC bioinformatics 11: 422

He X, Chang S, Zhang J et al. (2008) MethyCancer: the database of human DNA methylation and cancer. Nucleic Acids Res 36: D836–841

Hirayama A, Kami K, Sugimoto M et al. (2009) Quantitative metabolome profiling of colon and stomach cancer microenvironment by capillary electrophoresis time-of-flight mass spectrometry. Cancer research 69: 4918–4925

Hummel J, Selbig J, Walther D et al. (2007) The Golm Metabolome Database: a database for GC-MS based metabolite profiling. In: Nielsen J, Jewett M (eds) Metabolomics, vol 18. Topics in Current Genetics. Springer, Berlin Heidelberg, pp 75–95

Hunter S, Jones P, Mitchell A et al. (2012) InterPro in 2011: new developments in the family and domain prediction database. Nucleic Acids Res 40: D306–312

International-Cancer-Genome-Consortium, Hudson TJ, Anderson W et al. (2010) International network of cancer genome projects. Nature 464: 993–998

Irizarry RA, Ladd-Acosta C, Carvalho B et al. (2008) Comprehensive high-throughput arrays for relative methylation (CHARM). Genome research 18: 780–790

Kanehisa M, Goto S (2000) KEGG: kyoto encyclopedia of genes and genomes. Nucleic Acids Res 28: 27–30

Kanehisa M, Goto S, Sato Y et al. (2012) KEGG for integration and interpretation of large-scale molecular data sets. Nucleic Acids Res 40: D109–114

Kato K, Yamashita R, Matoba R et al. (2005) Cancer gene expression database (CGED): a database for gene expression profiling with accompanying clinical information of human cancer tissues. Nucleic Acids Res 33: D533–536

Kind T, Wohlgemuth G, Lee do Y et al. (2009) FiehnLib: mass spectral and retention index libraries for metabolomics based on quadrupole and time-of-flight gas chromatography/mass spectrometry. Analytical chemistry 81: 10038–10048

Koscielny G, Le Texier V, Gopalakrishnan C et al. (2009) ASTD: The Alternative Splicing and Transcript Diversity database. Genomics 93: 213–220

Langfelder P, Horvath S (2008) WGCNA: an R package for weighted correlation network analysis. BMC bioinformatics 9: 559

Lehne B, Schlitt T (2009) Protein-protein interaction databases: keeping up with growing interactomes. Human genomics 3: 291–297

Lewis BP, Burge CB, Bartel DP (2005) Conserved seed pairing, often flanked by adenosines, indicates that thousands of human genes are microRNA targets. Cell 120: 15–20

Li J, Duncan DT, Zhang B (2010) CanProVar: a human cancer proteome variation database. Human mutation 31: 219–228

Luo W, Brouwer C (2013) Pathview: an R/Bioconductor package for pathway-based data integration and visualization. Bioinformatics 29: 1830–1831

Maglott D, Ostell J, Pruitt KD et al. (2011) Entrez Gene: gene-centered information at NCBI. Nucleic Acids Res 39: D52–57

Martens L, Hermjakob H, Jones P et al. (2005) PRIDE: the proteomics identifications database. Proteomics 5: 3537–3545

McClellan J, King MC (2010) Genetic heterogeneity in human disease. Cell 141: 210–217

Metabolomics-Society (2014) Metabolomics Society

Miranda KC, Huynh T, Tay Y et al. (2006) A pattern-based method for the identification of MicroRNA binding sites and their corresponding heteroduplexes. Cell 126: 1203–1217

Murphy JA, Barrantes-Reynolds R, Kocherlakota R et al. (2004) The CDKN2A database: Integrating allelic variants with evolution, structure, function, and disease association. Human mutation 24: 296–304

Mutwil M, Øbro J, Willats WGT et al. (2008) GeneCAT—novel webtools that combine BLAST and co-expression analyses. Nucleic Acids Research 36: W320–W326

NCBI (1988) National Center for Biotechnology Information.

Novellino L, Castelli C, Parmiani G (2005) A listing of human tumor antigens recognized by T cells: March 2004 update. Cancer immunology, immunotherapy: CII 54: 187–207

Olivier M, Eeles R, Hollstein M et al. (2002) The IARC TP53 database: new online mutation analysis and recommendations to users. Human mutation 19: 607–614

Omenn GS, States DJ, Adamski M et al. (2005) Overview of the HUPO Plasma Proteome Project: results from the pilot phase with 35 collaborating laboratories and multiple analytical groups, generating a core dataset of 3020 proteins and a publicly-available database. Proteomics 5: 3226–3245

Ongenaert M, Van Neste L, De Meyer T et al. (2008) PubMeth: a cancer methylation database combining text-mining and expert annotation. Nucleic Acids Res 36: D842–846

Pani G, Galeotti T, Chiarugi P (2010) Metastasis: cancer cell's escape from oxidative stress. Cancer metastasis reviews 29: 351–378

Patterson MN, Hughes IA, Gottlieb B et al. (1994) The androgen receptor gene mutations database. Nucleic Acids Res 22: 3560–3562

Punta M, Coggill PC, Eberhardt RY et al. (2012) The Pfam protein families database. Nucleic Acids Res 40: D290–301

Rebhan M, Chalifa-Caspi V, Prilusky J et al. (1998) GeneCards: a novel functional genomics compendium with automated data mining and query reformulation support. Bioinformatics 14: 656–664

Rebhan M, ChalifaCaspi V, Prilusky J et al. (1997) GeneCards: Integrating information about genes, proteins and diseases. Trends Genet 13: 163–163

Reuter S, Gupta SC, Chaturvedi MM et al. (2010) Oxidative stress, inflammation, and cancer: how are they linked? Free radical biology & medicine 49: 1603–1616

Roberts A, Pimentel H, Trapnell C et al. (2011) Identification of novel transcripts in annotated genomes using RNA-Seq. Bioinformatics 27: 2325–2329

Robinson MD, McCarthy DJ, Smyth GK (2010) edgeR: a Bioconductor package for differential expression analysis of digital gene expression data. Bioinformatics 26: 139–140

Rojas I, Golebiewski M, Kania R et al. (2007) Storing and annotating of kinetic data. In silico biology 7: S37–44

Schetter AJ, Leung SY, Sohn JJ et al. (2008) MicroRNA expression profiles associated with prognosis and therapeutic outcome in colon adenocarcinoma. JAMA 299: 425–436

Schomburg I, Chang A, Ebeling C et al. (2004) BRENDA, the enzyme database: updates and major new developments. Nucleic Acids Res 32: D431–433

Shannon P, Markiel A, Ozier O et al. (2003) Cytoscape: a software environment for integrated models of biomolecular interaction networks. Genome research 13: 2498–2504

Smigielski EM, Sirotkin K, Ward M et al. (2000) dbSNP: a database of single nucleotide polymorphisms. Nucleic Acids Res 28: 352–355

Subramanian A, Tamayo P, Mootha VK et al. (2005) Gene set enrichment analysis: a knowledge-based approach for interpreting genome-wide expression profiles. Proceedings of the National Academy of Sciences of the United States of America 102: 15545–15550

Sun R, Love MI, Zemojtel T et al. (2012) Breakpointer: using local mapping artifacts to support sequence breakpoint discovery from single-end reads. Bioinformatics 28: 1024–1025

Szabo C, Masiello A, Ryan JF et al. (2000) The breast cancer information core: database design, structure, and scope. Human mutation 16: 123–131

TANTIGEN (2009) TANTIGEN: Tumor T cell Antigen Database.

The-Metabolomics-Innovation-Centre (2014) The Metabolomics Innovation Centre.

The-Nutrigenomics-Organization (2008) The Nutrigenomics Organization.

The-UniProt-Consortium (2014) Activities at the Universal Protein Resource (UniProt). Nucleic Acids Research 42: D191–D198

Wang J, Mullighan CG, Easton J et al. (2011) CREST maps somatic structural variation in cancer genomes with base-pair resolution. Nature methods 8: 652–654

Wishart DS, Jewison T, Guo AC et al. (2013) HMDB 3.0–The Human Metabolome Database in 2013. Nucleic Acids Res 41: D801–807

Wishart DS, Knox C, Guo AC et al. (2009) HMDB: a knowledgebase for the human metabolome. Nucleic Acids Res 37: D603–610

Wishart DS, Tzur D, Knox C et al. (2007) HMDB: the Human Metabolome Database. Nucleic Acids Res 35: D521–526

Wyman SK, Parkin RK, Mitchell PS et al. (2009) Repertoire of microRNAs in epithelial ovarian cancer as determined by next generation sequencing of small RNA cDNA libraries. PloS one 4: e5311

Xenarios I, Salwinski L, Duan XJ et al. (2002) DIP, the Database of Interacting Proteins: a research tool for studying cellular networks of protein interactions. Nucleic Acids Res 30: 303–305

Xiao F, Zuo Z, Cai G et al. (2009) miRecords: an integrated resource for microRNA-target interactions. Nucleic Acids Res 37: D105–110

Xu K, Cui J, Olman V et al. (2010) A comparative analysis of gene-expression data of multiple cancer types. PloS one 5: e13696

Yamada T, Letunic I, Okuda S et al. (2011) iPath2.0: interactive pathway explorer. Nucleic Acids Res 39: W412–415

Yanaihara N, Caplen N, Bowman E et al. (2006) Unique microRNA molecular profiles in lung cancer diagnosis and prognosis. Cancer Cell 9: 189–198

Zhou B, Xiao JF, Tuli L et al. (2012) LC-MS-based metabolomics. Molecular bioSystems 8: 470–481

Chapter 14
Understanding Cancer as an Evolving Complex System: Our Perspective

"… antagonistic coevolution is a cause of rapid and divergent evolution and likely to be a major driver of evolutionary changes within species" (Paterson et al. 2010).

"Now, here, you see, it takes all the running you can do, to keep in the same place", as the Queen told Alice in Lewis Carrol's novel, *Through the Looking-Glass, and What Alice Found There* (1871). Evolutionary biologist Leigh Van Valen considered that this statement captured the essence of his proposal that inter-species interactions are a key driving force of evolution and named it the "Red Queen Hypothesis" in his 1973 publication (Valen 1973). Paterson and colleagues recently provided an elegant demonstration that microbial evolution follows this Hypothesis, by showing how the dynamic equilibrium between two co-existing populations, one bacterial and one viral parasite, is regained through rapid evolution by one party once the other is genetically given a new competitive edge (Paterson et al. 2010). They went further to propose inter-species interactions as a **major** driving force of evolution. This hypothesis has been used as a guiding principle when writing this monograph.

14.1 What Is Cancer?

This is a topic that has been studied in many millions of scientific articles according to PubMed. Yet, cancer researchers and clinicians are still searching for an answer that can guide us to the root(s) of the problem, and hence can help to develop more effective treatment of this often deadly illness. Throughout this book, cancer is considered as a process of cellular survival in an increasingly more stressful and more demanding microenvironment, which co-evolves with the diseased cell population; and cell proliferation is a cancer's way of survival in the sense that when proliferation stops, the cells die.

Clearly, different cancers have selected distinct survival routes, hence giving rise to diverse phenotypes as discussed throughout this book, which is probably due to

© Springer Science+Business Media New York 2014
Y. Xu et al., *Cancer Bioinformatics*, DOI 10.1007/978-1-4939-1381-7_14

the variations across a variety of cancer-inducing microenvironments. With that said, the commonalities shared by different cancers, including the cancer hallmarks (Hanahan and Weinberg 2011), seem to be substantially more significant than their exhibited phenotypic diversities. This strongly suggests that different cancers have to overcome similar stresses and may have utilized the same or similar survival "guides", which are probably provided by the same or related cellular programs originally used for other purposes, rather than created opportunistically through blind selection of malfunctions or mutations in an arbitrary fashion as some may believe. Furthermore, we posit, based on the material presented throughout this book that proliferation may not be just *a* pathway to survival, but instead it may represent the most feasible way for survival when considering the types of stressors that cancers must overcome or escape from in a sustained manner.

14.2   What Stresses Must Cancer Cells Overcome?

The current knowledge suggests that neoplastic cells have different types of life-or-death stresses at different developmental stages. As discussed in Chap. 5, the initial stress is dominantly induced by chronic hypoxia and/or ROS accumulation, possibly as a result of chronic inflammation or inheritable genetic mutations, which leads to the continuous accumulation of glucose metabolites. Removal of these metabolites becomes a stress that the host cells must overcome, or die. We believe that this is the cancer-defining stress from the onset as cell division represents the most feasible way for the cells to alleviate the stress, i.e., by converting the accumulated glucose metabolites into DNA and lipids of the new daughter cells as discussed in Chap. 6. Clearly this offers only a temporary solution to the problem as the hypoxic environment will lead to glucose-metabolite accumulation again, and hence a vicious cycle.

It has been well established that as a cancer progresses, the ROS level, as well as the associated oxidative stress, continues to increase as discussed in Chap. 9, and severe ROS levels will lead to cell death (Dröge 2002). The increased ROS level may be the result of at least two key activities: cancer metabolism (possibly coupled with malfunctioning mitochondria) and repeated tumor angiogenesis, hence repeated re-oxygenation. This, we believe, is one of the main stresses that accompanies cancer development throughout, i.e., beginning with cell proliferation and ending with metastasis. It should be emphasized that the impact of ROS level changes can be significant as they can fundamentally alter the cellular redox biochemistry, arguably one of the most essential chemical aspects in a cell.

Another key stress during the development of cancer, once the glycolytic fermentation pathway becomes heavily used, is mediated by lactic acidosis, which kills normal cells and enables encroachment by cancer cells as discussed in Chap. 8. Cancer cells have adapted to this stress by activating multiple mechanisms encoded in human cells to neutralize the acid.

   While the initiation of a cancer is closely associated with the lack of O_2, the main challenge among other stresses that a metastatic cancer must overcome is the increased O_2 level when moving from a hypoxic environment, where the migrating cells have been residing for years, to one that is rich in blood and hence an O_2-rich environment as discussed in Chap. 11. The increased O_2 level in metastatic sites will lead to various types of damage to the metastatic cancer cells that are not equipped to handle this level of O_2, hence casting a severe stress on such cells.

   Avoiding the activation of apoptosis is a constant battle and thus another stressor for any cancer since its cells tend to accumulate a variety of abnormalities. In fact, these aberrations would trigger apoptosis in normal cells as discussed in Chap. 7.

   Immune response-induced stresses are another class of stressors that the neoplastic cell must overcome. Interestingly, as a cancer evolves, it seems capable of converting the less sophisticated innate immune cells such as macrophages to form conjugative relationships with cancer cells, while adapting themselves to weaken the attacks from the adaptive immune cells such as T-cells.

   There are of course other stresses associated with cancer development, including DNA damage and nutrient depletion-induced stresses. These, however, we believe are secondary compared to the above types of stress.

14.3   Stresses *Versus* Proliferation

Knowing the major stress types a cancer must overcome and the common characteristics shared by different cancers, one is tempted to speculate that cell division must represent a most effective pathway for the stressed cells to reduce their stress levels, and hence has been selected as a common method for survival by different cancers. Specifically, cell division represents a feasible and direct way for chronic hypoxic cells to rid themselves of the accumulated products resulting from the reprogrammed energy metabolism as discussed in Chap. 5, because of: (1) the natural link between the accumulated glucose metabolites (specifically G6P) and hyaluronic acid synthesis under hypoxic conditions, and (2) the link between hyaluronic acid fragments and tissue-repair signaling, hence cell proliferation and survival.

   Interestingly, cell proliferation also represents a way for neoplastic cells to overcome the lactic acidosis typically associated with a cancer environment since the encoded mechanisms for neutralizing the excess protons are available only to proliferating cells, but not to normal non-dividing cells. Currently, there are no data that can be used to assess the relative level of contribution between acidosis-induced stresses *versus* other stressors that have the potential to drive the observed cell proliferation, knowing that this may vary across different cancer types.

   As mentioned above, the cellular ROS level will continually increase as a cancer progresses; moreover, it has been established that ROS can induce cell proliferation (Sauer et al. 2001; Gough and Cotter 2011; Chiu and Dawes 2012). Hence, we speculate that cell division may represent a way to reduce the ROS level since each doubling will divide the ROS population in the mother cell between the two daughter cells in some fashion. To the best of our knowledge, no studies have reported how the ROS molecules are divided between the two daughter cells, but it is natural to assume that some type of division of the ROS population will take place between the two new cells, thus resulting in a reduced ROS level in each and alleviating, at least partially and temporarily, ROS-induced stress. From this, we posit that cell proliferation, possibly in conjunction with other conditions, provides an avenue to reduce the ROS-induced stress. If such an induction requires "other conditions," which are not generally available, some subpopulation may create the triggering conditions, consequently leading to the selection and expansion of this subpopulation by evolution and selection.

Knowing that the overall ROS level continues to increase, we further speculate that the rate of ROS reduction through cell division intrinsically could not maintain pace with the rate of ROS increase due to cancer metabolism and re-oxygenation, at least for some cancers. Once the ROS level exceeds a certain threshold, it will trigger the synthesis of hyaluronic acid as discussed in Chap. 10. This time, however, they will trigger the process of cancer metastasis. Following this argument one can suggest that ROS drives cell proliferation during the development of cancer and then drives the cancer to colonize other sites when it becomes too high.

While cell proliferation appears to be essential to the survival of the primary cancer or cancer-forming cells, it seems to be only a side-product, at least to a large extent, of other survival mechanisms in response to the increased O_2-induced stress in metastatic cancer. As discussed in Chap. 11, metastatic cancer does not require cell proliferation to overcome the increased O_2 level-induced stress. Instead, such cancer requires the increased cholesterol level to stay viable. However, the increased cholesterol level in an O_2-rich environment gives rise to oxidized cholesterols and steroidogenic products, which lead to accelerated growth of cancer as a side product rather than a way for stress reduction and hence survival. This may be an important reason why cancer drugs, designed for primary cancers, are less or not effective for metastatic cancers, that has led to the belief that metastatic cancer is a terminal illness. Since for primary cancer, reducing cell proliferation, as is the case with some of the current hormone-based drugs, means increasing stress to the cells and possibilities to ultimately kill them. In contrast, cell proliferation is not a channel for metastatic cancer cells to reduce stress. Hence, slowing down or stopping cell proliferation in metastatic cancer will not lead to increased stress to the cells and therefore will not have the same effect as on primary cancers. Importantly, one needs to understand: although both primary cancer and metastatic cancer proliferate, they do so for different reasons.

This realization, along with the discussion in Chap. 11, suggests that metastatic cancer may not necessarily be a terminal disease. This belief has been ingrained in us because of the many observations of ineffectiveness when using the same drugs to treat primary and metastatic cancers. The latter, however, is probably a different type of disease from the primary counterpart and, consequently, may require a different regimen of drugs for efficacious treatment. This could be, for example, in the form of a reduction in the intake or *de novo* synthesis of cholesterol, since oxidized cholesterol serves as a main driver of metastatic cancer as outlined in Chap. 11.

We suspect that the relationships between cell proliferation and reduction in immune- or apoptosis-induced stresses are very different from those discussed above. From our current understanding, cell proliferation probably does not reduce stress-levels induced by immune or apoptotic attacks; instead they seem to be related to the maintenance of tissue homeostasis, i.e., when neoplastic cells are killed by immune responses or apoptosis in a cancer tissue, signals may be released to initiate repair of the lost tissue, hence triggering new cell generation. Overall, cell proliferation in this case represents a way to correct for the eliminated cells rather than an essential pathway to survival as in the above situations.

14.4   Different Survival Routes by Different Cancers

The above discussion, while focused on general stresses for all or the majority of the solid-tumor cancers to overcome, does provide a useful framework for one to study the detailed stresses and survival pathways for individual cancer cases. This framework can help in guiding the search for "rate-limiting" factors along the survival pathway of a cancer, hence possibly leading to more effective drugs and treatment plans. Specifically, by understanding why a specific cancer type tends to utilize a certain survival pathway (in a more general sense than the survival pathways discussed in Chap. 7), one can possibly derive the main stress(es) that the underlying cells must overcome, based on the same type of statistical association analysis used throughout this book. For example, one can ask: *What stresses drive melanoma to grow so much faster than the vast majority of the other cancer types, when the cancer switches from the radial growth phase to the vertical growth phase?*

This problem can potentially be addressed through a comparison of the various micro-environmental factors such as the oxygen level, the ROS level, the acidity level and immune responses in melanoma samples in the vertical growth phase *versus* the earlier stages. This approach can also be readily extended with samples of other cancer types in an effort to identify those stressors, or combinations thereof, that are particularly high. A comparison of the results can provide a framework enabling a focus on those frequently occurring stressors. Analyses can then be conducted to determine statistical associations between the identified stressors and possible facilitators for the high proliferation rate, such as growth factor receptors, nuclear receptors or possibly other functional genes that can be linked to both the stressors and cell proliferation, possibly via some intermediate steps. Such association analyses between specific stressors and key survival steps, when applied in an iterative manner, have the potential to lead to the identification of a sequence of short-range connections, which together link the stressors to the observed proliferation.

When studying cancer as a process of survival, two key observations were made that may be useful for future studies: (1) cancer cells tend to use mechanisms encoded in human cells to overcome or escape from encountered stressors, rather than inventing new ones through a random selection of genomic mutations or other malfunctions; and (2) the selected responses to stressors tend to be relatively large cellular programs but not a single reaction or a small number of reactions, hence providing a substantive pathway, possibly with several sites for intervention.

For (1), the evolving cells basically create a micro-environmental condition that can trigger the proper response system to help overcome the stress. This pattern has been observed repeatedly when studying various topics throughout this book. One good example is that neoplastic cells overcome lactic acidosis through proliferation. This is important as acidosis will trigger apoptosis in a cell unless the cell is in a proliferative state. Only then can such cells activate encoded mechanisms for neutralizing acidity as discussed in Chap. 8. Clearly, this is the result of natural section of the subpopulation of the cells that are able to trigger the proper responses. When a cancer cannot create a triggering condition for the right response(s) for self-preservation, the stress level will increase, ultimately leading to the activation of a general stress-response system at the epigenomic level as discussed in Chap. 9. All possible responses encoded in the genome will be systematically attempted, and if one of the responses is appropriate, as judged by survival, the subpopulation that used this response will survive and future generations of the cell will have the effective stress response encoded in their epigenomes. Clearly adaption through utilization of such general stress-response mechanisms also makes the cancer more malignant.

For (2), knowing that the survival pathway tends to be well coordinated among its individual steps, with substantial commonalities shared by different samples of the same cancer type, it is only reasonable to speculate that each survival route is a well-designed pathway, already encoded in the human genome. The utilization of the tissue repair system by neoplastic cells for survival is a good example. The richness of the hyaluronic acid-based signaling system can provide different survival routes for different stress conditions, hence giving rise to different phenotypes. An alternative explanation could be that cancer cells select random mutations or biomolecules in abnormal functional states (e.g., proteins with accidentally oxidized residues by ROS) in a blind (unguided) fashion and converge on a combination of mutations or abnormally functioning molecules that ultimately leads to survival. The chance for this to happen is extremely small, which is clearly not in agreement with the observed commonalities shared by different cancer cases of the same type in different patients, as well as the frequencies of cancer occurrence in the world.

Within this framework, the key roles played by genomic mutations in sporadic cancers, as we posit, are facilitators to make the survival pathways proceed more efficiently and sustainably, which are defined by programs such as the tissue repair system. That is, the main role(s) of genetic mutations may be to substitute for ongoing cellular functions that are either constitutively activated or repressed through regulation or accidental modifications on specific molecules, hence making the execution of the same functions more efficient and more sustainable. An alternative is to create a new functionality through selection of a genomic mutation, which will be easily detected by the layers of the surveillance systems and the affected cells will be destroyed. This is particularly important in the early stage of cancer development when survival pathways are not activated (e.g., by the tissue repair system), the apoptosis system is still intact and the immune system is not impaired. Mutations in familial cancers provide a good example here as they tend to change the cellular conditions in a slow and subtle manner, such as slowly and gradually increasing the ROS level over an extended period of time as discussed in Chap. 5. Hence, we posit that mutations in sporadic cancers are generally manifested as facilitators. In contrast, germline mutations may dominantly act as drivers in familial cancers. They, however, would not function in the way of "driver mutations" as defined in the current literature, since they probably all lead to gradual ROS accumulation and possibly hypoxia, mimicking the initial conditions that drive sporadic cancers.

14.5 Tissue *Versus* Cell Level Problem?

Knowing that cell proliferation is essential to the survival of the diseased cells, one can probably surmise that the essence of a cancer is predominantly a tissue rather than a cell level problem. This is because the ability for a cell to divide is dominantly determined at the tissue level rather than at the cell level, at least for solid-tumor cancers, e.g., an activated oncogene could not start cell proliferation in a tissue environment, a principle that has been discussed frequently in this book.

14.6 Holistic *Versus* Reductionist Approaches

Studies of cancer at the molecular and cellular level, particularly using cell culture or immune-deficient animal models, have generated enormous amounts of data and hypotheses as discussed throughout this book. At the same time, they may have also generated some misconceptions about the complexity of the disease as a result of tackling this intrinsically difficult-to-decompose problem using approaches guided by reductionist thinking. For example, the extracellular matrix is an integral part of the cancer problem. To study cancer cells independent of their co-evolving extracellular matrix is somewhat analogous to studying the evolution of *Pseudomonas fluorescens* without considering its antagonist co-adapting viral parasite, phage $\Phi 2$ (Paterson et al. 2010), as discussed in Chap. 5. The same can be said about the immune system.

While cell-based studies on one hand simplify a complex cancer problem, they may concurrently introduce even greater complexity as cancer is an intricately coupled system among its components. For example, the result of activating gene X may be highly context-dependent, but this context-dependence information may not be clear to the experimenter based on the available knowledge of specific cell lines. For this reason, some of the context information may be unknowingly ignored. When assessing the impact of the activation of gene X in two distinct cell lines (originating from different tissue types) in the same culture medium, separately, one may obtain conflicting results and incorrectly conclude that the response

Fig. 14.1 An object
consisting of two
components, white and black
in the top part of the figure,
along with the density
distributions of *white points*
and *black points* in the lower
half of the figure

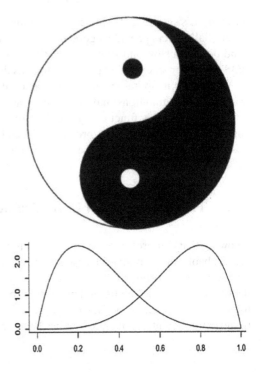

mechanism to the activation of the gene is different between the two cell types. Yet, the reality may be that the two cell lines have essentially the same response system but require different cofactors that are constitutively present in their original tissue environments, which are unknowingly lacking in the cell culture experiment. We suspect that the conflicting results observed on the functional roles of the human SPARC gene across different studies as discussed in Chap. 8 may fall into this category.

Figure 14.1 illustrates the idea using a simple mathematics problem where the relationship between the white and the black objects is very clear in two dimensional space, but when examining it in a reduced space, such as in one dimensional space (e.g., checking the density distributions of the white and black pixels), the relationship becomes more complex. This is essentially the approach that some cell-based studies have taken by examining the problem in a simplified setting (i.e., one dimensional) where the picture is quite clear in its full complexity (i.e., two dimensional space). One can imagine simple images like this in which the relationships could become arbitrarily more complex depending on how the two dimensional problem is projected to the one dimensional space. Therefore, studying a problem in an artificially designed environment (equivalent to dimension reduction in the example provided) could potentially make problems substantially more complex than they are.

The main point here is that cancer may potentially not be as complex as believed. The "complexity" may partially come from years of experience of tackling this intrinsically indecomposable problem, at least among key ingredients such as cells and their ECM, using highly reductionist approaches.

The above discussion, by no means, suggests that cancer is a simple problem. We hope that it will serve a purpose to encourage *omic* data analysts and computational biologists not to be intimidated by the fact that after decades of intensive studies by tens of thousands of cancer biologists worldwide, we are still searching for the root causes of cancer and the associated mechanisms. Our purpose is to attract computational biologists and bioinformaticians to undertake challenging cancer-related problems, for example those presented in this book, in the literature and personally generated, and take a more holistic approach to mine the available data and ask deep, penetrating questions. Of course, the basis for doing this is having a general knowledge of the biology of cancer as well as a state-of-the-art understanding about the specific problem one aims to investigate. It is hoped that the information presented herein may serve as a starting point for computational biologists to enter into this extremely important, interesting and challenging field of cancer research. The timing is perfect since a considerable amount of information has been generated on cancer biology over the past 100 years, and an enormous amount of *omic* data have been made publicly available, with more to come on a frequent basis. This wealth of data enables one to address fundamentally important cancer-biology questions. We whole-heartedly believe that computation-based holistic approaches, which complement the current thinking and tactics, are urgently needed in the field in order to make the next quantum leap in our understanding of cancer.

While a considerable amount of knowledge has been gained concerning cancer in the past 100 years, the reality is that our current understanding is somewhat fragmented. It, unfortunately, has not yet reached a level with a large framework or theory through which all the available information can be integrated to study this disease in its full complexity as an evolving tissue-level system. For this, our critical analysis must encompasses all of the indispensable players, namely diseased cells, stromal cells, immune cells and the extracellular matrices. The emerging knowledge about hyaluronic acid and fragments serving as a key signaling system used by cancer for its survival, as well as the epigenome as a general stress-response (*versus* condition-specific stress response) mechanism, may provide a framework for connecting many of what now appear to be disjointed facts. The information derivable from the large-scale *omic* data has the potential to provide the local connectivity information when assembling all the facts, like jigsaw puzzle pieces (Fig. 14.2), big and small, into one integrated system. This will allow cancer researchers and clinicians to identify the "weakest links" within a cancer system and guide the searches for effective drugs to treat individual cancer cases.

Fig. 14.2 A schematic illustration of integrating all of the knowledge learned in the past about cancer into one framework so systems-level studies can be carried out for identification of drivers, facilitators and weakest links of cancer evolution for substantially improved capabilities for treating the disease

References

Chiu J, Dawes IW (2012) Redox control of cell proliferation. Trends in cell biology 22: 592–601

Dröge W (2002) Free radicals in the physiological control of cell function. Physiological reviews 82: 47–95

Gough DR, Cotter TG (2011) Hydrogen peroxide: a Jekyll and Hyde signalling molecule. Cell death & disease 2: e213

Hanahan D, Weinberg RA (2011) Hallmarks of cancer: the next generation. Cell 144: 646–674

Paterson S, Vogwill T, Buckling A et al. (2010) Antagonistic coevolution accelerates molecular evolution. Nature 464: 275–278

Sauer H, Wartenberg M, Hescheler J (2001) Reactive oxygen species as intracellular messengers during cell growth and differentiation. Cell Physiol Biochem 11: 173–186

Valen LV (1973) A new evolutionary law. Evolutionary Theory 1: 1–30

Index

© Springer Science+Business Media New York 2014   363
Y. Xu et al., *Cancer Bioinformatics*, DOI 10.1007/978-1-4939-1381-7

Printed in the United States
By Bookmasters